AF772890

Ingo Glöckner

Fuzzy Quantifiers

Studies in Fuzziness and Soft Computing, Volume 193

Editor-in-chief
Prof. Janusz Kacprzyk
Systems Research Institute
Polish Academy of Sciences
ul. Newelska 6
01-447 Warsaw
Poland
E-mail: kacprzyk@ibspan.waw.pl

Further volumes of this series
can be found on our homepage:
springer.com

Vol. 178. Claude Ghaoui, Mitu Jain, Vivek Bannore, Lakhmi C. Jain (Eds.)
Knowledge-Based Virtual Education, 2005
ISBN 3-540-25045-X

Vol. 179. Mircea Negoita, Bernd Reusch (Eds.)
Real World Applications of Computational Intelligence, 2005
ISBN 3-540-25006-9

Vol. 180. Wesley Chu, Tsau Young Lin (Eds.)
Foundations and Advances in Data Mining, 2005
ISBN 3-540-25057-3

Vol. 181. Nadia Nedjah, Luiza de Macedo Mourelle
Fuzzy Systems Engineering, 2005
ISBN 3-540-25322-X

Vol. 182. John N. Mordeson, Kiran R. Bhutani, Azriel Rosenfeld
Fuzzy Group Theory, 2005
ISBN 3-540-25072-7

Vol. 183. Larry Bull, Tim Kovacs (Eds.)
Foundations of Learning Classifier Systems, 2005
ISBN 3-540-25073-5

Vol. 184. Barry G. Silverman, Ashlesha Jain, Ajita Ichalkaranje, Lakhmi C. Jain (Eds.)
Intelligent Paradigms for Healthcare Enterprises, 2005
ISBN 3-540-22903-5

Vol. 185. Spiros Sirmakessis (Ed.)
Knowledge Mining, 2005
ISBN 3-540-25070-0

Vol. 186. Radim Bělohlávek, Vilém Vychodil
Fuzzy Equational Logic, 2005
ISBN 3-540-26254-7

Vol. 187. Zhong Li, Wolfgang A. Halang, Guanrong Chen (Eds.)
Integration of Fuzzy Logic and Chaos Theory, 2006
ISBN 3-540-26899-5

Vol. 188. James J. Buckley, Leonard J. Jowers
Simulating Continuous Fuzzy Systems, 2006
ISBN 3-540-28455-9

Vol. 189. Hans Bandemer
Mathematics of Uncertainty, 2006
ISBN 3-540-28457-5

Vol. 190. Ying-ping Chen
Extending the Scalability of Linkage Learning Genetic Algorithms, 2006
ISBN 3-540-28459-1

Vol. 191. Martin V. Butz
Rule-Based Evolutionary Online Learning Systems, 2006
ISBN 3-540-25379-3

Vol. 192. Jose A. Lozano, Pedro Larrañaga, Iñaki Inza, Endika Bengotxea (Eds.)
Towards a New Evolutionary Computation, 2006
ISBN 3-540-29006-0

Vol. 193. Ingo Glöckner
Fuzzy Quantifiers: A Computational Theory, 2006
ISBN 3-540-29634-4

Ingo Glöckner

Fuzzy Quantifiers

A Computational Theory

Springer

Dr. Ingo Glöckner
FernUniversität in Hagen
LG Prakt. Inf. VII
Informatikzentrum
Universitätsstr. 1
58084 Hagen
Germany
E-mail: ingo.gloeckner@fernuni-hagen.de

Library of Congress Control Number: 2005936355

ISSN print edition: 1434-9922
ISSN electronic edition: 1860-0808
ISBN-10 3-540-29634-4 Springer Berlin Heidelberg New York
ISBN-13 978-3-540-29634-8 Springer Berlin Heidelberg New York

This work is subject to copyright. All rights are reserved, whether the whole or part of the material is concerned, specifically the rights of translation, reprinting, reuse of illustrations, recitation, broadcasting, reproduction on microfilm or in any other way, and storage in data banks. Duplication of this publication or parts thereof is permitted only under the provisions of the German Copyright Law of September 9, 1965, in its current version, and permission for use must always be obtained from Springer. Violations are liable for prosecution under the German Copyright Law.

Springer is a part of Springer Science+Business Media
springer.com
© Springer-Verlag Berlin Heidelberg 2006
Printed in The Netherlands

The use of general descriptive names, registered names, trademarks, etc. in this publication does not imply, even in the absence of a specific statement, that such names are exempt from the relevant protective laws and regulations and therefore free for general use.

Typesetting: by the author and TechBooks using a Springer LaTeX macro package

Printed on acid-free paper SPIN: 11430025 89/TechBooks 5 4 3 2 1 0

Preface

From a linguistic perspective, it is *quantification* which makes all the difference between "having no dollars" and "having a lot of dollars". And it is the meaning of the quantifier "most" which eventually decides if "Most Americans voted Kerry" or "Most Americans voted Bush" (as it stands). Natural language (NL) quantifiers like "all", "almost all", "many" etc. serve an important purpose because they permit us to speak about properties of collections, as opposed to describing specific individuals only; in technical terms, quantifiers are a 'second-order' construct. Thus the quantifying statement "Most Americans voted Bush" asserts that the set of voters of George W. Bush comprises the majority of Americans, while "Bush sneezes" only tells us something about a specific individual. By describing collections rather than individuals, quantifiers extend the expressive power of natural languages far beyond that of propositional logic and make them a universal communication medium.

Hence language heavily depends on quantifying constructions. These often involve fuzzy concepts like "tall", and they frequently refer to fuzzy quantities in agreement like "about ten", "almost all", "many" etc. In order to exploit this expressive power and make fuzzy quantification available to technical applications, a number of proposals have been made how to model fuzzy quantifiers in the framework of fuzzy set theory. These approaches usually reduce fuzzy quantification to a comparison of scalar or fuzzy cardinalities [197, 132]. Thus for understanding a quantifying proposition like "About ten persons are tall", we must know the degree to which we have j tall people in the given situation, and the degree to which the cardinal j qualifies as "about ten". However, the results of these methods can be implausible in certain cases [3, 35, 54], i.e. they can violate some basic linguistic expectations. These problems certainly hindered the spread of the 'traditional' approaches into commercial applications.

This monograph proposes a new solution which shows itself immune against the deficiencies of existing approaches to fuzzy quantification. The theory covers all aspects of fuzzy quantification which are necessary for a comprehensive treatment of the subject, ranging from abstract topics like the proper semantical description of fuzzy quantifiers to practical concerns like robustness of interpretations and efficient implementation. The novel theory of fuzzy quantification is

- a compatible extension of the Theory of Generalized Quantifiers [7, 9];
- not limited to absolute and proportional quantifiers;
- a genuine theory of fuzzy multi-place quantification;
- not limited to finite universes of quantification;
- not limited to quantitative (automorphism-invariant) examples of quantifiers;
- based on a rigid axiomatic foundation which enforces meaningful and coherent interpretations;
- fully compatible with the formation of negation, antonyms, duals, and other constructions of linguistic relevance.

To accomplish this, the theory introduces

- a formal framework for analysing fuzzy quantifiers;
- a system of formal postulates for those models of fuzzy quantification that are semantically conclusive;
- a number of theorems showing that these models are plausible from a linguistic point of view and also internally coherent;
- a discussion of several constructive principles along with an analysis of the generated models;
- a collection of algorithms for implementing fuzzy quantifiers in the models and examples for their application.

Specifically, this work approaches the semantics of fuzzy NL quantification by formalizing the essential linguistic expectations on plausible interpretations, and it also presents examples of models which conform to these semantical requirements.

The book further discloses a general strategy for implementing quantifiers in a model of interest, and it develops a number of supplementary methods which will optimize processing times. In order to illustrate these techniques, the basic procedure will then be detailed for three prototypical models, and the complete algorithms for implementing the relevant quantifiers in these models will be presented. Hence the proposed solution is not only theoretically appealing, but it also avails us with computational models of fuzzy quantification which permit its use in practical applications. The quantifiers covered by the new algorithms include the familiar absolute and proportional types known from existing work on fuzzy quantification, but they also comprise some other cases of linguistic relevance. In particular, the treatment of quantifiers of exception ("all except k") and cardinal comparatives ("many more than") is innovative in the fuzzy sets framework.

The material presented in the book will be of interest to those working at the crossroads of natural language processing and fuzzy set theory. The range of applications comprises fuzzy querying interfaces, where fuzzy quantifiers support more flexible queries to information systems; systems for linguistic data summarization, where fuzzy quantifiers are used to build succinct linguistic summaries of given databases; decision-support systems, where fuzzy quantifiers implement multi-criteria decisionmaking, and many others. In all of these applications, the 'old' models of fuzzy quantification can easily be replaced with the improved models developed here. I hope that by removing the 'plausibility barrier' faced by earlier techniques and rendering possible a more reliable use of fuzzy quantifiers, the new theory of fuzzy NL quantification will contribute to a break-through of these techniques into commercial applications.*

Hagen
September 2005

Ingo Glöckner

* The monograph is based on the author's PhD thesis titled *"Fuzzy Quantifiers in Natural Language: Semantics and Computational Models"* [52] which was published by Der Andere Verlag, Osnabrück (now Tönning), Germany under ISBN 3-89959-205-0. Kind permission of Der Andere Verlag is acknowledged to republish this improved version of the manuscript.

Overview of the Book

The book is organized as follows. The *first chapter* presents a general introduction into the history and main issues of fuzzy natural language quantification. To this end, the basic characteristics of linguistic quantifiers are first reviewed in order to distill some requirements on a principled theory of linguistic quantification. Following this, we consider the problem of linguistic vagueness and its modelling in terms of continuous membership grades, which is fundamental to fuzzy set theory. Zadeh's traditional framework for fuzzy quantification [197, 199] is then explained. It is argued that the framework is too narrow for a comprehensive description of fuzzy NL quantification, and the particular approaches that evolved from it are shown to be inconsistent with the linguistic data.

The *second chapter* introduces a novel framework for fuzzy quantification in which the linguistic phenomenon can be studied with the desired comprehensiveness and formal rigor. The framework comprises: fuzzy quantifiers, which serve as the operational model or target representation of the quantifiers of interest; semi-fuzzy quantifiers, which avail us with a uniform and practical specification format; and finally QFMs (quantifier fuzzification mechanisms), which establish the link between specifications and operational quantifiers. The basic representations underlying the proposed framework are directly modelled after the generalized quantifiers familiar in linguistics [7, 9]. Compared to the two-valued linguistic concept, semi-fuzzy quantifiers add approximate quantifiers like "almost all", while fuzzy quantifiers further admit fuzzy arguments ("tall", "young" etc.). Thus, semi-fuzzy quantifiers and fuzzy quantifiers are the apparent extension of the original generalized quantifiers to Type III and Type IV quantifications in the sense of Liu and Kerre [108]. The organization of my approach into specification and operational layers greatly simplifies its application in practice. The commitments intrinsic to this analysis of fuzzy quantification are discussed at the end of the chapter, which also judges the actual coverage of the proposed approach compared to the phenomenon of linguistic quantification in its full breadth.

The book presents a total of 275 theorems and it would have been impossible to include, or even sketch, the proofs of all of them here. In order to keep the size of this work within limits and also to improve its readability, the theorems cited in this work have been proven separately in a series of technical reports [53, 55, 56, 57, 58, 59].[†]

Appendix B lists a theorem reference chart which connects every theorem cited here with its original proofs published in these reports.

[†] These reports are available from `http://pi7.fernuni-hagen.de/gloeckner` or directly from the author, email: `iglockner@web.de`.

Contents

Overview of the Book IX

1 An Introduction to Fuzzy Quantification: Origins and Basic Concepts 1
- 1.1 Motivation and Chapter Overview 1
- 1.2 Logical Quantifiers vs. Linguistic Quantifiers 4
- 1.3 The Vagueness of Language 13
- 1.4 Fuzzy Set Theory: A Model of Linguistic Vagueness 18
- 1.5 The Case for Fuzzy Quantification 20
- 1.6 The Origins of Fuzzy Quantification 23
- 1.7 Issues in Fuzzy Quantification 25
- 1.8 The Modelling Problem 27
- 1.9 The Traditional Modelling Framework 29
- 1.10 Chapter Summary: The Need for a New Framework 50

2 A Framework for Fuzzy Quantification 55
- 2.1 Motivation and Chapter Overview 55
- 2.2 Fuzzy Quantifiers 65
- 2.3 The Dilemma of Fuzzy Quantifiers 68
- 2.4 Semi-fuzzy Quantifiers 70
- 2.5 Quantifier Fuzzification Mechanisms 74
- 2.6 The Quantification Framework Assumption 75
- 2.7 A Note on Nullary Quantifiers 78
- 2.8 A Note on Undefined Interpretation 79
- 2.9 Chapter Summary 80

3 The Axiomatic Class of Plausible Models 85
- 3.1 Motivation and Chapter Overview 85
- 3.2 Correct Generalization 86
- 3.3 Membership Assessment 88
- 3.4 The Induced Propositional Logic 90

3.5 The Aristotelian Square ... 92
3.6 Unions of Argument Sets ... 96
3.7 Monotonicity in the Arguments ... 97
3.8 The Induced Extension Principle ... 101
3.9 Determiner Fuzzification Schemes: The DFS Axioms ... 106
3.10 Chapter Summary ... 107

4 Semantic Properties of the Models ... 111
4.1 Motivation and Chapter Overview ... 111
4.2 Correct Generalisation ... 111
4.3 Properties of the Induced Truth Functions ... 112
4.4 A Different View of the Induced Propositional Logic ... 115
4.5 Argument Permutations ... 116
4.6 Cylindrical Extensions ... 118
4.7 Negation and Antonyms ... 119
4.8 Symmetrical Difference ... 121
4.9 Intersections of Arguments ... 122
4.10 Argument Insertion ... 123
4.11 Monotonicity Properties ... 126
4.12 Properties of the Induced Extension Principle ... 129
4.13 Quantitativity ... 131
4.14 Extensionality ... 133
4.15 Contextuality ... 134
4.16 Semantics of the Standard Quantifiers ... 136
4.17 Fuzzy Inverse Images ... 140
4.18 The Semantics of Fuzzy Multi-Place Quantification ... 141
4.19 Chapter Summary ... 144

5 Special Subclasses of Models ... 149
5.1 Motivation and Chapter Overview ... 149
5.2 Models Which Induce the Standard Negation ... 150
5.3 Models Which Induce the Standard Disjunction ... 154
5.4 The Standard Models of Fuzzy Quantification ... 155
5.5 Axiomatisation of the Standard Models ... 157
5.6 Chapter Summary ... 158

6 Special Semantical Properties and Theoretical Limits ... 161
6.1 Motivation and Chapter Overview ... 161
6.2 Continuity Conditions ... 162
6.3 Propagation of Fuzziness ... 163
6.4 Conjunctions and Disjunctions of Quantifiers ... 165
6.5 Multiple Occurrences of Variables ... 165
6.6 Convex Quantifiers ... 166
6.7 Conservativity ... 170
6.8 Fuzzy Argument Insertion ... 172
6.9 Chapter Summary ... 173

7 Models Defined in Terms of Three-Valued Cuts and Fuzzy-Median Aggregation ... 179
7.1 Motivation and Chapter Overview ... 179
7.2 Alpha-Cuts are not Suited to Define Models ... 180
7.3 From Two-Valued Logic to Kleene's Logic, and Beyond ... 183
7.4 Adaptation of the Mechanism to Quantifiers ... 184
7.5 Three-Valued Cuts of Fuzzy Subsets ... 187
7.6 The Fuzzy-Median Based Aggregation Mechanism ... 190
7.7 The Unrestricted Class of $\mathcal{M}_{\mathcal{B}}$-QFMs ... 191
7.8 Characterisation of $\mathcal{M}_{\mathcal{B}}$-DFSes ... 194
7.9 A Simplified Construction Based on Mappings $\mathcal{B}' : \mathbb{H} \longrightarrow \mathbf{I}$ ... 197
7.10 Characterisation of the Models in Terms of Conditions on $\mathcal{B}'$ ... 198
7.11 Examples of $\mathcal{M}_{\mathcal{B}}$-Models ... 199
7.12 Properties of $\mathcal{M}_{\mathcal{B}}$-Models ... 201
7.13 A Standard Model with Unique Properties: $\mathcal{M}_{CX}$ ... 206
7.14 Chapter Summary ... 216

8 Models Defined in Terms of Upper and Lower Bounds on Three-Valued Cuts ... 221
8.1 Motivation and Chapter Overview ... 221
8.2 The Unrestricted Class of $\mathcal{F}_{\xi}$-QFMs ... 223
8.3 Characterisation of $\mathcal{F}_{\xi}$-Models ... 224
8.4 Examples of $\mathcal{F}_{\xi}$-Models ... 226
8.5 Properties of the $\mathcal{F}_{\xi}$-Models ... 227
8.6 Chapter Summary ... 234

9 The Full Class of Models Defined in Terms of Three-Valued Cuts ... 237
9.1 Motivation and Chapter Overview ... 237
9.2 The Unrestricted Class of $\mathcal{F}_{\Omega}$-QFMs ... 237
9.3 Characterisation of the $\mathcal{F}_{\Omega}$-Models ... 240
9.4 The Unrestricted Class of $\mathcal{F}_{\omega}$-QFMs ... 243
9.5 The Classes of $\mathcal{F}_{\omega}$-Models and $\mathcal{F}_{\Omega}$-Models Coincide ... 243
9.6 Characterisation of the $\mathcal{F}_{\omega}$-Models ... 244
9.7 Examples of $\mathcal{F}_{\Omega}$-Models ... 246
9.8 Properties of $\mathcal{F}_{\Omega}$-Models ... 249
9.9 Chapter Summary ... 255

10 The Class of Models Based on the Extension Principle ... 259
10.1 Motivation and Chapter Overview ... 259
10.2 A Similarity Measure on Fuzzy Arguments ... 259
10.3 The Unrestricted Class of $\mathcal{F}_{\psi}$-QFMs ... 261
10.4 Characterisation of the $\mathcal{F}_{\psi}$-Models ... 264
10.5 The Classes of $\mathcal{F}_{\psi}$-Models and $\mathcal{F}_{\omega}$-Models Coincide ... 266
10.6 The Unrestricted Class of $\mathcal{F}_{\varphi}$-QFMs ... 268

10.7 The Classes of $\mathcal{F}_\varphi$-Models and $\mathcal{F}_\psi$-Models Coincide 270
10.8 Characterisation of the $\mathcal{F}_\varphi$-Models 271
10.9 An Alternative Measure of Argument Similarity 272
10.10 Chapter Summary 274

11 Implementation of Quantifiers in the Models 277
11.1 Motivation and Chapter Overview 277
11.2 'Simple' Quantifiers 278
11.3 Direct Implementation of Special Quantifiers in $\mathcal{M}_{CX}$ 282
11.4 The Core Algorithms for General Quantifiers 287
11.5 Refinement for Quantitative Quantifiers 293
11.6 Computation of Cardinality Bounds 309
11.7 Implementation of Unary Quantifiers 313
11.8 Implementation of Absolute Quantifiers and Quantifiers of Exception 318
11.9 Implementation of Proportional Quantifiers 321
11.10 Implementation of Cardinal Comparatives 332
11.11 Chapter Summary and Application Examples 344

12 Multiple Variable Binding and Branching Quantification ... 355
12.1 Motivation and Chapter Overview 355
12.2 Lindström Quantifiers 357
12.3 (Semi-)fuzzy L-Quantifiers and a Suitable Notion of QFMs ... 359
12.4 Constructions on L-Quantifiers 360
12.5 The Class of Plausible L-Models 365
12.6 Nesting of Quantifiers 368
12.7 Application to the Modelling of Branching Quantification 371
12.8 Chapter Summary 372

13 Discussion 375
13.1 A Novel Theory of Fuzzy Quantification 375
13.2 Comparison to Earlier Work on Fuzzy Quantification 383
13.3 Future Perspectives 387

A An Evaluation Framework for Approaches to Fuzzy Quantification 391
A.1 Motivation and Chapter Overview 391
A.2 The Evaluation Framework 392
A.3 The Sigma-Count Approach 395
A.4 The OWA Approach 400
A.5 The FG-Count Approach 406
A.6 The FE-Count Approach 412
A.7 Chapter Summary 414

B Theorem Reference Chart 417

References . . . 425

Index . . . 437

List of Symbols . . . 449

List of Figures . . . 457

List of Tables . . . 459

these criteria in turn. Borderline cases, to begin with, are cases in which we find ourselves unable or 'unwilling' to decide whether or not the property of interest applies. Thus a borderline bald person is neither clearly bald nor clearly not bald. It is important to understand that the unclarity associated with borderline cases is not due to a lack of information, because a borderline bald person will remain a borderline case even if we gather further information, e.g. by inspecting his scalp with a magnifying-glass. Similarly, deciding if a borderline tall person is tall will not be simplified if we measure the person's height with the utmost precision. For example, 1.84 metres vs. 1.84000001 metres is not likely to make a difference. Consequently, assertions like *"Bill is bald"* or *"Tim is tall"* (assuming they are borderline cases), must be considered neither true nor false, but rather 'indeterminate' or 'in-between'. This means that vague predicates violate the law of 'tertium non datur', which suggests an extension of the two-valued system of truth values. Vague predicates further lack sharp boundaries. For example, there is no exact point on a scale of ages which separates "young" persons from those no longer young. Rather we have a *fuzzy boundary*, i.e. a blurred area of gradual transition. In general, then, this means that a vague predicate F not only has boundary cases; there is not even a clear separation between the clear cases and the borderline cases. It rather appears that in the boundary area, there is a gradual shift of F-ness which runs the gamut from clear positives to clear negatives. Finally vague predicates are susceptible to Sorites paradoxes. According to Keefe and Smith [90, p. 9/10],

> A paradigm Sorites set-up for the predicate F is a sequence of objects x_i, such that the two premises
> (1) Fx_1
> (2) For all i, if Fx_i, then Fx_{i+1}
> both appear true, but, for some suitable large n, the putative conclusion
> (3) Fx_n
> seems false.

For example, (1) a person of age ten is young; (2) a young person remains young when aging by 1 millisecond; (3) a ninety-year old is young. By instantiating the general scheme, similar examples can be constructed for arbitrary vague predicates.

In the examples given so far, we preferred 'simple' predicates which mainly depend on a single dimension – like "tall" (height), "young" (age) or "bald" (which was supposed to depend on the number of hair for simplicity). Obviously, multidimensional predicates like "big", or predicates like "beautiful" or "important", where the relevant dimensions are not even clear, also exhibit vagueness. In addition, the vagueness of natural language is not restricted to common nouns like "heap" or "importance" and adjectives like "tall" or "young". In contrast, vagueness is ubiquitous in NL and can be found right across all syntactic categories. For examples, verbs like "hurry", adverbs like "slowly", modifiers like "very" and quantifiers like "many" or "almost all" are all vague. At this point, let us elucidate the classificatory difference between vagueness and the other sources of imperfection in NL.

- *Underspecificity*, to begin with, is *"a matter of being less than adequately informative for the purpose in hand"*, examples being *"Someone said something"* and *"an integer greater than thirty"* [90, p. 5]. These examples demonstrate that underspecificity per se has nothing to do with fuzzy boundaries, borderline cases, and the Sorites paradox.
- *Ambiguity* refers to one-to-many relationships between words and word senses. "Liver", for example, can denote someone who lives, as in "loose liver", but it can also refer to the part of the body, as in "liver complaint". The individual word senses of an ambiguous term can either be vague or not, just like ordinary word senses. Hence, ambiguity as such has nothing to do with vagueness.
- *Context dependence* describes the variability in the meaning of NL terms which can be attributed to a change in external factors. (We have already met with this phenomenon when discussing quantifiers like "many"). For example, a small basketball player may well exceed the height of a tall equestrian. Obviously, many vague predicates are context-dependent. Relational adjectives like "tall", "young", "expensive" in particular, depend on a standard of comparison which is usually not stated explicitly, and must therefore be resolved from the context. However, as pointed out by Keefe and Smith [90, p. 6]:

  > *"Fix on a context which can be made as definite as you like (in particular, choose a specific comparison class): "tall" will remain vague, with borderline cases and fuzzy boundaries, and the sorites paradox will retain its force. This indicates that we are unlikely to understand vagueness or solve the paradox by concentrating on context-dependence."*

In communication, the imperfection of NL results in some uncertainty regarding the information conveyed. It is therefore instructive to relate the above notions of underspecificity, ambiguity, context-dependence and vagueness to the 'classical' model of uncertainty, i.e. probability theory. It appears that underspecificity, ambiguity and context-dependence all introduce a number of *alternatives* regarding the intended interpretation. In the case of underspecificity, these alternatives result from the use of overly general terms; in the case of ambiguity, the alternatives express different word senses; and in the case of context-dependence, we have the option of choosing different comparison classes. In order to decide between these alternatives, we cannot do better but rely on experience and thus, expectations. This indicates that probability theory, which is concerned with the formalization of expectations, might provide a suitable framework for discussing these phenomena. Vagueness, however, is not primarily an epistemic question and caused by a lack of information. For example, if someone is borderline tall, then *"no amount of further information about his exact height (and the heights of others) could help us decide whether he* is *tall*" [90, p. 2]. Consequently, a probabilistic model which fills in expectations of the missing pieces of information will not help here. Moreover, the probabilistic model assumes a clear distinction between positive and negative outcomes of an event. Vagueness, however, means a lack of such a definite or sharp distinction, thus undermining the very preconditions of probabilistic modelling. In addition, vagueness does not seem to involve any kind of "events".

Compared to the precision achieved in logic, the phenomenon of vagueness appears as a deficiency or 'imperfection' at first sight. Nevertheless, vagueness appears to be a universal characteristic of all natural languages. As opposed to artificial languages designed on the logician's desk, natural languages had to stand the test of actual language use; these languages were never meant to be 'ideal', and they had to be practical, robust and flexible in the first place. While artificial systems of logic essentially depend on everything to be made 100% precise, this requirement is unkown to NL, which is tolerant against vague and imprecise information. The demand for accuracy intrinsic to classical logic might create a burden of 'precisification' and formalization which makes its application impractical or even not feasible in some cases. Moreover, the vagueness of NL accounts for the imprecision of our senses. Not surprisingly, then, perceptual predicates like "red" which directly refer to perceptual categories are typically vague. Wright [169] considers such predicates 'tolerant', because there is *"a notion of degree of change too small to make any difference"* to their applicability. Natural languages profit from this tolerance, which lets them absorb a certain degree of the imprecision of our senses. It should be of high interest to add such features to logical languages as well in order to achieve similar flexibility and robustness.

The models of vagueness described in the literature can be classified into epistemic approaches, supervaluationist approaches, and finally models based on many-valued logics, usually either three-valued or continuous-valued. The latter proposals are also known as 'degree theories of vagueness'.
From the *epistemic* point of view [24, 25, 151, 168], borderline propositions are either true or false, but they are unknowingly so. Thus, vague predicates are not inherently different from exact predicates; we are only ignorant of their precise boundaries. Obviously, this epistemic view, which denies the very phenomenon of vagueness, does not contribute much to the goal of utilizing vagueness for robust computer programs. When trying to model vague predicates in a two-valued logic, we will be forced to introduce an artificial precise boundary, which gives a wrong impression of total accuracy, and also results in a substantial loss of information. The *supervaluationist approach* attempts to avoid the bias made by committing to such an artificial boundary:

> *"Replacing vagueness by precision would involve fixing a sharp boundary between the positive and negative extensions and thereby deciding which way to classify each of the borderline cases. However, adopting any one of these ways of making a vague predicate precise – any one of "precisification" or*

> *"sharpening" – would be arbitrary. For there are* many, *equally good, sharpenings. According to supervaluationism, our treatment of vague predicates should take account of* all *of them"* Keefe and Smith [90, p. 24].

Based on the resulting alternatives, the truth of a proposition is then determined by the following *principle of supervaluation*:

> *"A sentence is true iff it is true on all precisifications, false iff false on all precisifications, and neither true nor false otherwise"* Keefe and Smith [90, p. 24].

This 'third case' admitted in supervaluationism is often not considered a genuine truth value on a par with "true" and "false", but rather a truth value gap [47], i.e. no truth value at all, or a truth value 'glut' [14], i.e. both true and false. However, if the third case is declared a genuine third truth value, say $\frac{1}{2}$, this naturally takes us to the modelling of vagueness in terms of a three-valued logic, as proposed by Tye [164].
In a theoretical set-up which only supports three cases ("true"; "false"; and "gap", "glut", or third truth value), we are forced to introduce a sharp boundary between the clear positives and the borderline cases (similarly between borderline cases and clear negatives). This appears unnatural and contradicts the principle of 'tolerance' mentioned above. Considering the vague predicate "tall", for example, there is a strong intuition that changing the height of a person by 0.1 mm cannot decide if someone is clearly tall or not clearly tall; 1 millisecond of elapsed time will not decide if someone is clearly young or not clearly young, etc. This observation suggests the use of a continuous-valued model, in which the various shades of F-ness of a vague predicate F can be represented by real numbers ranging from 0 (completely false) to 1 (fully true). The resulting *degree theories* of vagueness obviously account for borderline cases (which result in intermediate truth values) and fuzzy boundaries (which can be modelled as a smooth transition between the extreme cases of 0 and 1). The Sorites paradox can now be resolved as follows: The inductive premise (2), "If Fx_i then Fx_{i+1}", is not completely true, but only very nearly true. Although there is never a substantial drop in degrees of truth between consecutive x_i's, repeated application of (2) will accumulate these effects. Consequently, the truth of Fx_i will decrease as i becomes larger, until it approaches complete falsity, as witnessed by the falsity of the putative conclusion Fx_n. Thus, the Sorites paradox resolves nicely, and the continuous-valued model gives a satisfactory account of all three characteristics of vagueness.
In the literature, there are different proposals for degree theories, and defences of using a continuous-valued model [14, 46, 68, 105, 111, 188]. However, only one of these approaches has acquired practical significance outside philosophical debates, and is highly visible both in scientific research and commercial applications.

than the universe of discourse. Implicit quantifications, too, are often approximate by nature:

f. "often", "rarely", "recently", "mostly", "almost always",...
(approximate temporal specification)
g. "almost everywhere", "hardly anywhere", "partly",...
(approximate spatial specification)
h. "usually", "typically",...
(dispositional quantification)
i. "in almost every way", "in most respects",...
(quantification over properties, ways of doing things)

As mentioned above, there are examples of implicit quantification in dispositional or habitual propositions which do not involve a visible quantifier at all, like the earlier *"Slimness is attractive"*. Zadeh suggests that these cases be treated as (covert) uses of vague quantifiers like "usually" [199, 200].

It is rather instructive, and also of great importance to the analysis of fuzzy quantifiers adopted in this book, to relate the above cases I.–IV. to the 'fields of quantification' of Liu and Kerre [108, p. 2], who classify the possible instances of fuzzy quantification into four basic categories. The classification is presented for unrestricted quantification only, i.e. for a unary quantifier Q and a single argument A, as in *"There are Q A's"*, *"Q things are A"*, or $QxA(x)$ in a logical notation. Noticing that both the quantifier Q and predicate A can either be crisp or fuzzy, Liu and Kerre obtain the following table of possible combinations:

Table 1.3. The fields of quantification identified by Liu and Kerre [108, p. 2]

	A crisp	A fuzzy
Q crisp	I	II
Q fuzzy	III	IV

In this work, we are mainly concerned with multi-place quantifiers. For quantifiers which accept several arguments, the classification must be generalized as follows.

Type I: the quantifier is precise and all arguments are crisp;
Type II: the quantifier is precise, but fuzzy arguments are admitted;
Type III: the quantifier may be fuzzy, but all arguments are crisp;
Type IV: both the quantifier and its arguments are allowed to be fuzzy.

Let us now return to the examples introduced at the beginning of this section. It is clear that in the above example I., we have a Type I quantification, in case II., a Type II quantification, etc. However, the two classifications are not identical. In examples I.–IV., there is a dichotomy between precise and approximate quantifiers; in particular, approximate quantifiers cannot be precise. In the classification of Liu and Kerre, by contrast, the term 'fuzzy', which refers to fuzzy sets, includes both the crisp and genuinely fuzzy cases. Consequently Type I quantifications are a special case of both Type II and Type III quantifications, which in turn are special cases of Type IV quantifications. Generally speaking, a two-valued modelling is only adequate for Type I quantifications. The most general form, i.e. Type IV quantification, not only includes the remaining cases; it is also most relevant from the perspective of applications. Consequently, achieving a proper treatment of Type IV quantifications is the main objective of every model of fuzzy quantification.

1.6 The Origins of Fuzzy Quantification

It was L.A. Zadeh who brought fuzzy NL quantifiers to scientific attention by writing the pioneering papers on the modelling of these quantifiers with methods from fuzzy set theory. In a series of papers starting in the mid-1970's [189, 190, 191, 192, 193, 194, 195, 196, 197], Zadeh developed the fundamentals of possibility theory[4] [189, 191] which served as the basis for his proposal on natural language semantics, the meaning representation language PRUF [192]. These research efforts were aimed at the development of a theory of approximate reasoning [189, 190, 192], in which the knowledge about the variables is represented in terms of possibility distributions, from which the reasoning process constructs further distributions or linguistic truth values. In these publications Zadeh exposes his first ideas about fuzzy quantifiers, however only in short passages, e.g. [193, p. 166-168], and usually in an exemplary way. In fact, there were no systematic inquiries into the subject before 1983, when Zadeh published the first treatise solely devoted to fuzzy quantifiers. In [197], he introduces the basic distinction of quantifiers of the first and the second kinds (i.e. absolute and proportional), and he develops a framework for modelling such quantifiers in terms of fuzzy numbers. In addition, several generalizations of the familiar notion of cardinality to fuzzy sets are considered, and it is shown how these cardinality measures can be utilized for implementing fuzzy quantification. Finally, Zadeh also makes a proposal for syllogistic reasoning with fuzzy quantifiers. It is this seminal publication which established fuzzy quantification as a special branch of fuzzy set theory. Some other publications of Zadeh on the subject, which are also influential, comprise: a theory of commonsense knowledge [198], which emphasizes the role of fuzzy NL quantifiers; a further refinement of fuzzy reasoning with fuzzy quantifiers [199]; Zadeh's quantificational model of dispositions [200]; and the discussion of fuzzy quantifiers in connection with uncertainty management in expert systems [201].

---

[4] Roughly speaking, possibility theory is concerned with the representation and processing of imprecise information expressed by fuzzy propositions. For example, "Marcel is young" conveys some information about Marcel's age, which can be described by a possibility distribution.

Again, there are some precursors to Zadeh's work on fuzzy quantification. In *many-valued logic*, the research into quantifiers which accept many-valued arguments was launched in 1939 when J.B. Rosser [138] presented the first treatise on the subject. Other important contributions to quantification in many-valued logics, which antedate Zadeh's work, were made by Rosser/Turquette [139, 140, 141] and Rescher [136, 137]. The treatment of quantifiers in many-valued logic, however, is usually restricted to Type II quantifications only, i.e. generalizations of precise quantifiers applied to many-valued arguments. In principle, some of these methods, notably the proposal of Rescher [136], are also capable of expressing many-valued Type IV quantifications. In fact, Rescher [137, p. 201] remarks that he considers the generalization of orthodox $\forall$ and $\exists$ a 'moot question', giving to understand that he has more interesting cases in mind. He envisions quantifications referring to the semantical status of many-valued propositions, which is made accessible from the object language. Rescher's examples (for a three-valued logic) include the following quantifiers:

Table 1.4. Rescher's quantifiers

Quantifier	**Meaning**
$(\exists^I x)Px$	"P has borderline cases"
$(\forall^T x)Px$	"Everything is a clear positive of P"
$(M^T x)Px$	"Most things are clear positives of P"
$(M^I x)Px$	"Most things are borderline cases of P"

The basic idea of accessing the degree of determinacy (i.e. vagueness or fuzziness) through quantification is indeed an interesting one, and we will return to Rescher's examples later. At this point, it is sufficient to observe that the quantifiers $\exists^I, \forall^T, M^T$ and M^I, although applied to three-valued arguments, always result in a two-valued, precise quantification. Thus, Rescher's examples, too, are not an anticipation of Type IV quantifications, which were first considered by Zadeh.

Obviously, natural language quantifiers are also of intrinsic interest to linguists, and there is indeed a linguistic school of 'generalized quantifiers' which emerged at roughly the same time when Zadeh presented his early ideas on fuzzy quantifiers. The linguistic theory, whose beginnings are Barwise [5, 6] and Barwise/Cooper [7], will be discussed at length later on. At this point, it is sufficient to remark that the linguistic model always starts from crisp arguments and precise quantifiers; approximate cases like "many" are either not considered at all [92] or forced into the precise framework and 'modelled' as precise quantifiers [7, 69]. Barwise and Cooper [7] are well aware of the vagueness inherent to these quantifiers, and suggest an extension of their basic analysis to a three-valued model; however, they do not elaborate this idea further. Zadeh, it appears, is familiar with the linguistic account of NL quantification, which he mentions in the introduction to his 1983 publication [197, p. 149]. Nevertheless, there is no visible influence of the linguistic analysis on his proposal. To sum up, Zadeh was indeed the first to recognize that quantifying constructions in NL usually express Type IV quantifications, and the basic notion of a fuzzy quantifier he introduces incorporates both fuzziness in quantifiers and in their arguments. His 1983 publication marks the beginning of specialized research into fuzzy quantification and today, there are substantial contributions covering diverse aspects of the subject. Before turning to the technical details of Zadeh's proposal, we shall identify the main directions of research in the evolving field. We will then discuss Zadeh's basic concept of a fuzzy quantifier and the framework for fuzzy quantification involving such quantifiers that he proposes.

1.7 Issues in Fuzzy Quantification

From a methodological perspective, we can discern three main issues in fuzzy quantification: interpretation, reasoning and summarization. These three areas differ both in their goals, and in the aspect of fuzzy quantification being investigated.

Interpretation.

The most basic problem to be solved is that of precisely describing the meaning of fuzzy quantifiers, i.e. the *modelling problem.* Making available such descriptions is a prerequisite of implementing fuzzy quantifiers on computers. Obviously the chosen interpretations must match the expected meaning of the NL quantifiers of interest in order to avoid misunderstanding. The computer models should therefore mimick the use of these quantifiers in ordinary language. In order to develop such interpretations, we need a formalism which lets us express the semantics of fuzzy quantifiers with sufficient detail. In this framework, the relationship between linguistic target quantifiers and their mathematical interpretations must then be analysed. In other words, a methodology must be developed which lets us identify the mathematical model of an NL quantifier of interest by describing its expected behaviour. Given a fuzzy quantifying proposition like *"Many spacious cars are expensive"*, we can then use the models of "most", "spacious", "cars" and "expensive" to construct the interpretation of the expression as a whole. As witnessed by the example, which is based on an approximate quantifier "many" applied to the fuzzy argument of "spacious cars", this methodology must be developed for general Type IV quantifications. Only then will it become possible to model the interesting cases, like the above example involving gradual evaluations. For example, Type IV quantifications are also necessary to rank a collection of cars according to the quality criterion *"Few important parts are made from plastic"*. Similar criteria are very common in everyday reasoning and typically used when we have to decide among several options. A good deal of the literature on fuzzy quantifiers is concerned with the issue of interpretation, although most of these publications also introduce some prototypical application. Examples comprise the works of Zadeh [197], Ralescu [132, 133] and Yager [179, 185]. See Liu and Kerre [108], Barro et al. and Delgado et al. [34] for overviews of the approaches described in the literature. A more detailed discussion of interpretations for fuzzy quantifiers and the various facets of the 'modelling problem' will be given in the next section.

Reasoning

Another class of papers is concerned with the manipulation of expressions involving fuzzy quantifiers. In this case, the goal is that of deducing further knowledge from a set of facts and rules involving fuzzy quantifiers. As opposed to interpretation, this reasoning process will usually not demand complete knowledge of the given situation, i.e. there is some uncertainty concerning the exact values of some of the variables. However, even partial or imperfect knowledge of a situation is often sufficient to draw valuable and reliable conclusions; this is evident in our everyday problem solving. In computer models of this form of 'approximate reasoning' or 'reasoning under uncertainty', such partial knowledge can be expressed, for example, in terms of facts and rules involving fuzzy predicates (or possibility distributions of linguistic variables) and fuzzy quantifiers. Ideally, the axiom schemes and inference rules used by the reasoning system should permit the derivation of new truths (valid formulas) from given ones, and the assumed calculus should also cover the full space of logical consequences of the base knowledge. The classical procedure to achieve this is to define a semantical notion of entailment and then make sure that the calculus parallel semantic entailment, i.e. the calculus should be both correct and complete. However, this route was not taken in Zadeh's pioneering work [197, 199, 200]. Important contributions of other authors comprise [40, 131, 152, 161, 162, 176, 177], to name just a few. These approaches are often more particular about the semantical justification of their reasoning schemes. Thöne [161], for example, shows that it is possible to draw precise conclusions in the presence of uncertainty. A survey of reasoning with fuzzy quantifiers is presented in [109].

Summarisation

The third main issue in fuzzy quantification is concerned with the generation of quantifying expressions which best describe a given situation. This problematic is different from the generation of quantifying expressions in a fuzzy reasoning system, where the system's limited knowledge of the situation is further elaborated by applying the calculus rules. In contrast, a system for data summarisation is usually supposed to have complete knowledge of the situation of interest, usually stated in terms of elementary facts. The goal is to construct a succinct description of the situation which captures the important characteristics of the data. In order to express these descriptions, the system is equipped with a repertoire of NL concepts (represented by fuzzy subsets) and of fuzzy quantifiers, which introduce the possible quantities in agreement like "few", "many", "almost all" etc. Typically, the situation to be described comprises a large number of individuals. This is why the extraction of the quantifying description can be viewed as a process of summary generation. The generated descriptions are particularly useful because they can be rephrased in ordinary language. A typical example of such a summary is "Much sales of components is with a high commission", see [83, p. 31]. As witnessed by the example, the process results in a concise linguistic summary which is often more informative than a list of plain facts. Owing to these advantages, fuzzy quantifiers have become the preferred tool in *linguistic data summarization* [82, 83, 135, 186], the fundamentals of which were developed by Yager [171, 182]. In general there are many possible ways of summarizing a given situation. Research has therefore focussed on the development of heuristic criteria which will guide the summary generation to the most promising solutions. The fundamental requirement on a candidate summary is of course its *truthfulness*, i.e. it should properly describe the situation at hand. Truth or 'validity' is not sufficient to identify the optimal summary, though. Yager [171, 182] therefore adds a measure of informativeness. Subsequent research has identified further validity criteria, which decide on the quality of the generated summary. Kacprzyk and Strykowski [83, p. 30], for example, present a system which relies on the following indicators: The degree of imprecision/fuzziness of the summary, its degree of covering the data, the degree of 'appropriateness', and finally the length of the summary. The system has been used for the linguistic summarization of sales data at a computer retailer [82, 83].

1.8 The Modelling Problem

This book will focus exclusively on the issue of interpretation, or more precisely, the *modelling problem* of achieving a meaningful interpretation for NL quantifiers with methods from fuzzy set theory. There is good reason for concentrating on this problem of improving the quality of interpretations. First of all, the modelling problem is fundamental, i.e. none of the above research directions in fuzzy quantification is really independent of its solution.

To demonstrate that the modelling problem is fundamental, let us first discuss its relevance to *reasoning with fuzzy quantifiers*; data summarization and other applications will be considered in turn. In order to carry out reasoning with fuzzy quantifiers, one needs a calculus which specifies the 'admissible moves'. This calculus should parallel the semantics of the logical language, i.e. it should be correct and complete in the ideal case. The semantics of the logical language, however, not only specifies an interpretation for predicates like "tall"; it is also concerned with the interpretation of the other non-logical symbols and fuzzy quantifiers in particular. This means that the chosen in-

skeleton to the modelling problem for a certain class of NL quantifiers which must then be instantiated by concrete interpretation mechanisms. The majority of later approaches to fuzzy quantification described in the literature have adopted Zadeh's ideas. Thus Zadeh's proposal can be said to constitute the *traditional framework for fuzzy quantification*. In presenting his framework, Zadeh [197, 199] demarcates a class of quantifiers to be treated; he proposes a mathematical model for these quantifiers as well as a system of specifications for such quantifiers; he evolves a general strategy for determining the target operators starting from these descriptions, and he also makes two proposals for concrete models of fuzzy quantification (interpretation mechanisms). We shall consider these components in turn. The subject of study, to begin with, are absolute and proportional quantifiers, or quantifiers of the 'first kind' and 'second kind' in Zadeh's terminology [197, p. 149]:

> *"We shall employ the class labels 'fuzzy quantifiers of the first kind' and 'fuzzy quantifiers of the second kind' to refer to absolute and relative counts, respectively, with the understanding that a particular quantifier, e.g.* many, *may be employed in either sense, depending on the context."*

Turning to concrete instances of such quantifiers, Zadeh notes that [197, p. 149]:

> *"Common examples of quantifiers of the first kind are:* several, few, many, not very many, approximately five, close to ten, much larger than ten, a large number, *etc. while those of the second kind are:* most, many, a large fraction, often, once in a while, much of, *etc."*

It should be apparent from these examples that Zadeh adopts a broad notion of quantifiers here, i.e. he considers both quantifiers on the universe of discourse (like "several" or "most") as well as quantifiers defined on other base sets (like "often" and "once in a while"). However, Zadeh does not distinguish quantifiers on the universe of discourse from quantifiers on other base sets. To be sure, he uses the labels 'explicit' vs. 'implicit' quantification, which we introduced to differentiate between these cases, but they bear a different meaning in Zadeh's work. Following Zadeh, a quantifier is only considered implicit if there is no visible quantifying element at all, as is the case for dispositional propositions. By contrast, we include all linguistic constructions which need quantification for their interpretation, but do not involve a quantifier proper on the NL surface. In any case, we can assert that Zadeh's approach is concerned with absolute and proportional quantifiers in a wide sense which also embraces examples like "often", "usually" etc. In a subsequent paper, however, Zadeh identifies quantifiers of the first and second kinds with one-place and two-place quantifiers in general [199, p. 757]:

> *"It is useful to classify fuzzy quantifiers into quantifiers of the first kind, second kind, third kind, etc., depending on the arity of the second-order fuzzy predicate which the quantifier represents."*

In order to eliminate this potential source of confusion, we will prefer the class labels 'absolute' and 'proportional' here, rather than quantifiers of the 'first' and 'second' kinds. We will use cardinals $n = 1$ or $n = 2$ to denote the arity of the quantifiers and thus to discern the unrestricted and restricted uses. It is apparent from Zadeh's modelling examples in [197] that he, too, is concerned with both cases. The unrestricted use then corresponds to a quantifier restricted by a crisp predicate, as in *"Few men are wise"*. Here, *"men"* can be viewed as supplying the domain in which the unrestricted statement *"There are few wise"* can then be interpreted. To sum up, Zadeh's framework targets at the modelling of the unrestricted and restricted use of absolute and proportional quantifiers. Zadeh also mentions quantifiers of the third kind, exemplified by "many more than" [199, p. 757] or by likelihood ratios and the certainty factors used in expert systems [197, p. 149], but the framework is only developed for the first two kinds.

Now that the scope of Zadeh's approach to fuzzy quantification has been explained, we will discuss Zadeh's solution to the modelling problem for the chosen quantifiers. Let us first consider the repository of modelling devices, i.e. the target interpretations for NL quantifiers of the indicated types that Zadeh proposes. It is apparent from the above characterization of Type IV quantifications that a fuzzy quantifier must accept one or more fuzzy arguments and result in a membership degree which signifies the desired outcome of quantification. Thus, a fuzzy quantifiers is essentially a *fuzzy second-order relation*, see [199, p. 756/757]. In his examples, Zadeh only considers unary quantifiers and quantifiers involving two arguments, although he mentions the possibility of more general cases [198, p. 149]. We will therefore focus on quantifiers of arities $n = 1$ or $n = 2$ in the following. Hence let $E \neq \varnothing$ be some base set. A *unary fuzzy quantifier* $\widetilde{Q}$ on E, then, is a mapping which associates a gradual quantification result $\widetilde{Q}(X) \in [0, 1]$ to each choice of the fuzzy argument $X \in \widetilde{\mathcal{P}}(E)$. Some examples are:

$$\widetilde{Q}_1(X) = \mu_X(e) \quad \text{for some fixed } e \in E$$

$$\widetilde{Q}_2(X) = \sup\{\mu_X(e) : e \in E\}$$

$$\widetilde{Q}_3(X) = \mu_{[j]} \quad \text{where } \mu_{[j]} \text{ is the } j\text{-th largest membership grade of } X,\ E \text{ finite}$$

$$\widetilde{Q}_4(X) = \frac{\sum_{e \in E} \mu_X(e)}{|E|}, \qquad E \text{ finite.}$$

The extension to fuzzy two-place quantifiers should be obvious: a *binary fuzzy quantifier* on E is a mapping which associates a gradual quantification result $\widetilde{Q}(X_1, X_2)$ in the unit range $[0, 1]$ to each choice of the fuzzy arguments $X_1, X_2 \in \widetilde{\mathcal{P}}(E)$. For some first examples, consider the two-place fuzzy quantifiers defined as follows:

$$\widetilde{Q}_5(X_1, X_2) = \widetilde{Q}_2(X_1 \cap X_2)$$
$$\widetilde{Q}_6(X_1, X_2) = \widetilde{Q}_3(X_1 \cap X_2), \quad E \text{ finite}$$
$$\widetilde{Q}_7(X_1, X_2) = \widetilde{Q}_3(X_1 \setminus X_2), \quad E \text{ finite}$$
$$\widetilde{Q}_8(X_1, X_2) = \frac{\sum_{e \in E} \min(\mu_{X_1}(e), \mu_{X_2}(e))}{\sum_{e \in E} \mu_{X_1}(e)}, \quad E \text{ finite.}$$

The above definition of unary and binary fuzzy quantifiers obviously accounts for fuzzy quantification of the intended kinds. We have one or two arguments, which are allowed to be fuzzy, and the quantifier determines a gradual result. Thus, the proposed concept of fuzzy quantifiers is capable of expressing the desired Type IV quantifications for the unrestricted and retricted uses of absolute and proportional quantifiers. Let us now consider the relationship of these modelling devices and the natural language quantifiers to be modelled. For example, are the quantifiers $\widetilde{Q}_1$ to $\widetilde{Q}_8$ introduced above realized in natural language and if so, which linguistic quantifiers do they express? And conversely (this is the more important form of the modelling problem): Is there a systematic way of interpreting NL quantifiers in terms of such fuzzy quantifiers, and how can a particular choice of interpretation be justified? Thiele [158] has successfully analyzed this relationship for the 'classical' examples of universal and existential quantifiers. However, there are no attempts yet to directly describe this relationship for a broader class of quantifiers. It is more promising to devise a way for describing NL quantifiers and to introduce interpretation mechanisms (the particular 'models', or 'approaches' to fuzzy quantification) which decide on the interpretation of these descriptions. Hence let us consider the specifications of quantifiers and the mechanisms for interpretation that Zadeh envisions.

Zadeh's approach rests on two fundamental ideas, one of which relates to the proposed specifications, while the other clarifies the role of the interpretation mechanisms. Zadeh's first idea is concerned with the representation of fuzzy quantifiers. As pointed out in [199, p. 756],

> *"...the concept of a fuzzy quantifier is related in an essential way to the concept of cardinality – or, more generally, the concept of measure – of fuzzy sets. More specifically, a fuzzy quantifier may be viewed as a fuzzy characterization of the absolute or relative cardinality of a collection of fuzzy sets.*

Hence Zadeh utilizes that the considered simple quantifiers can be expressed in terms of the cardinality of their argument or on the relative share of two cardinalities: Absolute quantifiers like *"About fifty Y_1's are Y_2's"* depend on $|Y_1 \cap Y_2|$ while proportional quantifiers like *"Most Y_1's are Y_2's"* depend on $|Y_1 \cap Y_2|/|Y_1|$. In this way, it is possible to replace the second-order notion of a fuzzy quantifier with a first-order representation as a fuzzy subset of the real line (for absolute quantifiers) or of the unit interval (for proportional quantifiers):

$$\widetilde{Q}(Y) = \mu_Q(|Y|) \qquad \text{for absolute quantifiers}$$
$$\widetilde{Q}(Y_1, Y_2) = \mu_Q(|Y_1 \cap Y_2|/|Y_1|) \qquad \text{for proportional quantifiers}$$

Notes 1.1.

- This reduction is admissible in the crisp case, when the cardinalities $|Y_1 \cap Y_2|$ and $|Y_1|$ are well-defined. We shall see later that Zadeh envisions a similar, cardinality-based evaluation in the fuzzy case as well. However, it is not necessarily so that fuzzy quantification must always reduce to a fuzzy cardinality measure.
- We have followed Zadeh in considering the unrestricted use of absolute quantifiers and the two-place use of proportional quantifiers as fundamental; it should be obvious how to adapt these equations for restricted absolute and unrestricted proportional quantification, i.e. $\widetilde{Q}(Y_1, Y_2) = \mu_Q(|Y_1 \cap Y_2|)$ and $\widetilde{Q}(Y) = \mu_Q(Y/|E|)$, resp.
- In principle, a membership function $\mu_Q : \mathbb{N} \longrightarrow \mathbf{I}$ should be sufficient for specifying absolute quantifiers which refer to cardinal numbers, after all. Some of the later models of fuzzy quantification, however, will demand a specification for arbitrary real numbers.
- In the above equation for proportional quantifiers, the case that $Y_1 = \varnothing$ is silently ignored in the literature. The use of an additional constant $v_0 \in \mathbf{I}$ would be necessary in order to fix a unique quantification result $\widetilde{Q}(\varnothing, Y_2) = v_0$ in this case as well.

To sum up, Zadeh proposes a reduction of second-order fuzzy quantifiers to first-order specifications. In terms of these simplified descriptions, then, an absolute quantifier can be described by a membership function $\mu_Q : \mathbb{R}^+ \longrightarrow \mathbf{I}$, while a proportional quantifier can be represented by a membership function $\mu_Q : \mathbf{I} \longrightarrow \mathbf{I}$. A possible choice of μ_Q for the proportional quantifier "almost all" is shown in Fig. 1.1. In this case, we have a membership function $\mu_{\textbf{almost all}}$ defined by

$$\mu_{\textbf{almost all}}(x) = S(x, 0.7, 0.9) \tag{1.1}$$

for all $x \in \mathbf{I}$, where S is Zadeh's S-function defined by

$$S(x,\, \alpha,\, \gamma) = \begin{cases} 0 & : \; x \le \alpha \\ 2 \cdot \left(\dfrac{x-\alpha}{\gamma-\alpha}\right)^2 & : \; \alpha < x \le \dfrac{\alpha+\gamma}{2} \\ 1 - 2 \cdot \left(\dfrac{x-\gamma}{\gamma-\alpha}\right)^2 & : \; \dfrac{\alpha+\gamma}{2} < x \le \gamma \\ 1 & : \; x > \gamma \end{cases} \tag{1.2}$$

for all x, α, $\gamma \in \mathbf{I}$. Zadeh [197, p. 150] further proposes to view these first-order representations as fuzzy numbers:

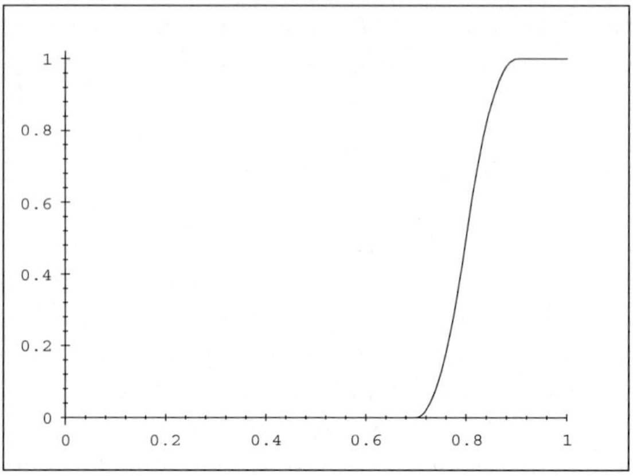

Fig. 1.1. A possible definition of $\mu_{\textbf{almost all}}$

> *"More generally, we shall view a fuzzy quantifier as a fuzzy number which provides a fuzzy characterization of the absolute or relative cardinality of one or more fuzzy or nonfuzzy sets."*

However, there might be some convexity implications of fuzzy numbers (depending on the chosen notion of fuzzy number) which are violated by artificial quantifiers like "an even number of". Therefore we will avoid the term "fuzzy number" in connection with Zadeh's first-order representations.

In the book, the following unambiguous notation will be used to discern the second-order notion of fuzzy quantifiers and their first-order descriptions. The second-order fuzzy quantifiers are symbolized as mappings $\widetilde{Q} : \widetilde{\mathcal{P}}(E) \longrightarrow \mathbf{I}$ or $\widetilde{Q} : \widetilde{\mathcal{P}}(E)^2 \longrightarrow \mathbf{I}$, i.e. the μ_Q-notation will not be used for them and they are further marked by the tilde $\widetilde{}$ in order to signify that they refer to Type IV quantifications and thus accept fuzzy arguments. The first order-descriptions, on the other hand, will always be written in the μ_Q-notation, which then denotes a mapping $\mu_Q : \mathbb{R}^+ \longrightarrow \mathbf{I}$ or $\mu_Q : \mathbf{I} \longrightarrow \mathbf{I}$, depending on the type of the quantifier of interest (i.e. absolute or proportional). Hence $\widetilde{\textbf{almost all}}$ is a second-order fuzzy quantifier $\widetilde{\textbf{almost all}} : \widetilde{\mathcal{P}}(E)^2 \longrightarrow \mathbf{I}$, while $\mu_{\textbf{almost all}}$ refers to the membership function of its first order description, $\mu_{\textbf{almost all}} : \mathbf{I} \longrightarrow \mathbf{I}$. In order to eliminate a possible source of confusion, we shall introduce a different terminology for $\widetilde{Q}$ (the second-order fuzzy quantifier) and μ_Q (the membership function of the first-order representation). For the sake of making a classificatory distinction, the term 'fuzzy quantifier' will be reserved for the second-order construct $\widetilde{Q}$. The first-order descriptions $\mu_Q : \mathbf{I} \longrightarrow \mathbf{I}$ or $\mu_Q : \mathbf{I} \longrightarrow \mathbf{I}$, by contrast, will be called *fuzzy linguistic quantifiers*. It should be pointed out that Zadeh and his followers identify the above cases and generally use 'fuzzy quantifiers' and 'fuzzy linguistic quantifiers' interchangeably. In this book, however, fuzzy quantifiers and fuzzy linguistic quantifiers will be technical terms with different meanings as explained above.

In addition, when talking about *approaches based on fuzzy linguistic quantifiers*, we will refer to those approaches which operate on these μ_Q's in order to determine the interpretation of NL quantifiers. In this way, the terminology of the traditional account of fuzzy quantification (which rests on first-order μ_Q's or 'fuzzy linguistic quantifiers') is clearly separated from the novel approach exposed here, which introduces a very different framework.

At this point, the repository of modelling devices as well as suitable specifications of such quantifiers have been introduced (i.e. unary or binary fuzzy quantifiers as the operations, and fuzzy linguistic quantifiers as representations). It should now be pretty obvious how an interpretation mechanism $\mathcal{Z}$ must look like which closes the gap between the first order representations μ_Q and corresponding operational quantifiers $\widetilde{Q} = \mathcal{Z}(\mu_Q)$. In principle, then, the interpretation mechanism is expected to map specifications of absolute quantifiers $\mu_Q : \mathbb{R}^+ \longrightarrow \mathbf{I}$ to unary or binary quantifiers $\mathcal{Z}^{(1)}_{\mathrm{abs}}(\mu_Q) : \widetilde{\mathcal{P}}(E) \longrightarrow \mathbf{I}$ and $\mathcal{Z}^{(2)}_{\mathrm{abs}}(\mu_Q) : \widetilde{\mathcal{P}}(E)^2 \longrightarrow \mathbf{I}$, in order to express the unrestricted and restricted use of the quantifier for a given base set $E \neq \varnothing$. A similar definition is also needed for specifications of proportional quantifiers $\mu_Q : \mathbf{I} \longrightarrow \mathbf{I}$, which should be mapped to unary quantifiers $\mathcal{Z}^{(1)}_{\mathrm{prp}}(\mu_Q) : \widetilde{\mathcal{P}}(E) \longrightarrow \mathbf{I}$, for modelling the unrestricted use, and $\mathcal{Z}^{(2)}_{\mathrm{prp}}(\mu_Q) : \widetilde{\mathcal{P}}(E)^2 \longrightarrow \mathbf{I}$ to express restricted quantification. By introducing this $\mathcal{Z}^{(i)}_{\mathrm{abs}}$ and $\mathcal{Z}^{(i)}_{\mathrm{prp}}$ notation, we will be able to describe the approaches in Zadeh's tradition in a uniform way. (See Chap. A in the appendix for details).

Zadeh not only proposes the use of first-order representations and fuzzy cardinality measures for modelling fuzzy quantification, though. He also explains how to design such interpretation mechanisms. His second fundamental idea on fuzzy quantification, then, is concerned with the functioning of a model of fuzzy quantification. Specifically, Zadeh [197, p. 159] proposes to determine the quantification results from the fuzzy linguistic quantifier μ_Q and the given arguments through a computational metaphor which reduces fuzzy quantification to a comparison of fuzzy cardinalities or ratios of cardinalities. He thus suggests to view the following propositions as semantically equivalent:

$$\text{There are } Q\ A\text{'s} \Leftrightarrow \mathrm{Count}(A) \text{ is } Q$$
$$Q\ A\text{'s are } B\text{'s} \Leftrightarrow \mathrm{Prop}(B|A) \text{ is } Q.$$

Hence in order to evaluate an absolute quantifying statement like *"There are about eighty X's"*, one needs a scalar or fuzzy measure of the cardinality of fuzzy sets, which determines a quantity Count(X). The resulting quantity must then be compared to the fuzzy linguistic quantifier, i.e. to the given μ_Q, and this comparison will determine the numerical score of the final quantification result, symbolized $\mathcal{Z}^{(1)}_{\mathrm{abs}}(\mu_Q)(X)$ in the above notation. Turning to proportional quantifiers, a proposition like *"Almost all X_1's are X_2's"* is believed to depend on some quantity Prop($X_2|X_1$) denoting the proportion, or relative share of X_1's which are X_2's. Again, this quantity can either be a scalar number or a fuzzy subset in the unit interval $\mathbf{I} = [0, 1]$. By comparing the resulting quantity to the given μ_Q in some way, one then obtains the quantification result, symbolized $\mathcal{Z}^{(2)}_{\mathrm{prp}}(\mu_Q)(X_1, X_2)$ in the proposed notation. It should be apparent from this description that Zadeh's ideas leave two degrees of freedom for developing approaches to fuzzy quantification: (a) the choice of the measure of fuzzy cardinality and relative cardinality, i.e. the definitions to substitute for $\mathrm{Count}(X)$ and $\mathrm{Prop}(X_2|X_1)$; and (b), the way in which the comparison of these (absolute or relative) cardinalities is accomplished.

Before turning to the individual approaches which instantiate this framework, let us clarify a few more differences between Zadeh's terminology and the classifications used in this work. Zadeh's usage of the terms 'universe of discourse' and 'base set' differs from the way they are used in this book. What we will call the 'universe of discourse', i.e. the dedicated base set which supplies the individuals that we can talk about, has no special name in Zadeh's system, and is identified with other base sets, i.e. collections over which a quantification can possibly range (points in time or in space etc.). Zadeh uses the term 'universe of discourse', rather than 'base set', to denote all of these collections. In Zadeh's writings [197, p. 159], the term 'base set' is reserved for the qualifying first argument A in a pattern like *"Most A's are B's"*, and it therefore corresponds to the 'restriction' of a quantifier in the terminology of this book. Zadeh has no special term for denoting the second argument B, which we call the 'scope' of the quantifier. To be sure, Zadeh also uses the word 'scope'. In his terminology, however, the scope of a (binary) quantifier is the argument tuple (B, A), where $A, B \in \widetilde{\mathcal{P}}(E)$ are given choices of the variables in the pattern "Q A's are B's", see [197, p. 47].

1.9.1 A Survey of Existing Approaches

The abstract framework for fuzzy quantification described above introduces specifications and target operations for absolute and proportional quantifiers, and it clarifies the role of interpretation methods. In order to be useful in practice, however, the framework must be populated with concrete examples of possible interpretation mechanisms, and Zadeh [197] makes two instructive proposals. The first method described in [197], which we will call the *Σ-count approach* in this monograph, is based on a scalar measure of the cardinality of fuzzy sets and of fractions of such cardinalities. This scalar measure, known as the Σ-count or 'power' of a fuzzy set [36], serves to implement the cardinality comparisons in "$\mathrm{Count}(X)$ is Q" and "$\mathrm{Prop}(Y|X)$ is Q" to which Zadeh attempts to reduce all Type IV quantifications. Zadeh's second proposal, called the *FG-count approach* in this monograph, no longer supposes that the cardinality of a fuzzy set can be represented by a single scalar number. By contrast, the cardinality of a fuzzy set will now be represented by a fuzzy subset of the cardinal numbers, the so-called FG-count [197].

Other authors have usually adopted Zadeh's classification of absolute and relative quantifiers, and his basic framework for representing and interpreting quantifiers of these types. Most authors also share his assumption that Type IV quantifications be reducible to a comparison of (absolute or relative) cardinalities. Thus, the approaches described in the literature mainly differ in the measure of fuzzy cardinality used and in the way that the required comparison of fuzzy cardinalities is accomplished. Some of these new proposals are directly motivated from negative evidence against an earlier technique; the modified method then targets at an improvement for the critical cases. An example is Ralescu's model of fuzzy quantification [132] which is presented as an improvement compared to Yager's FG-count-based method [174].

Other proposals attempt a generalization of existing methods, e.g. by replacing min and max with general fuzzy conjunctions and disjunctions [33, 35] or by making use of general implications [3, p. 15, eq. (3)]. Finally, there are also some approaches based on conceptually different techniques [179, 181, 183, 185].

Let us now consider the main approaches. Yager [174] presented a model of fuzzy quantification suited for nondecreasing quantifiers, which is essentially based on the FG-count. In [175], Yager proposes a generalization of his method to arbitrary fuzzy conjunctions and disjunctions. In addition, Yager extends the method towards binary proportional quantifiers, thus incorporating importances [180, p.72]. His method, although not introduced in this way, is a generalization of the basic FG-count model (see Sect. A.5).
Ralescu [132] attempts to improve upon Yager's model with an alternative proposal for interpreting fuzzy quantifiers motivated by a possibilistic argument. The method not only considers the degree to which a given fuzzy set has at least k elements, but also the degree to which the remaining elements are *not* contained in it. We will call Ralescu's proposal the *FE-count approach* here because the method can be reconstructed in terms of the FE-count, a measure of fuzzy cardinality described by Zadeh [197]. Each of the three basic methods mentioned so far – i.e. the Σ-count, FG-count and FE-count approaches – adopts its particular notion of fuzzy cardinality, which serves as the basis for performing cardinality comparisons. Thus, these approaches exemplify Zadeh's basic idea of reducing Type IV quantifications to a gradual comparison "Count(X) is Q" or "Prop($Y|X$) is Q".

A different approach was proposed by Yager [179], who models fuzzy quantifiers as Ordered Weighted Averaging (OWA) operators. Thus, fuzzy quantification is treated as a special case of aggregation problem. The basic OWA approach, which is restricted to unary proportional quantifiers based on nondecreasing choices of μ_Q, has later been extended towards nonincreasing and unimodal quantifiers [184]. In addition, Yager has developed several methods for incorporating importances [179, 181, 183]. It should be remarked that the OWA approach, although originally not defined in terms of a cardinality measure, also fits into the general framework because it can be reformulated as a function of a fuzzy cardinality, see note on p. 47. Another reduction to a fuzzy cardinality measure is described in [35]. The three approaches mentioned earlier as well as the OWA approach, have been most important to the theoretical development of fuzzy quantification and they are most frequently found in applications (in particular the Σ-count and OWA method are very popular). Consequently, these four methods will be considered the *main approaches* to fuzzy quantification, and they will be discussed at some more length in the remainder of this chapter. Apart from these well-known methods, there are several minor variants and generalizations which also belong in Zadeh's general framework and share his basic assumptions.

To begin with, there are several variants of the OWA approach which differ in their formalization of two-place quantification, or 'importance qualification' in Yager's terminology. Here we will focus on Yager's original proposal [179]; two alternative techniques to include weights are described in [181, 183]. A recent proposal of Yager, which is also concerned with importance qualification in an OWA setting, is mentioned in [186]. Yager's models of two-place quantification make essential use of a coefficient derived from the given quantifier, the so-called 'degree of orness' (see Sect. A.4 in the appendix for details). Others have tried to develop improved methods for restricted proportional quantification which are also parametrized by the degree of orness. Vila et al. [165], for example, propose a convex combination (i.e. linear interpolation) of restricted universal and existential quantification, where the degree of orness is used as the interpolation coefficient. Delgado, Sánchez and Vila [33] propose an integral-based method which incorporates both the FG-count and OWA approaches, depending on the choice of fuzzy conjunction and disjunction.[5] The method is extended towards restricted proportional quantification, again, by employing an aggregation operator parametrized by the degree of orness. The same authors later presented a general cardinality-based method which also encompasses the FG-count and OWA approaches as special cases [35]. This proposal certainly marks an important theoretical advance because it explains these techniques from a unifying, cardinality-based perspective. In addition, the 'core' approaches are extended to arbitrary quantifiers (without special assumptions on monotonicity), and a general analysis of relative cardinalities and binary proportional quantification is also provided. Finally, Barro et al. [3] consider a generalization of Yager's [180] inclusion approach towards general fuzzy implications (but they do not recommend their method due to some intrinsic problems which it shares with Yager's proposal).

Due to the conceptual vicinity of fuzzy quantification and cardinality assessments, another point of departure is the improvement or generalization of the underlying cardinality measures, from which novel methods for fuzzy quantification can then be derived. The work of Delgado et al. [35] mentioned above falls into this category. Historically, the discussion of cardinality measures for fuzzy sets starts with Zadeh's proposal of a fuzzy cardinality in [193], and Blanchard's [15] subsequent research into a suitable notion of fuzzy cardinality. Dubois and Prade [42] improve on Zadeh's original proposal in [193] (which may generate non-convex fuzzy cardinalities) and on the FE-

[5] the method is also described in English in Barro et al. [3, p. 19-21].

count measure (which discerns less structure than possible) by presenting an alternative measure of fuzzy cardinality with improved formal properties. In addition, Dubois and Prade develop a methodology for cardinality comparisons of fuzzy sets, which establishes a meaningful interpretation to statements like "A has more elements than B" when A, B are fuzzy sets. However, they seem to view this as an application of cardinality measures, not as a special kind of fuzzy quantification. Some further publications on the cardinality of fuzzy sets have been contributed by Ralescu [134] and Wygralak [170].

Finally let us consider those approaches to fuzzy quantification which depart from the general picture framed by Zadeh. Prade [127], to begin with, presents a fuzzy pattern-matching approach to the evaluation of fuzzy quantifiers which can be applied to arbitrary absolute quantifiers $\mu_Q : \mathbb{R}^+ \longrightarrow \mathbf{I}$ without special monotonicity requirements. The approach does not generate a unique quantification result $\mathcal{Z}(\mu_Q)(X) \in \mathbf{I}$, though, but rather determines interval-valued interpretations $[N, \Pi]$, where $N, \Pi \in \mathbf{I}$. As shown by Bosc and Lietard [21, p. 11], the result interval contracts into a unique scalar for monotonic quantifiers; the approach then coincides with the FG-count approach. Bosc and Lietard [21, 22] further show that the basic FG-count approach can be reduced to a Sugeno integral, whereas the basic OWA approach can be expressed as a Choquet integral. This means that these 'core' approaches can be explained by a comprehensive theory of fuzzy measures and integrals. Moreover, the integral-based reformulation is no longer restricted to quantifiers expressed in terms of cardinalities; obviously, it can handle arbitrary fuzzy measures (i.e. nondecreasing unary quantifiers). However, Bosc and Lietard abstract from the problem of supporting non-monotonic and multi-place quantifiers.

Yager [173, p. 196] presents a general method which does not adopt Zadeh's proposal of reducing Type IV quantifications to a comparison of fuzzy cardinalities. He explains the method as follows:

> *"One commonly used approach of accomplishing this is called the substitution approach ... In this approach, we try to represent a quantified statement by an equivalent logical sentence involving atoms which are instances of the predicate evaluated at the elements in D."*

(Here D is the domain of discourse, i.e. the base set). Yager attributes this method to Suppes 1957 [155]. But, Rescher [137, p. 203] also mentions the method in his standard reference on many-valued logic, and he believes it was introduced earlier by Wittgenstein. In fact, as noted by Bocheński [17, p. 15, p. 349], the method has been rediscovered over the centuries, and its beginnings can be attributed to Albert of Saxony (1316-1390), who first proposed a reduction of existential and universal propositions to disjunctions and conjunctions, cf. [17, (34.07), p. 234]. From a practical perspective, the substitution approach is not yet a concrete method but rather a framework for methods because a number of choices must be made when applying the method to continuous-valued quantifiers. (for a concrete example of a conforming model, see Th. 7.87 on p. 208 below).

Another line of research independent of Zadeh's framework was pursued by H. Thiele [156, 157, 158]. Starting from a definition of 'general fuzzy quantifiers', i.e. mappings $\widetilde{Q} : \widetilde{\mathcal{P}}(E) \longrightarrow \mathbf{I}$, Thiele develops various concepts useful for classifying such quantifiers. These methods permit him to characterize the general classes of fuzzy universal and existential quantifiers, called T- and S-quantifiers, respectively. Later Thiele also discussed 'median quantifiers' [159]. These are likely not realized in language, however, and they will be of no importance in this monograph. Another specialized class of quantifiers, so-called 'implicational' quantifiers, has been analysed by Hájek and Kohout [77] within their checklist-paradigm for defining fuzzy truth functions.

To sum up, a variety of interpretation mechanisms or more generally, methods for fuzzy quantification, have been proposed which all target at the proper modelling of these quantifiers in conformance with their meaning in ordinary language; a survey of approaches is also given in [4, 34, 108]. However, there is no consensus about the proper choice, i.e. no clear favourite which best answers the semantical expectations. Thus all models exist in parallel and the Σ-count, OWA and FG-count approaches are often met in theoretical treatise and in applications (only the FE-count approach is hardly ever used – with a good cause, as we shall see). In other words, no single model has emerged which clearly outperforms the other approaches. Quite the opposite, it appears that each method comes with its own internal deficiencies, and notes on the counter-intuitive behavior of these approaches are scattered across the literature [2, 4, 35, 42, 54, 132, 134, 184].

Zadeh's first proposal, the Σ-count approach, was criticized by Yager [184], who presents a counter-example which documents the unwanted 'accumulative' behaviour of the method when there are a lot of small (but non-zero) membership grades. Barro et al. [4, p. 93] and Glöckner [54, p. 12-15] have further criticism. And Zadeh himself warns that his approach be incompatible with the formation of antonyms [197, p. 167]. Ralescu [132] criticizes Yager [174] and hence the basic FG-count approach, demonstrating that the method yields an implausible semantics when the quantifier is non-monotonic or unimodal, like the example "a few". Although Yager excludes these cases from consideration [174, p. 639], Ralescu's example is instructive in illustrating the limits of the FG-count model. Arnould and Ralescu [2] also argue against Yager [174] whose model is essentially the FG-count approach. However, the probabilistic model underlying their criticism makes an unusal methodological commitment. Yager's inclusion approach [180], which extends the basic FG-count model towards importance qualification, was criticized by Glöckner [54, p. 21-25], see Sect. A.5 below. He also presents a counter-example against Ralescu's proposal [132], which reveals that the model assigns an implausible interpretation to the existential quantifier [54, p. 25/26]. Dubois and Prade [42] put further evidence against the FG-count and FE-count approaches by criticizing the underlying cardinality measures for fuzzy sets. Glöckner [54,

p. 16-20], Delgado, Sánchez and Vila [35, p. 32] and Barro, Bugarín, Cariñena and Díaz-Hermida [4, p. 95] present counter-examples against the three variants of the OWA approach for restricted proportional quantification described in [179], [181] and [183], respectively. Among other points of criticism, these extensions of the OWA method to importance qualification fail to be monotonic (i.e. inequalities between quantifiers are not preserved), and the approach cannot faithfully represent its target class of operators even for crisp arguments. The quantifier interpolation method of Vila et al. [165] fails for similar reasons, as shown by Barro et al. [4, p. 96] and Delgado et al. [35, p. 32]. Turning to the proposals of Delgado, Sánchez and Vila, their integral-based method [33] shows non-monotonic behaviour and other incorrect results even for crisp arguments, as reported by Barro et al. [4, p. 96] with regard to restricted proportional quantification. The cardinality-based method of Delgado et al. [35], which offers a unifying perspective on both the FG-count and OWA approaches, has lost much of its original appeal since Barro et al. [4, p. 97] subjected it to a critical inspection. Specifically, the resulting interpretations can show a discontinuous dependency on the membership grades of the involved fuzzy sets, and the approach is neither compatible with negation, antonyms, nor dualisation of quantifiers. Finally, Barro et al. [3, p. 15, eq. (3)] discourage the use of their own extension of Yager's inclusion approach; apparently, it only serves to inspect a possible direction of research and demonstrate that it should not be pursued further. In sum, then, we have a scattered picture of rivaling approaches, all of which appear to struggle with their own problems.

We shall now discuss the four main approaches at some more length (further technical details and supplementary information can be found in Appendix A). In order to provide a common point of reference for the various counter-examples that will be presented, a fixed scenario has been chosen. Thus, all examples will originate from a domain of meteorological images which describe 'cloudiness situations'. These images form part of the document base of an experimental retrieval system for multimedia weather documents [62, 63]. This system had to be capable of ranking satellite images according to accumulative criteria like "Almost all of Southern Germany is cloudy". Consequently, we will now be concerned with an *image ranking task*, where the images of interest are first evaluated for their compatibility with a given quantifying criterion, and the computed numerical scores are then used to rearrange the images into a meaningful order, which should parallel the perceived quality of match. The fuzzy sets (or images, in this case) involved in this process are depicted in Fig. 1.2. Here, the base set E comprises the pixel coordinates which form the shape of Germany. There are two images to be considered. X_1 is a fuzzy subset of E which represents Southern Germany. The region of white pixels fully belongs to Southern Germany, and the grey pixels are those which belong to Southern Germany to some degree. X_2 is a fuzzy subset which represents a cloudiness situation. White pixels are fully cloudy, grey pixels are cloudy to some degree. (In the right image, the contours of Germany have been added in order to facilitate interpretation.

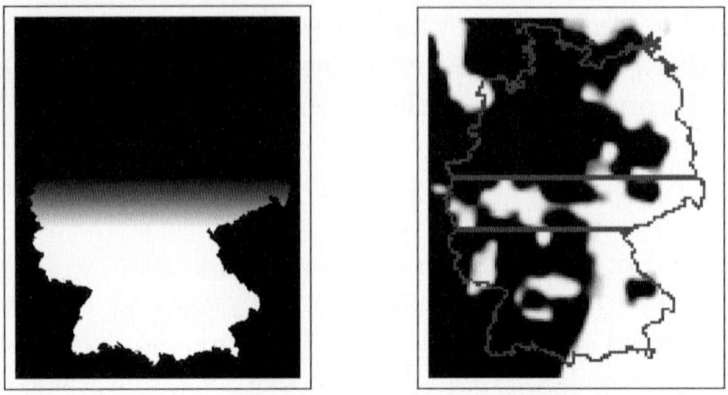

Fig. 1.2. Running example of a base set E, the pixels forming Germany. **Left.** Fuzzy subset X_1: Southern Germany. **Right.** Fuzzy subset X_2: Cloudy

The lower part fully belongs to Southern Germany, the upper part belongs to Germany, but not to Southern Germany, and the middle part contains the intermediate cases.) The images X_1 (southern Germany) and X_2 (representing a cloudiness situation) constitute the fuzzy arguments to which a binary proportional quantifier $\widetilde{Q}$ will be applied, which determines a quantification result $\widetilde{Q}(X_1, X_2) \in \mathbf{I}$. Given a suitable choice of the quantifier, $\widetilde{Q}(X_1, X_2)$ will express the degree to which "Almost all of southern Germany is cloudy" in the given situation. The simple scenario so defined provides a testbed for approaches to fuzzy quantification, because each method $\mathcal{Z}$ will determine its specific choice of quantifier $\widetilde{Q} = \mathcal{Z}^{(2)}_{\mathrm{prp}}(\mu_Q)$. The specific properties of $\mathcal{Z}$ will express in the numerical scores $\tau = \mathcal{Z}^{(2)}_{\mathrm{prp}}(\mu_Q)(X_1, X_2)$ which it assigns to the example images. In many cases, an implausible model will not only show up in the numerical scores, but also in the computed ranking of the images, which differs from the expected order by relevance.

1.9.2 The Sigma-Count Approach

Let us first review the *Σ-count approach* proposed by Zadeh [197], a simple model of fuzzy quantification derived from a scalar cardinality measure for fuzzy sets. The model rests on the computation of *Σ-counts*, also known as the *power* of a fuzzy set, which are defined as the sum of all membership grades, thus

$$\Sigma\text{-Count}(X) = \sum_{e \in E} \mu_X(e)\,.$$

Zadeh's basic idea is that Σ-Count(X) *"may be used as a* numerical summary *of the fuzzy cardinality of a fuzzy set"* [193, p. 167], and can thus be substituted for the ordinary cardinality of crisp sets in order to evaluate quantifying expressions. For unrestricted absolute statements *"There are Q X 's"*, then, we get the associated score $\mu_Q(\Sigma\text{-Count}(X))$, based on the given $\mu_Q : \mathbb{R}^+ \longrightarrow \mathbf{I}$

and $X \in \widetilde{\mathcal{P}}(E)$. The so-called *relative Σ-count* [189] is used for modelling proportional quantification, i.e. the interpretation is based on the relative share of those X_1 that are X_2. For a statement *"Q X_1's are X_2's"* involving a proportional quantifier, we then obtain the score

$$\tau = \mu_Q \left(\frac{\Sigma\text{-Count}(X_1 \cap X_2)}{\Sigma\text{-Count}(X_1)} \right)$$

where $\mu_Q : \mathbf{I} \longrightarrow \mathbf{I}$ and $X_1, X_2 \in \widetilde{\mathcal{P}}(E)$. It is apparent from these formulas that the Σ-count approach depends only on very basic arithmetics, which makes it easy to implement and thus attractive for applications. However, there are serious concerns regarding its ability to catch the actual meaning of NL quantifiers. The first line of criticism is concerned with the known deficiency that the Σ-Count approach *accumulate* membership grades, see Ralescu [132, 134] and Yager [184]. Hence suppose $E = \{\text{John}, \text{Lucas}\}$ is a set of persons and **bald** the fuzzy subset defined by

$$\mu_{\mathbf{bald}}(\text{John}) = 0.5, \quad \mu_{\mathbf{bald}}(\text{Lucas}) = 0.5\,.$$

The Σ-count approach then judges the statement *"Exactly one person is bald"* as being fully true, which is clearly inacceptable. In general terms, the problem is that a number of smaller membership grades can possibly become indiscernible from a single large membership grade. The images depicted in Fig. 1.3 further elucidate this mismatch with the expected semantics of quantifiers in ordinary language. In the center result image there is about ten percent cloud coverage of Southern Germany, and the condition *"About ten percent of Southern Germany are cloudy"* correctly evaluates true in the Σ-count approach. In the right image, however, *all* of Southern Germany is almost not cloudy (i.e. cloudy to the low degree of 10%), which clearly does not mean that ten percent of Southern Germany are cloudy.[6] In fact, Zadeh anticipated such problems, and therefore remarked that *"for some applications, it is necessary to eliminate from the count those elements of F whose grade of membership falls below a specified threshold"* [193, p. 167]. However, these thresholds must be considered an ad-hoc device which does not offer explanatory value.

The treatment of Type II quantifications in the Σ-count approach is also counterintuitive. To see this, consider an NL quantifier of the precise type, e.g. "more than 30 percent". Such quantifiers require a membership function μ_Q which is also two-valued.[7] But if μ_Q is two-valued, then the Σ-count approach produces a 'degenerate' model in terms of a fuzzy quantifier which only

[6] This is apparent because everyone who claims that 10 percent of Southern Germany be cloudy in the right image, must also be able to tell *which* 10% of Southern Germany are cloudy. However, there is no reasonable choice for that because all pixels which belong to Southern Germany are almost not cloudy.

[7] As we shall see below in Sect. A.3 which discusses the issue in some more depth, it is not permissible to replace the two-valued mapping μ_Q with a smooth choice of quantifier in order to avoid these effects, because the intended semantics of the quantifier would be lost.

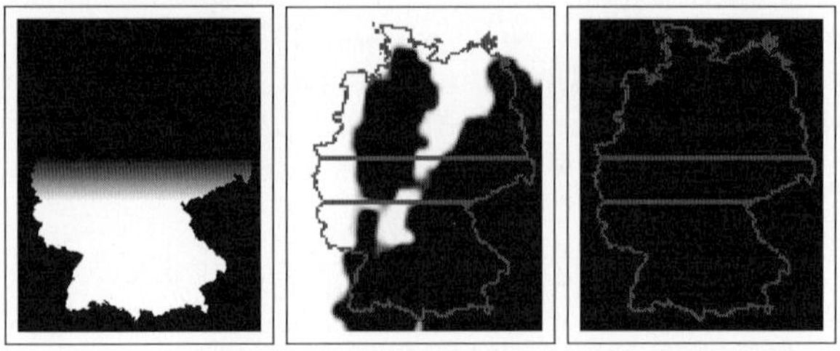

Fig. 1.3. The Σ-count approach accumulates 'small' membership grades in an undesirable way. Results of "About 10 percent of Southern Germany are cloudy". Left: Southern Germany. Result for center image: 1 (plausible). Result for right image: 1 (inadequate)

returns the crisp results 0 or 1 (i.e., fully false or fully true). In particular, the resulting interpretations are discontinuous mappings and very sensitive to slight changes in the membership degrees of the fuzzy sets supplied to the quantifier. The example in Fig. 1.4 illustrates this extreme brittleness of the Σ-count approach. Here, the center image and the right image depict very similar cloudiness situations. However, the results obtained from the Σ-count approach are radically different (0 vs. 1). Obviously, this sensitivity to minor variations in the membership degrees is inacceptable for many real-world applications which depend on sensory data and are thus faced with noise, limited accuracy of sensors, quantization errors etc.[8] To sum up, the Σ-count approach is appealing at first sight due to its simple definition, which makes it easy to implement. However, the known deficiencies of the approach do not encourage its use for applications. Its accumulating effects and brittleness can already be observed in the simple case of one-place quantification. This means that the Σ-Count approach does not possess a reasonable 'core' which might provide a starting point for a plausible extension to a broader class of

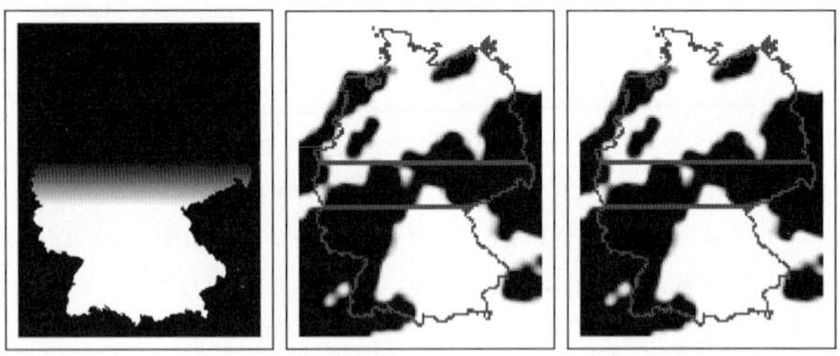

Fig. 1.4. Results of computing "At least 60 percent of Southern Germany are cloudy" with the Σ-count approach. Left: Southern Germany. Result for center image: 0, result for right image: 1

[8] Zadeh's proposal of adding thresholds will produce discontinuous behaviour even for genuine fuzzy quantifiers with smooth membership functions.

quantifiers. For these reasons, the Σ-count approach will not be considered further in this monograph.

1.9.3 The OWA Approach

Next we shall review Yager's OWA approach to fuzzy quantification [179, 180], which is based on ordered weighted averaging (OWA) operators. The basic OWA approach is only concerned with proportional quantifiers, which are further required to be 'regular nondecreasing', i.e. $\mu_Q : \mathbf{I} \longrightarrow \mathbf{I}$ has $\mu_Q(0) = 0$, $\mu_Q(1) = 1$ and $\mu_Q(x) \le \mu_Q(y)$ whenever $x \le y$. Now let $m = |E|$ be the number of elements in the given domain. Then $\mu_Q(j/m) - \mu_Q((j-1)/m)$ is the quantifier's increment if the number of elements $j-1$ is increased to j, or proportionally from $(j-1)/m$ to j/m. By $\mu_{[j]}(X)$ we denote the j-th greatest membership grade of X (including duplicates), i.e. the j-th element in the ordered sequence of membership values. Intuitively, $\mu_{[j]}(X)$ expresses the degree to which X has at least j elements. Using this notation, Yager's rule for unrestricted proportional quantification then assigns the following score to "Q things are X", viz

$$\tau = \sum_{j=1}^{m} (\mu_Q(j/m) - \mu_Q((j-1)/m)) \cdot \mu_{[j]}(X)\,, \tag{1.3}$$

where $\mu_Q : \mathbf{I} \longrightarrow \mathbf{I}$ must be regular nondecreasing and $X \in \widetilde{\mathcal{P}}(E)$. This 'core' OWA approach is defined for unrestricted (one-place) quantification only. Recognizing the need to incorporate importances and thus support two-place quantification as well, Yager [179, p. 190], [180] suggests to transform the fuzzy arguments X_1, X_2 supplied to a two-place quantifier into a single fuzzy set Z to which the original rule for unary quantification is then applied. A restricted quantifying statement "Q X_1's are X_2's" is then assigned the score of the unrestricted statement "There are Q Z's", which can already be handled by (1.3). The fuzzy set Z is defined by

$$\mu_Z(e) = \max(\mu_{X_1}(e), 1 - \mathrm{orness}(\mu_Q)) \cdot \mu_{X_2}(e)^{\max(\mu_{X_1}(e), \mathrm{orness}(\mu_Q))}$$

see equation (A.2) in the appendix for the definition of $\mathrm{orness}(\mu_Q)$, the so-called 'degree of orness' of μ_Q. This extension to restricted quantification makes it possible to treat cases like *"Almost all young are poor"*, which require two arguments **young**, **poor** $\in \widetilde{\mathcal{P}}(E)$ to be taken into consideration. There is negative evidence regarding its semantical plausibility, however, because the proposed formula assigns incorrect models to any 'genuine' proportional quantifiers (see pp. 402+ below). In fact, the only two cases which are handled successfully are the logical quantifiers "all" and "exists". Here we shall confine ourselves to a suggestive example which demonstrates that the theoretical defects are also potentially detrimental to applications. Hence consider the situation depicted in Fig. 1.5. The two results in the figure were obtained from

OWA operators with importance qualification. In this case, the task was to assess the truth of the proposition *"At least 60 percent of Southern Germany are cloudy"*, which is obviously the case in the center image, but not in the right image. The OWA approach, though, ranks the right image higher than the center image. As witnessed by Fig. 1.5, the results of the OWA approach show an undesirable dependency on cloudiness grades in regions III and IV, which do not belong to Southern Germany at all. This kind of behaviour runs counter to the intuitive meaning of the NL quantifier "at least 60 percent".

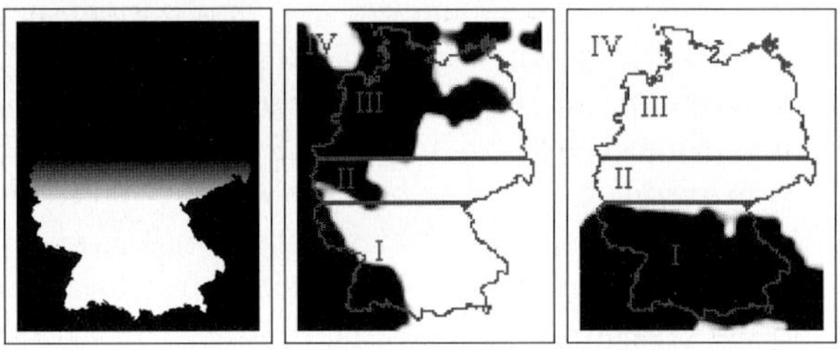

Fig. 1.5. Application of the OWA-approach. Results of "At least 60 percent of Southern Germany are cloudy". Left: Southern Germany. Result for center image: OWA: 0.1, desirable outcome: 1. Result for right image: OWA: 0.6, desirable: 0

Additional difficulties arise when the core OWA approach is extended beyond regular nondecreasing quantifiers. To see this, consider the quantifier "less than 60 percent", which does not belong to the regular nondecreasing type. In order to make the OWA approach applicable to such quantifiers, we must resort to equivalent NL paraphrases. For example, we can use "at least 60 percent" and evaluate *"Less than 60 percent of the* X_1*'s are* X_2*'s"* by negating the result of *"At least 60 percent of the* X_1*'s are* X_2*'s"*, which means a reduction to the regular nondecreasing type. Alternatively, we can use *"More than 40 percent of the* X_1*'s are* X_2*'s"* and evaluate *"Less than 60 percent of the* X_1*'s are* X_2*'s"* by computing *"More than 40 percent of the* X_1*'s are not* X_2*'s"*. Intuitively, we should expect that the final results of these computations will coincide because they stem from equivalent paraphrases. Yager's proposal for binary proportional quantification treats both cases differently, though. A counter-example is shown in Table 1.5. Similar remarks apply to the modelling of unimodal and other more complex quantifiers, see pp. 405+ for a thorough discussion. To sum up, the basic OWA approach for regular nondecreasing one-place quantifiers appears to be rather well-behaved, at least no counter-examples are known. Thus the 'core' approach (which is too limited for important applications of quantifiers) will be considered a potential starting point for future extensions. These must target at covering non-monotonic and two-place quantifiers (importance qualification) as well. As mentioned above, earlier proposals to affect this extension do not achieve the interpretation quality of the basic method. As was shown

Table 1.5. "Less than 60 percent of Southern Germany are cloudy". Results of possible renderings when applying the OWA approach to the images in Fig. 1.5

Quantifier	**center image**	**right image**
not at least 60 percent	0.9	0.4
more than 40 percent not	0.4	0
desired result	0	1

by Barro et al. [3] and Delgado et al. [35, p. 32], this negative evidence also generalizes to Yager's alternative methods for extending the OWA approach to two-place quantification.

1.9.4 The FG-Count Approach

Apart from the Σ-count approach, Zadeh [197] proposes another model of fuzzy quantification; see also Yager [174] who uses the same basic formulas in the context of rule-based expert systems. The *FG-count approach* tries to improve upon the use of Σ-counts by replacing the original scalar measure with a fuzzy measure of the cardinality of fuzzy sets. Hence let FG-count(X) denote the fuzzy set of cardinals defined by

$$\mu_{\text{FG-count}(X)}(j) = \mu_{[j]}(X) \tag{1.4}$$

for all $j \in \mathbb{N}$. (Here and in the following it is convenient to stipulate that $\mu_{[0]}(X) = 1$ and $\mu_{[j]}(X) = 0$ for $j > |E|$). Intuitively, FG-count(X) captures the degree to which X has at least j elements (for all possible j). The FG-count incorporates a richer base of information compared to the Σ-count, which now becomes a summary of the FG-count defined by

$$\Sigma\text{-Count}(X) = \sum_{j=0}^{\infty} \mu_{\text{FG-count}(X)}(j) - 1\,,$$

see Zadeh [197, p. 157].[9] The new measure of fuzzy cardinality can be used to derive a corresponding model of fuzzy quantification. In the 'basic' FG-count approach, the score of an absolute quantifying statement "There are Q X's" becomes

[9] It should be apparent from (1.3) and (1.4) that the OWA-approach, too, can be expressed in terms of the fuzzy cardinality measure FG-count(X). The OWA score then becomes

$$\tau = \sum_{j=1}^{m} (\mu_Q(j/m) - \mu_Q((j-1)/m)) \cdot \mu_{\text{FG-count}(X)}(j)\,.$$

This demonstrates that the OWA approach, although not originally introduced in this way, is in fact a cardinality-based model of fuzzy quantification.

$$\tau = \max\{\min(\mu_Q(j), \mu_{\text{FG-count}(X)}(j)) : j = 0, \ldots, |E|\}$$

where $\mu_Q : \mathbb{R}^+ \longrightarrow \mathbf{I}$ must be nondecreasing and $X \in \widetilde{\mathcal{P}}(E)$. The basic FG-count approach so defined only covers nondecreasing quantifiers, and it does not incorporate importances. A possible extension towards the 'hard case' of two-place, weighted quantification has been described by Yager [180, p.72].[10] A proportional quantifying statement "Q X_1's are X_2's" is then assigned the score,

$$\tau = \max\left\{\min\left(\mu_Q\left(\frac{\sum_{v\in S}\mu_{X_1}(v)}{\sum_{v\in E}\mu_{X_1}(v)}\right), H_S\right) : S \in \mathcal{P}(E)\right\}$$
$$H_S = \min\{\max(1 - \mu_{X_1}(v), \mu_{X_2}(v)) : v \in S\},$$

where $\mu_Q : \mathbf{I} \longrightarrow \mathbf{I}$ is again assumed to be nondecreasing and $X_1, X_2 \in \widetilde{\mathcal{P}}(E)$. In order to get some impression of the resulting interpretations, we consider the situation depicted in Fig. 1.6. There are no clouds at all in (the support of) Southern Germany#1, so we should expect that "At least five percent of Southern Germany are cloudy" be completely false. The actual result, 0.55, thus reveals a defect of the proposed model. The example further demonstrates that the resulting operators can be discontinuous: If we replace Southern Germany#1 with the slightly different Southern Germany#2, then the result jumps to 0.95, although there are still no clouds in the region of interest. Hence some interpretations obtained from the two-place formula are very sensitive to noise, which discourages the use of the model in practical applications. Implausible results will also be obtained when quantifications like *"Less than 30 percent of Southern Germany are cloudy"*, which cannot be interpreted directly due to their monotonicity pattern, are reduced to NL paraphrases involving nondecreasing quantifiers: (a) *"It is not the case that at least 30 percent of Southern Germany are cloudy"*; and (b) *"More than 70 percent of Southern Germany are not cloudy"*. These NL paraphrases are equivalent and should therefore be interchangeable. The proposed extension of the FG-count approach to two-place quantification, though, produces incoherent results in both cases. To sum up, the 'core' FG-count approach appears to be well-behaved and does not have any overt flaws. But to be useful in practice, it must be extended beyond nondecreasing quantifiers, and support for restricted quantification must be added. As witnessed by the above examples, existing attempts at such an extension have not been successful.

1.9.5 The FE-Count Approach

Finally we review the method proposed by Ralescu [132]. This method is rather similar to Yager's proposal [174], i.e. to the basic FG-count approach.

[10] Zadeh's original proposal for two-place proportional quantification in the FG-count setting had to be excluded from consideration. The relationship between the above method for two-place quantification and the basic FG-count approach is explained in Sect. A.5 in the appendix.

(a) Southern Germany#1 (b) Southern Germany#2 (c) cloudy

Fig. 1.6. Application of the FG-count approach. Results of *"At least 5 percent of Southern Germany are cloudy"*. Result for Southern Germany#1: 0.55, Result for Southern Germany#2: 0.95. The desired result: 0

However, Ralescu attempts to overcome the restriction to nondecreasing quantifiers by replacing the $\mu_{[j]}(X)$ terms with the expression $\mu_{[j]}(X) \wedge \neg \mu_{[j+1]}(X)$ which not only considers the degree to which X contains at least j elements, i.e. $\mu_{[j]}(X)$, but also the degree to which X has less than $j+1$ elements, i.e. $1 - \mu_{[j+1]}(X)$. Thus a sentence of the form *"There are Q X's"* which involves unrestricted absolute quantification will now be assigned the interpretation

$$\tau = \max\{\min\{\mu_Q(j), \mu_{[j]}(X), 1 - \mu_{[j+1]}(X)\} : j = 0, \ldots, |E|\},$$

where $\mu_Q : \mathbb{R}^+ \longrightarrow \mathbf{I}$ and $X \in \widetilde{\mathcal{P}}(E)$.
The method will be called 'FE-count approach' here because it makes implicit use of the FE-count, a measure of fuzzy cardinality introduced by Zadeh [197]. The FE-count of a fuzzy set X is the fuzzy set FE-count$(X) \in \widetilde{\mathcal{P}}(\mathbb{N})$ with membership function $\mu_{\text{FE-count}(X)}(j) = \min(\mu_{[j]}(X), 1 - \mu_{[j+1]}(X))$. This means that Ralescu's model can also be written as

$$\tau = \max\{\min(\mu_Q(j), \mu_{\text{FE-count}(X)}(j)) : j = 0, \ldots, |E|\}.$$

This reformulation demonstrates that Ralescu's method is a fuzzy-cardinality based approach to fuzzy quantification which rests on the FE-count as the cardinality measure.

A generalization of the basic FE-count method to the interpretation of two-place and proportional quantifiers has not been described in the literature. Let us now consider the example shown in Fig. 1.7. The crisp image region shown in left image is clearly nonempty, because it contains several 'white' pixels with full degree of membership. The fuzzy image region displayed in the right image is certainly nonempty to the same degree of 1.0, because it embraces the crisp nonempty image region shown on the left. The FE-count approach, however, only rates the left image as nonempty to the desired degree of 1.0. Although the right image displays a larger region, the result drops to 0.5, which is clearly inacceptable. This means that the FE-count model shows unexpected results even in the simplest case of one-place quantification based

tially a matter of discovering the proper cardinality measures and comparison operators, see Barro et al. [3, p. 2]. However, as pointed out in the comparison of logical and linguistic quantifiers on p. 10, not all linguistic quantifiers can be reduced to cardinality assessments, because not all of these quantifiers treat every element in the domain alike and are thus 'quantitative'. In order to avoid this a priori limitation, a comprehensive solution to fuzzy quantification must be formulated *independently of any cardinality measure*.

It is likely due to the internal problems of existing approaches that applications of fuzzy quantifiers, like fuzzy database querying, tend to support more than one interpretation method. Kacprzyk and Zadrożny's fuzzy querying interface to Microsoft ACCESS [84], for example, supports both the Σ-count and OWA models. And Bosc and Pivert [23], when discussing an extension of the classical SQL language with fuzzy retrieval features, suggest the use of even three methods, i.e. Σ-counts, OWA operators/Choquet integral, and finally FG-counts/Sugeno integral. Thus, the burden of selecting the model which works best in the given situation is shifted to the user of these systems. However, it must be doubted that the users of the database can decide on the proper interpretation when even the system designers feel unable to do that.

When viewing the development of fuzzy quantification in retrospect, one recognizes that research in this area has lost much of its original impetus: In the beginning, i.e. starting with Zadeh's pioneering 1983 publication, there was rapid progress; it did not take longer than 1988 that the main approaches had been proposed. Following that came the current period, which is marked by advances in applications like reasoning with fuzzy quantifiers, multi-criteria decisionmaking, fuzzy querying of databases or linguistic data summarization. However, there was a slow-down on the methodical side and no substantial progress in the modelling of fuzzy quantifiers. The seeming stability of this area, and the spread of fuzzy quantifiers into experimental applications, does not imply that the research efforts have resulted in a theoretically satisfactory analysis. The popular methods still show the same problematic behaviour and the few recent proposals for interpretation methods [33, 34, 35, 165] are also linguistically inconsistent [3, 35]. This general slow-down of progress in the theoretical analysis of fuzzy quantification hints at a wrong way of thinking about fuzzy quantification.

A novel framework which targets at a reliable semantics for fuzzy quantifiers will need to reconcile fuzzy set theory and linguistics. Zadeh himself concedes that his approach be 'different' from (i.e. incompatible with) the linguistic model [197, p. 149]. However, it appears that few people have reflected the consequences of this departure from the competent scientific discipline. In fact, it is not even possible to view the two-valued quantifiers usually considered by linguists [7, 9, 92] as a special case of Zadeh's fuzzy linguistic quantifiers. In order to cover a substantial class of linguistic quantifiers, the novel framework must give up the use of first-order representations and the reduction to cardinality comparisons. The known approaches to fuzzy quantification are motivated by concepts of fuzzy set theory; they have not been designed and systematically evaluated from a linguistic perspective. What the numerous counterexamples really demonstrate is that we may not accept a bona fide assumption of linguistic plausibility. Quite the opposite: Linguistic adequacy must be central to the very design of the interpretation methods, and it must be explicitly guaranteed by appropriate means. In the monograph, then, we will be concerned with the development of a new framework to fuzzy quantification which is much closer to linguistics. Special attention will be paid to the formalization of criteria which capture the intuitive expectations on meaningful interpretations of quantifiers. Based on this analysis of the essential plausibility criteria, we can then formulate a theory of natural language quantification which rests on an axiomatic foundation.

We then obtain the apparent definitions,

$$\forall(Y) = 1 \Leftrightarrow Y = E \qquad \text{and} \qquad \exists(Y) = 1 \Leftrightarrow Y \neq \varnothing \tag{2.1}$$

for all $Y \in \mathcal{P}(E)$. It is instructive to consider the relationship between unary $\forall$ and the two-place quantifier "all", in order to clarify the distinction between the restricted and unrestricted use of NL quantifiers that was already mentioned in the introduction:

- In the case that the NL quantifier, which is implicitly two-place, is employed in a quantifying expression which involves two overt arguments, this corresponds to the restricted use of the quantifier. *"All bachelors are men"*, for example, finds an interpretation in terms of a two-place quantification $\mathbf{all}(\mathbf{bachelors}, \mathbf{men})$. The restricted use of the quantifier corresponds to the natural way in which quantifiers are applied in NL.
- In some cases, it is also possible to model quantifying NL statements by one-place quantification, although the involved NL quantifier regularly takes two arguments. This is the type of quantification is called unrestricted because the full domain E is substituted for the unspecified first argument, thus filling in the restriction of the quantifier. The logical quantifier $\forall : \mathcal{P}(E) \longrightarrow \mathbf{2}$ for example, which models the unrestricted use of the NL quantifier "all", can be expressed in terms of two-place **all** by

  $$\forall(Y) = \mathbf{all}(E, Y) \qquad \text{for all } Y \in \mathcal{P}(E).$$

  The example illustrates how the full domain E provides the missing restriction (first argument) required by the two-place base quantifier. In other words, the unrestricted quantification in $\forall(Y)$ means that *"All elements of the domain are Y"* or simply, *"All things are Y"*.

Clearly the unrestricted use, which is ubiquitous in predicate logic, is of minor importance to the description of natural language. We will see later in Sect. 6.7, p. 171 that two-place, restricted quantification can be reduced to the unrestricted use of quantifiers in certain cases. However, such reduction is only possible for certain types of quantifiers, and it does not generalize to the fuzzy case.

Apart from the above examples of absolute quantifiers which depend on absolute counts, and quantifiers of exception like "all except k", there are many other types of NL quantifiers that deserve interest.[3] The most prominent class is certainly that of proportional quantifiers, which depend on the relative share (or ratio) of Y_1's that are Y_2's. Assuming for convenience that the base set E be finite, we can introduce some generic examples from which a broad range of NL quantifiers can be derived.

[3] For a more extensive discussion of absolute quantifiers and quantifiers of exception also covering implementation issues, see Sect. 11.8.

Definition 2.5.
Let $E \neq \varnothing$ be some base set. We define

$$
\begin{aligned}
&[\mathbf{rate} \geq r](Y_1, Y_2) = 1 \Leftrightarrow |Y_1 \cap Y_2| \geq r\,|Y_1| \\
&[\mathbf{rate} > r](Y_1, Y_2) = 1 \Leftrightarrow |Y_1 \cap Y_2| > r\,|Y_1| \\
&[\mathbf{rate} \leq r](Y_1, Y_2) = 1 \Leftrightarrow |Y_1 \cap Y_2| \leq r\,|Y_1| \\
&[\mathbf{rate} < r](Y_1, Y_2) = 1 \Leftrightarrow |Y_1 \cap Y_2| < r\,|Y_1| \\
&[\mathbf{rate} = r](Y_1, Y_2) = 1 \Leftrightarrow |Y_1 \cap Y_2| = r\,|Y_1| \\
&[r_1 \leq \mathbf{rate} \leq r_2] = 1 \Leftrightarrow r_1|Y_1| \leq |Y_1 \cap Y_2| \leq r_2|Y_1|
\end{aligned}
$$

for $r \in \mathbf{I}$, $Y_1, Y_2 \in \mathcal{P}(E)$.

In terms of these quantifiers, a statement like *"At least 70 percent of the Y_1's are Y_2's"* can now be modelled by

$$\textbf{at least 70 percent}(Y_1, Y_2) = [\mathbf{rate} \geq 0.7](Y_1, Y_2)\,,$$

and *"More than 60 percent of the Y_1's are Y_2's"* can be modelled by

$$\textbf{more than 60 percent}(Y_1, Y_2) = [\mathbf{rate} > 0.6](Y_1, Y_2)\,.$$

In particular, the quantifier "most" becomes

$$\mathbf{most} = [\mathbf{rate} > 0.5]\,. \tag{2.2}$$

This definition of "most" covers its technical sense, as in *"Most Americans voted for Bush"*, where one vote can result in an all-or-nothing decision. In other contexts, a gradual modelling would be more appropriate. By instantiating the generic quantifiers $[\mathbf{rate} \leq r]$ and $[\mathbf{rate} < r]$ in the apparent way, we can also cover natural language quantifiers of the type "at most r percent" and "less than r percent". Similarly, a statement like *"Exactly 10 percent of the Y_1's are Y_2's"* can now be interpreted as

$$\textbf{exactly 10 percent}(Y_1, Y_2) = [\mathbf{rate} = 0.1](Y_1, Y_2)\,.$$

Finally, the parametric quantifier $[r_1 \leq \mathbf{rate} \leq r_2]$ is useful for modelling a statement like *"Between 10 and 30 percent of the Y_1's are Y_2's"*, which then becomes

$$\textbf{between 10 and 30 percent}(Y_1, Y_2) = [0.1 \leq \mathbf{rate} \leq 0.3](Y_1, Y_2)\,.$$

A more detailed analysis of proportional quantifiers including approximate examples is given below in Sect. 11.9.

Next we shall discuss a couple of further quantifiers taken from different linguistic classes: quantifiers of comparison, definite quantifiers, proper names (which can naturally be modelled through quantification), composite quantifiers, and others. As to quantifiers of comparison, consider a statement like

2.3 The Dilemma of Fuzzy Quantifiers

The proposed notion of fuzzy quantifiers establishes a class of operators which incorporate the fuzziness of language. We have thus solved one part of the modelling problem by introducing the required operational quantifiers. The modelling problem has another facet, though. We still need to elucidate the relationship between the proposed modelling constructs (fuzzy quantifiers) and the NL quantifiers of interest. For *crisp* subsets of E, we have clear intuitions about the proper definition of linguistic quantifiers. For example, the definition of $\forall : \mathcal{P}(E) \longrightarrow \mathbf{2}$ in (2.1) is certainly uncontroversial. When turning to the fuzzy case, however, a plenty of possible options enters the scene (only to mention the wealth of conceivable fuzzy conjunctions that exist even in the simple propositional case). Thus, how can we: (a), determine a well-motivated choice of fuzzy quantifier which properly models a given NL quantifier and (b), defend this choice against all possible alternatives, which might appear equally plausible at first sight?

Hence let us return to the simple universal quantifier $\forall$, and make an attempt to extrapolate its meaning to the fuzzy case. Equation (2.3) above introduces a fuzzy analogue of this quantifier. However, what about the following alternative stipulations, which only constitute a small sample of the possible options:

$$\widetilde{\forall}'(X) = \frac{\sum_{e \in E} \mu_X(e)}{|E|}$$

$$\widetilde{\forall}''(X) = \begin{cases} 1 & : \quad X = E \\ 0 & : \quad \text{else} \end{cases}$$

$$\widetilde{\forall}'''(X) = \inf\left\{ \prod_{i=1}^{m} \mu_X(a_i) : A = \{a_1, \ldots, a_m\} \in \mathcal{P}(E) \text{ finite, } a_i \neq a_j \text{ if } i \neq j \right\}.$$

Some of these quantifiers apparently do not qualify as plausible models of $\forall$ in the fuzzy case, while others are not so easy to judge. The first alternative $\widetilde{\forall}'$ for example can be rejected right away, because it differs from two-valued $\forall$ on crisp arguments. The second example also violates some intuitive postulates. For example, producing crisp interpretations only does not seem very natural for a fuzzy quantifier. Moreover, the fuzzy quantifier $\widetilde{\forall}''$ shows an undesirable discontinuity in the membership grades $\mu_X(e)$ which makes it too brittle for applications. In addition, $\widetilde{\forall}''$ does not conform to Thiele's analysis of fuzzy universal quantifiers, see [156, 157, 158] and Sect. 4.16 below. As to the third example, $\widetilde{\forall}'''$ appears to be acceptable as a fuzzy model of universal quantification. In fact, both the original definition of $\widetilde{\forall}$ and the alternative $\widetilde{\forall}'''$ comply with Thiele's analysis, and must thus be considered admissible choices of fuzzy universal quantifiers.

The examples demonstrate that for elementary quantifiers, some basic intuitions persist when admitting fuzzy arguments. However, these are not necessarily strong enough to establish a unique preferred choice. The case-by-case verification that a proposed fuzzy quantifier $\widetilde{Q}$ make a plausible model of the given NL quantifier soon becomes too tedious when considering a larger number of quantifiers. Moreover, the manual procedure makes it very difficult for the stipulated correspondences to achieve the desired systematicity and consistency. To ensure coherence, such direct correspondence assertions should be avoided altogether. We need a general mechanism for establishing the desired correspondences especially to treat unclear cases where we cannot justify a particular interpretation intellectually.

The scarcity of intuitions concerning the proper modelling of NL quantifiers in the fuzzy case becomes evident once we move from the simplest example $\forall$ to other common quantifiers like "at least 10 percent". Given a finite base set E, we can readily define a corresponding two-valued quantifier $\textbf{at least 10 percent} : \mathcal{P}(E)^2 \longrightarrow \mathbf{2}$ by $\textbf{at least 10 percent} = [\textbf{rate} \geq 0.1]$, i.e.

$$\textbf{at least 10 percent}(Y_1, Y_2) = \begin{cases} 1 & : \quad |Y_1 \cap Y_2| \geq |Y_1|/10 \\ 0 & : \quad \text{else} \end{cases} \tag{2.7}$$

for all $Y_1, Y_2 \in \mathcal{P}(E)$. The equation reveals the dependency of the quantifier on the cardinality $|\bullet|$ of crisp sets. Now let us try to define a corresponding fuzzy quantifier $\widetilde{\textbf{at least 10 percent}} : \widetilde{\mathcal{P}}(E)^2 \longrightarrow \mathbf{I}$. In this case the arguments X_1, X_2 in

$$\widetilde{\textbf{at least 10 percent}}(X_1, X_2)$$

can be arbitrary fuzzy subsets $X_1, X_2 \in \widetilde{\mathcal{P}}(E)$. The familiar cardinality measure for ordinary sets, which made possible the above definition (2.7) of the two-valued quantifier, is not applicable to these fuzzy arguments, i.e. it cannot be used to define the fuzzy quantifier. And there is no reliable and uncontroversial notion of the cardinality of fuzzy sets, which we can substitute for $|\bullet|$ in the fuzzy case. To demonstrate this, let us consider the Σ-count as a replacement for $|\bullet|$. We then obtain the following model of "at least 10 percent",

$$\widetilde{\textbf{at least 10 percent}}(X_1, X_2) = \begin{cases} 1 : \ \Sigma\text{-Count}(Y_1 \cap Y_2) \geq \Sigma\text{-Count}(Y_1)/10 \\ 0 : \ \text{else} \end{cases}$$

This proposal is representative of approaches to fuzzy quantification which simply replace the crisp cardinality in (2.7) with some generalized cardinality for fuzzy sets. This parallelism with the formula for crisp sets might look suggestive, but it does not guarantee plausible interpretations. The proposed interpretation of "at least 10 percent" for example, can show abrupt changes in response to slight variations of the membership grades. Moreover the general criticism of the Σ-count approach exemplified by Fig. 1.3, also applies to the quantifier "at least 10 percent".

These difficulties in defining "at least 10 percent" for fuzzy arguments hint at an intriguing problem because the class of fuzzy quantifiers is certainly rich

Based on this concept, we can now express the following *quantification framework assumption* which is implied by the proposed framework:

> **Quantification framework assumption (QFA):**
> *If two base quantifiers of interest (i.e. NL quantifiers to be defined directly) have distinct interpretations $\widetilde{Q} \neq \widetilde{Q}'$ as fuzzy quantifiers, then they are already distinct on crisp arguments, i.e. $\mathcal{U}(\widetilde{Q}) \neq \mathcal{U}(\widetilde{Q}')$.*

The QFA condition ensures the applicability of the QFM framework because we can then represent $\widetilde{Q}, \widetilde{Q}'$ by $Q = \mathcal{U}(\widetilde{Q})$ and $Q' = \mathcal{U}(\widetilde{Q}')$, without compromising the existence of a QFM $\mathcal{F}$ which takes Q to $\widetilde{Q} = \mathcal{F}(Q)$ and Q' to $\widetilde{Q}' = \mathcal{F}(Q')$. If the QFA is violated by $\widetilde{Q}$ and $\widetilde{Q}'$, however, then it is impossible for any QFM to separate the quantifiers, because $\mathcal{U}(\widetilde{Q}) = \mathcal{U}(\widetilde{Q}')$ entails that the same interpretation $\mathcal{F}(\mathcal{U}(\widetilde{Q})) = \mathcal{F}(\mathcal{U}(\widetilde{Q}'))$ be assigned to both quantifiers.

The QFA expresses the linguistic postulate (universal principle) that there is no pair of base quantifiers in any natural language which coincide on two-valued arguments, but differ for fuzzy arguments. It should be pointed out that the linguistic theory of quantification, TGQ, depends on the same assumption, because it restricts attention to two-valued arguments only. In doing this, it is assumed that crisp arguments will reveal sufficient structure to discuss the phenomenon of interest. Starting from its basic notion of a generalized quantifier, TGQ has investigated various aspects of NL quantification like semantic universals, algebraic properties, expressive power and logical definability, as well as purely linguistic matters like 'there-insertion' or 'negative polarity' items.[7] The success of TGQ in describing these phenomena demonstrates that the relevant properties of NL quantifiers are indeed visible on the level of crisp arguments (i.e. of two-valued quantifiers and semi-fuzzy quantifiers).

At first sight, the fundamental assumption underlying the quantification framework appears uncritical because it makes a very elementary requirement. However, the condition had to be restricted to 'base quantifiers' because it is indeed possible to construct a few examples which must be treated with different methods. Hence let us discuss the notion of base or 'simplex' quantifier $\widetilde{Q}$, which must be directly expressed in terms of semi-fuzzy quantifiers, i.e. as $\widetilde{Q} = \mathcal{F}(Q)$.

If we treat all conceivable linguistic quantifiers as base quantifiers, then it is clear in advance that the QFA will be violated. This is because a continuous-valued logic cannot validate all axioms of Boolean algebra on formal grounds. Natural languages, however, permit us to express arbitrary Boolean combinations of a quantifier's arguments. Starting from a pair of Boolean expressions which are equivalent in the two-valued, but not in the fuzzy case, we can then construct a pair of quantifiers whose argument structure parallels these Boolean expressions. These quantifiers will coincide for crisp arguments, but

[7] see e.g. [7] (universals), [9] (algebraic properties), [91] (expressiveness), [10] (definability) and [79] (linguistic issues).

not necessarily for fuzzy arguments, thus violating the QFA criterion. A simple example is $\widetilde{Q}(X) = \widetilde{\forall}(X \cup \neg X)$, or *"Everything is X or not X"*, versus $\widetilde{Q}'(X) = 1$ *const*, which paraphrases a constantly true proposition like *"Everything belongs to the base set"*, which by cylindrical extension (i.e. adding a vacuous variable) becomes a unary quantifier. The standard choice of fuzzy universal quantifier (infimum) and the standard set of connectives then results in $\widetilde{Q} \neq \widetilde{Q}'$, although $\mathcal{U}(\widetilde{Q}) = \mathcal{U}(\widetilde{Q}') = 1$ coincide.
The example demonstrates that $\widetilde{Q}(X) = \widetilde{\forall}(X \cup \neg X)$ is not a base quantifier, but rather a non-simplex quantifier whose interpretation must be *constructed* from the interpretations of available base quantifiers (in this case, from $\widetilde{\forall} = \mathcal{F}(\forall)$).

Another instructive example of non-simplex quantifiers are the many-valued quantifiers of Rescher [137, p. 199], see also p. 24 in the introduction. By identifying all intermediate truth values $z \in (0,1)$ with the third truth value of Rescher's original definition, we extend Rescher's quantifiers $\exists^I$, $\forall^T$, M^T and M^I to unary fuzzy quantifiers defined for all fuzzy arguments. Rescher's quantifiers can also be realized linguistically, as shown by the following examples from ordinary English:

1. Many borderline-cases of bald men use hair restorers.
2. All clear cases of insanity are kept under strict observation.
3. All stationary patients are clear cases of dementia.

In order to demonstrate that Rescher's quantifiers are non-simplex, let us consider the quantifier $\widetilde{\exists}^I : \widetilde{\mathcal{P}}(E) \longrightarrow \mathbf{I}$ defined by $\widetilde{\exists}^I(X) = 1$ if $\mu_X(e) \in (0,1)$ for some $e \in E$, and $\widetilde{\exists}^I(X) = 0$ otherwise. The Type IV quantification $\widetilde{\exists}^I(P)$ then expresses the linguistic pattern *"There are borderline cases of P's"*. In order to check the validity of the QFA criterion in this case, let us suppose that $\widetilde{Q} = \widetilde{\exists}^I$, and the constantly false quantifier $\widetilde{Q}'(X) = 0$ which corresponds to a false proposition like *"The base set is empty"*, both be base quantifiers. We then observe that $\mathcal{U}(\widetilde{\exists}^I) = \mathcal{U}(\widetilde{Q}') = 0$ is the constantly false quantifier because for crisp arguments, one generally finds no borderline cases. Hence, there are two distinct quantifiers $\widetilde{Q} = \widetilde{\exists}^I$ and $\widetilde{Q}' = 0$ expressible in NL which coincide for crisp arguments. This demonstrates that $\widetilde{\exists}^I$ is not a base quantifier, because its interpretation must be different from the constantly false quantifier.

Linguists have postulated a universal principle of *variety* which states that 'trivial' or non-informative quantifiers like $\mathcal{U}(\widetilde{\exists}^I) = 0$ never occur as basic quantifiers of a natural language [9, p. 452]. This provides further evidence that $\widetilde{\exists}^I$ must be considered a constructed, rather than simplex quantifier.

Obviously, this does not mean that $\widetilde{\exists}^I$ cannot be handled by the framework at all, it just means that it cannot be expressed directly in the form $\widetilde{\exists}^I = \mathcal{F}(Q)$. In order to reduce $\widetilde{\exists}^I$ to the interpretation of a base quantifier, we simply decompose it into *"There are R's"*, where the proposition R abbreviates "borderline case of P". Thus, "borderline case of" is now viewed as part of the argument and no longer attached to the quantifier. Let $h : \mathbf{I} \longrightarrow \mathbf{I}$ be the mapping defined by $h(0) = h(1) = 0$ and $h(z) = 1$ otherwise. Then R can be computed from P by applying the 'linguistic hedge' h, i.e. $\mu_R(e) = h(\mu_P(e))$ for all $e \in E$. We can then express $\widetilde{\exists}^I$ as $\widetilde{\exists}^I(P) = \mathcal{F}(\exists)(R)$, where $\exists$ is the usual existential quantifier.

Generally speaking, it appears that quantifying constructions which explicitly refer to the fuzziness, vagueness or determinacy of their arguments violate the QFA and must thus be viewed as constructed quantifiers. In particular, linguistic hedges can only be applied to the arguments of a fuzzy quantifier but not semi-fuzzy quantifiers because modifications of membership grades are meaningless for crisp arguments. Thus quantifiers constructed from hedging are *not* base quantifiers. However, hedging quantifiers like those of Rescher are pretty exotic cases and all common quantifiers like "most", "many" etc. can be treated as simplex.

2.7 A Note on Nullary Quantifiers

Before closing the chapter, we will discuss the status of nullary quantifiers (this section) and the treatment of undefined quantifications in the proposed framework (next section). The chosen definitions of semi-fuzzy and fuzzy quantifiers admit the case of quantifiers with arity $n = 0$. These nullary quantifiers are artificial constructs (or boundary cases of quantifiers) not observed in natural language. In fact, every NL quantifier must at least support one argument slot, which accepts the interpretation of the verb phrase. For a minimal example, consider *"Joan sleeps"*, where "Joan" is mapped to a unary quantifier $\mathbf{joan} : \mathcal{P}(E) \longrightarrow \mathbf{2}$ and "sleep" denotes a set $\mathbf{sleep} \in \mathcal{P}(E)$. The example thus translates into $\mathbf{joan}(\mathbf{sleeps})$, which determines the resulting truth value from the known identity of "Joan" and the known set of sleeping people. This minimal example illustrates that the slot for the interpretation of the verb phrase is mandatory in natural language, i.e. we always have arities $n \geq 1$. Nevertheless, it can be pretty useful for theoretical investigations to have nullary quantifiers available. For example, they naturally arise from repeated argument insertions, to be discussed in Sect. 4.10. Moreover, admitting nullary quantifiers will permit a succinct presentation of the axiom system for plausible models in Def. 3.28, which will refer to nullary quantifiers in its first and most fundamental axiom (Z-1).

In order to elucidate the concept of nullary quantifiers and motivate the later notation for these quantifiers, we must recall some very basic mathematics. If A is some set and $n \in \mathbb{N}$, then A^n is commonly taken to denote set of mappings from $n = \{0, \ldots, n-1\}$ to A, i.e. $A^n = A^{\{0, \ldots, n-1\}}$. Consequently, $A^0 = A^\varnothing = \{\varnothing\}$. A nullary semi-fuzzy quantifier $Q : \mathcal{P}(E)^0 \longrightarrow \mathbf{I}$, i.e. $Q : \{\varnothing\} \longrightarrow \mathbf{I}$, is therefore uniquely determined by the value $Q(\varnothing) \in \mathbf{I}$ which it assumes for the *empty tuple*, written as $\varnothing$. In other words, there is a one-to-one correspondence between nullary semi-fuzzy quantifiers $Q : \mathcal{P}(E)^0 \longrightarrow \mathbf{I}$

and elements $Q(\varnothing) = y \in \mathbf{I}$ in the unit interval. We can therefore speak of nullary quantifiers as 'constant quantifiers'. To give an example, the quantifier $Q : \{\varnothing\} \longrightarrow \mathbf{I}$ defined by $Q(\varnothing) = \frac{1}{2}$, is a nullary semi-fuzzy quantifier, and $\mathbf{true}, \mathbf{false} : \{\varnothing\} \longrightarrow \mathbf{I}$, defined by $\mathbf{true}(\varnothing) = 1$ and $\mathbf{false}(\varnothing) = 0$, are of course also constant semi-fuzzy quantifiers.
Similar considerations apply to nullary *fuzzy* quantifiers $\widetilde{Q} : \widetilde{\mathcal{P}}(E)^0 \longrightarrow \mathbf{I}$. In this case, $\widetilde{\mathcal{P}}(E)^0 = \widetilde{\mathcal{P}}(E)^\varnothing = \{\varnothing\}$, which again establishes a one-to-one correspondence between nullary or 'constant' fuzzy quantifiers $\widetilde{Q} : \{\varnothing\} \longrightarrow \mathbf{I}$, and elements $\widetilde{Q}(\varnothing) \in \mathbf{I}$. In particular, the set $\mathbf{I}^{(\mathcal{P}(E)^0)} = \mathbf{I}^{\{\varnothing\}}$ of constant semi-fuzzy quantifiers over E coincides with the set of constant fuzzy quantifiers $\mathbf{I}^{(\widetilde{\mathcal{P}}(E)^0)} = \mathbf{I}^{\{\varnothing\}}$, for any considered base set $E \neq \varnothing$.

2.8 A Note on Undefined Interpretation

In TGQ, the concept of two-valued quantifiers is often extended by admitting undefined interpretations. Thus quantifiers like the definite "the" already mentioned above, need not denote in certain cases. To achieve this kind of modelling, the definition of two-valued quantifiers could be weakened to partial (incompletely defined) mappings. Barwise and Cooper have suggested a three-valued modelling, where the 'undefined' or 'undecided' case is represented by a third truth value [7, p. 171]. In this book, the base concept of two-valued quantifiers has *not* been extended by undefined interpretations, though. One reason for this decision is that the formal apparatus of TGQ is fully developed only for total quantifiers. For example, van Benthem [9, 10] explicitly restricts attention to the simple case of totally defined quantifiers. Other authors like Keenan and Stavi [92, p. 277] or Hamm [69] avoid undefined interpretations altogether and even stick to the Russelian analysis of definite quantifiers. Barwise and Cooper [7, p. 171] suggest that Kleene's three-valued logic might provide the blueprint for a refinement to undefined interpretations, but they do not detail this proposal.[8]

From the perspective of fuzzy quantification, the refinement to partial or three-valued quantifiers is still too coarse. The proposed notions of semi-fuzzy and fuzzy quantifiers, by contrast, can model all possible shades of a quantification result between 0 (totally false) and 1 (fully true). In this continuous-valued setting, the 'undefined' or 'undecided' result can then be represented by the intermediate value $\frac{1}{2}$. In the following chapters, all relevant concepts of TGQ will be generalized to semi-fuzzy and fuzzy quantifiers. The basic concept of a generalized quantifier (i.e. totally defined, two-valued quantifiers) will be sufficient to introduce these concepts, which will then reformulated for the continuous-valued case. (As a by-product of this, the concepts of TGQ will also be fully developed for three-valued evaluations).

---

[8] Interestingly, all standard models of fuzzy quantification developed below will comply with this strategy, see p. 156.

To sum up, we only need two-valued quantifiers here for motivating continuous-valued abstractions, and the central concepts of semi-fuzzy and fuzzy quantifiers already account for three-valued interpretations. Consequently, the proposed analysis would not profit from the additional use of partial mappings as a second mechanism for expressing indeterminate quantification results.

2.9 Chapter Summary

In the introduction, a framework for fuzzy quantification was described as comprising the following parts: A description of its scope, which fixes the range of considered quantifiers; a definition of operational target quantifiers; a definition of specifications for such quantifiers; and finally models of fuzzy quantification which connect the specifications of quantifiers to their operational forms. Noticing that the traditional framework for fuzzy quantification does not achieve a coherent interpretation for a broad class of NL quantifiers, the present chapter was concerned with the development of a new theoretical skeleton which avoids these problems. The novel framework was supposed to cover all quantifiers of linguistic relevance. Recalling the characteristics of linguistic quantifiers discussed in the introduction, this means that it had to support general multi-place quantifiers including non-monotonic and non-quantitative examples. The most useful point of departure for devising such a framework we considered the theory of generalized quantifiers (TGQ), which offers a suitable notion of (crisp) generalized quantifiers that only needs to be extended towards fuzzy quantification.
To this end, we first defined a suitable notion of *fuzzy quantifiers* which support fuzzy arguments and gradual quantification results. These are the target operations used for interpreting quantifying NL expressions in the new framework. However, it is difficult to establish a correspondence between NL quantifiers and matching fuzzy quantifiers because the familiar crisp cardinality can not be used to define a fuzzy quantifier. In order to simplify the specification of the target quantifier, the framework therefore introduces the novel concept of a *semi-fuzzy quantifier*. Semi-fuzzy quantifiers support gradual quantification results, but they only target at Type III quantifications in the sense of Liu and Kerre [108, p. 2] where all arguments are crisp. The novel quantifiers offer a uniform representation which is independent of particular types of quantifiers. The absolute quantifiers "about ten" and the proportional quantifier "most", for example, both map to two-place semi-fuzzy quantifiers $Q : \mathcal{P}(E)^2 \longrightarrow \mathbf{I}$.
Semi-fuzzy quantifiers include the classical generalized quantifiers of TGQ and thus embed all quantifiers of linguistic relevance. Owing to the restriction to crisp arguments, semi-fuzzy quantifiers admit suggestive definitions in terms of crisp cardinalities. This is of crucial importance to their usefulness as a specification medium. By taking an intermediate course between two-valued

quantifiers (which are too weak) and fuzzy quantifiers (which are too expressive for specification purposes), semi-fuzzy quantifiers solve the 'dilemma of fuzzy quantifers' concerning the conflict between expressive power and ease of specification.

It is the responsibility of a model of fuzzy quantification to describe the relationship between specifications (semi-fuzzy quantifiers) and the associated fuzzy quantifiers (target operations). The framework was therefore completed by introducing the concept of a *quantifier fuzzification mechanism* (QFM) which expresses these correspondences. The notion of QFM is very general and comprises all conceivable assignments of fuzzy quantifiers to the given semi-fuzzy quantifiers. Hence, it will be the main concern of fuzzy quantification theory to precisely describe those QFMs which achieve a plausible interpretation of linguistic quantifiers (and of course, it must also present concrete examples of such models). To identify the intended models, the theory must develop formal criteria for those interpretations of fuzzy quantifiers which are semantically conclusive. These criteria can be expressed as requirements on the admissible fuzzification mechanisms because all details of the mapping from specifications to target quantifiers are encapsulated by the QFMs.

The proposed framework spans both the generalized quantifiers of TGQ and the Type IV quantifications of fuzzy set theory. Due to its vicinity to the linguistic theory, the later translation of the conceptual apparatus of TGQ to fuzzy quantifiers will become a straightforward task. As to the coverage of quantificational phenomena, the framework includes a variety of extensional quantifiers which are permitted to be multi-place and non-quantitative (i.e. not necessarily definable in terms of cardinalities). In addition, there is no a priory limitation to finite domains.

When discussing the quantification framework assumption, we already met a few 'exotic' quantifiers which cannot be reduced to a semi-fuzzy quantifier because they then become meaningless. The prime example are Rescher's quantifiers or more generally, complex quantifiers built from a hedging construction. In the following, we will discuss some other quantificational phenomena not yet covered by the proposed framework. It should be remarked in advance that these cases are also not handled by the linguistic theory and of course, they cannot be treated by existing work on fuzzy quantification. The necessary extensions to cover these cases are sometimes obvious from the usual treatment of these phenomena in logic and linguistics. However, we must concentrate on the core issues of fuzzy NL quantification in this work, and the phenomena not treated do not seem to be intrinsically connected to linguistic quantification as such.

Firstly, the proposed framework does not account for *higher-order quantifiers*. As opposed to a first-order proposition which describes individuals, higher-order propositions refer to sets of individuals, sets of such sets etc. The corresponding higher-order logics, and even the full theory of finite types [143, 27, 1], were found to be valuable tools for linguistic description. To keep things simple, however, the current framework only formalizes first-order quantification. Most linguistic quantifiers are first-order but there are also some examples of higher-order quantification. As remarked by Keenan and Stavi, these express in 'higher-order' NPs like "two sets of dishes" or "a pride of lions" [92, p. 256].

The basic framework further abstracts from so-called *branching quantification*, in which the linear sequence of nested quantifiers is replaced with a partial order. The corresponding branching quantifiers can then operate in parallel and independently of each other [74, 166]. As already mentioned, the relevance of branching quantification to the analysis of language has been a matter of linguistic debate [130, 76, 6, 167] and Barwise [6] made a convincing point that branching quantification is needed to describe certain reciprocal constructions. To support such cases, Chap. 12 will extend the basic framework by fuzzy branching quantifiers. This is also a first step towards fuzzy higher-order quantification because branching quantification, as noted by Barwise, is just a human language technique to conceal the use of certain higher-order constructions.

The quantifiers of TGQ are *extensional*, i.e. they operate on sets of individuals (extensions) rather than 'meanings' or 'intensions' (usually represented as mappings from states of affairs into extensions), so there is no indexing by some parameter which corresponds to states of affairs. As pointed out by Barwise and Cooper [7, p. 203], they deliberately restricted themselves to an extensional fragment of English because they wanted to focus on the central issues. For similar reasons, the proposed framework for fuzzy quantification also assumes extensional quantifiers. The following characterization of extensional determiners due to Keenan and Stavi also generalizes to fuzzy quantifiers:

> *"To say that a det d is extensional is to say, for example, that whenever the doc[t]ors and the lawyers are the same individuals then d* doctors *and d* lawyers *have the same properties, e.g., d* doctors attended the meeting *necessarily has the same truth value as d* lawyers attended the meeting. *Thus* every *is extensional, since if the doctors and the lawyers are the same then, every doctor attended iff every lawyer did"* [92, p. 257].

A logic built 'on top' of these quantifiers[9] will then be extensional as well, according to the usual 'principle of extensionality' [50, p. 5],

$$\chi \leftrightarrow \chi' \models \varphi \leftrightarrow [\chi'/\chi]\varphi$$

i.e. substitutivity of expressions with identical reference. Following Frege [48], we will discern the *sense* of an NL expression ('Sinn' in Frege's terms, i.e. meaning, intension) and its *reference* (or 'Bedeutung' in Frege's terms, i.e. extension). Quine [129] has shown that there are certain positions in linguistic expressions called 'opaque contexts' in which the substitution of coextensional terms is no longer admissible but rather alters the overall meaning. Thus

[9] such a logic with generalized quantifiers has been developed by Barwise and Feferman [8, Ch. 2].

a complete model of linguistic quantification will likely have to deal with intensional arguments. And the quantifier itself might also be intensional, like "all alleged" [9, p. 448] or "an undisclosed number of" [92, p. 257]:

> *"It is not hard to imagine a situation in which the doctors and the lawyers are the same and but [sic]* an undisclosed number of lawyers attended the meeting *is true and* an undisclosed number of doctors attended the meeting *is false. (Imagine a meeting of medical personnel. The chairman announces the number of doctors in attendance but not the number of lawyers)"* [92, p. 257].

This means that the principle of extensionality fails in this case. However, as stated in Gamut [50, p. 228], *"there are not many really intensional NPs or determiners"*. In addition, it is relatively clear how to integrate intensional features into a given system of logic. The standard method of accomplishing this was developed by S. Kripke [102, 101], i.e. one must add a primitive notion of a possible state of affairs (or 'possible world' in modal logic), an accessibility relation on states, and define the semantics in terms of these states and their accessibility. It should be possible to incorporate intensional features into the proposed framework along the same lines. However, this direction will not be pursued further here in order to keep things simple.

Finally the Theory of Generalized Quantifiers, and the proposed extensions to fuzzy quantification as well, specifically target at 'count NPs' [92, p. 256] or 'discrete quantifiers' [9, p. 448] like "a lot of" or "many" rather than so-called 'mass quantifiers' or 'continuous quantifiers' like "much" or "little". However, mass quantifiers are important as well, especially for modelling temporal or spatial quantification. In this case, the underlying regions in time and space are possibly best modelled as atomless masses. Hence a comprehensive model will need to support some form of 'mass quantification'. The framework provides a good starting point for developing a theory of mass quantification, because its core concepts are defined for base sets of infinite cardinality as well. It is not the goal of this book to develop a theory of fuzzy mass quantification, however. To this end, it would certainly be necessary to incorporate additional elements from the theory of fuzzy measures.

P in the sense that whenever $P(Q)$ holds for a semi-fuzzy quantifier Q, we also have $P'(\mathcal{F}(Q))$, i.e. the corresponding property holds for the fuzzy quantifier $\mathcal{F}(Q)$ associated with Q. A plausible model of fuzzy quantification should preserve all properties of linguistic relevance. For example, if Q shows a certain monotonicity property, then $\mathcal{F}(Q)$ will also be monotonic.

Moreover, a plausible model should also preserve important *relationships* between quantifiers. The prime example are functional relationships between quantifiers which are established by certain *constructions* (like dualisation or negation). Hence if C is a construction on semi-fuzzy quantifiers which builds a new semi-fuzzy quantifier $C(Q)$ given Q, and C' is the corresponding construction on fuzzy quantifiers, then we can require that $\mathcal{F}$ be compatible with the construction in the sense that we always have

$$\mathcal{F}(C(Q)) = C'(\mathcal{F}(Q)),$$

i.e. it does not matter whether we first apply the construction and then 'fuzzify' using $\mathcal{F}$, or whether we first apply $\mathcal{F}$ and then perform the construction. Compatibility with such constructions corresponds to the well-known mathematical concept of a homomorphism (structure-preserving mapping).

The particular properties of quantifiers, relationships and constructions to be considered will usually be adopted from the linguistic theory of natural language quantification, TGQ. The reformulation of these notions will usually be easy due to the similarity between semi-fuzzy or fuzzy quantifiers and the original two-valued quantifiers.

The criteria that evolve from this procedure form a catalogue which is open to refinement: It can be developed incrementally and will not collapse entirely should individual criteria need to be altered. It is of course hoped that a careful analysis of meaningful interpretations and their formalization into axiomatic requirements will ultimately converge and form a stable body of criteria which identifies the class of plausible models. In this way, the research into semantical characteristics of NL quantifiers will make it possible to develop a theory of fuzzy quantification which rests on an axiomatic foundation.

Postulating another ad-hoc rule for interpretation (i.e. mathematical formula, algorithm) will only achieve modest progress, by contrast – should the formula prove defective, not too much insight would have been gained. For this reason, we will clearly separate the theoretical analysis (*What are plausible models?*) and the later identification and implementation of conforming models in this monograph.

3.2 Correct Generalization

Perhaps the most elementary condition on a quantifier fuzzification mechanism is that it properly generalize the original semi-fuzzy quantifier. We can express this succinctly if we recall the notion of an underlying semi-fuzzy quantifier $\mathcal{U}(\widetilde{Q})$ which simply restricts the fuzzy quantifier $\widetilde{Q}$ to crisp arguments, see Def. 2.11. It is natural to assume that a model of fuzzy quantification satisfy

$$\mathcal{U}(\mathcal{F}(Q)) = Q\,, \tag{3.1}$$

for all choices of semi-fuzzy quantifier Q, which means that $\mathcal{F}(Q)$ properly generalizes Q in the sense that

$$\mathcal{F}(Q)(Y_1, \ldots, Y_n) = Q(Y_1, \ldots, Y_n)$$

whenever all arguments are crisp, i.e. given that $Y_1, \ldots, Y_n \in \mathcal{P}(E)$.

The requirement of correct generalization is absolutely fundamental because it decides on the internal coherence of the QFM. The criterion expresses the 'downward compatibility' of the model with respect to to the original specification in terms of crisp arguments. In other words, we need correct generalisation to properly set up the 'fuzzification pattern' (i.e. use of a fuzzification mechanism) which underlies the quantification framework [59, p. 419+]. Due to this enabling role for the proposed framework, the above equation (3.1) will constitute the first requirement imposed on meaningful models of fuzzy quantification. We should note in advance, however, that the corresponding criterion (Z-1) in the final axiom system (stated in Def. 3.28) will be restricted nullary and unary quantifiers ($n \leq 1$). This serves the purpose of keeping the axioms simple. When taken together, these axioms will of course entail the original criterion, and thus guarantee that equality (3.1) holds for quantifiers of arbitrary arities $n \in \mathbb{N}$.

Let us now consider some examples involving such quantifiers of arities $n \leq 1$. As concerns unary (one-place) quantifiers, let us assume that $E \neq \varnothing$ is a given domain of persons. We shall further assume that there are some men among the persons in E, but no children. Hence the extensions of "men" and "children" in E, $\mathbf{men}, \mathbf{children} \in \mathcal{P}(E)$, satisfy $\mathbf{men} \neq \varnothing$ and $\mathbf{children} = \varnothing$, respectively. Now consider the existential quantifier $\exists : \mathcal{P}(E) \longrightarrow \mathbf{2}$, which determines the quantification results of $\exists(\mathbf{men}) = 1$ and $\exists(\mathbf{children}) = 0$. We certainly expect the fuzzification process to preserve these interpretations, and application of the associated fuzzy quantifier $\mathcal{F}(\exists) : \widetilde{\mathcal{P}}(E) \longrightarrow \mathbf{I}$ should result in $\mathcal{F}(\exists)(\mathbf{men}) = 1$ and $\mathcal{F}(\exists)(\mathbf{children}) = 0$.

Let us now turn to the case of nullary quantifiers ($n = 0$). It has already been remarked in Sect. 2.7 that the sets of nullary semi-fuzzy quantifiers and nullary fuzzy quantifiers coincide, regardless of the chosen base set $E \neq \varnothing$. Hence for every nullary semi-fuzzy $Q : \mathcal{P}(E)^0 \longrightarrow \mathbf{I}$, the considered quantifier *itself* already qualifies as a fuzzy quantifier, and there is nothing to generalize or to fuzzify. It is therefore essential that these quantifiers be mapped to themselves by every plausible choice of a QFM. Now returning to the above equation (3.1), it is easily observed that this is precisely what $\mathcal{U}(\mathcal{F}(Q)) = Q$ states in the case of nullary Q. This is because both the nullary semi-fuzzy quantifier Q, as well as the nullary fuzzy quantifier $\mathcal{F}(Q)$, are uniquely determined by their value obtained at the empty tuple $\varnothing$. By Def. 2.11, then, we obtain from $\mathcal{U}(\mathcal{F}(Q)) = Q$ and the fact that the empty tuple is crisp that these values coincide, i.e. $\mathcal{F}(Q)(\varnothing) = \mathcal{U}(\mathcal{F}(Q))(\varnothing) = Q(\varnothing)$. Hence Q and $\mathcal{F}(Q)$ coincide, as desired. To give an example, consider the nullary quantifier **true** defined by $\mathbf{true}(\varnothing) = 1$. A conforming choice of $\mathcal{F}$ should result in $\mathcal{F}(\mathbf{true}) = \mathbf{true}$, in order to catch the semantics of the constant quantifier. This completes the discussion of the first requirement which must be fulfilled by all models of fuzzy quantification.

3.3 Membership Assessment

Membership assessment (crisp or fuzzy) can be modelled through quantification. For an element e of the given base set, we can define a two-valued quantifier π_e which checks if e is present in its argument. Similarly, we can define a fuzzy quantifier $\widetilde{\pi}_e$ which returns the degree to which e is contained in its argument. It is natural to require that the crisp quantifier π_e be mapped to $\widetilde{\pi}_e$, which plays the same role in the fuzzy case.

In order to express this precisely on the formal level, we will now introduce this special type of semi-fuzzy quantifiers (and a corresponding type of fuzzy quantifiers) that allow us to treat membership assessments as a special case of quantification. Let us first recall the concept of a *characteristic function* and agree on the following notation.

Definition 3.1.
Suppose E is a given set and $A \in \mathcal{P}(E)$ a (crisp) subset of E. By $\chi_A : E \longrightarrow \mathbf{2}$ we denote the characteristic function *of A, i.e. the mapping defined by*

$$\chi_A(e) = \begin{cases} 1 & : & e \in A \\ 0 & : & e \notin A \end{cases}$$

for all $e \in E$.

Building on this concept, let us state the following definition of projection quantifiers, which covers the case of crisp arguments:

Definition 3.2.
Suppose E is a base set and $e \in E$. The projection quantifier $\pi_e : \mathcal{P}(E) \longrightarrow \mathbf{2}$ *is defined by*

$$\pi_e(Y) = \chi_Y(e)\,,$$

for all $Y \in \mathcal{P}(E)$.

Notes 3.3.

- The use of the term 'projection quantifier' is motivated by the observation that we can view π_e as the projection of the 'E-tuple' $\chi_A : E \longrightarrow \mathbf{2}$ onto that component $\chi_A(e)$, which is indexed by e.

- To present an example, suppose $E = \{\text{Joan, Lucas, Mary}\}$ is a set of persons and **married** $\in \mathcal{P}(E)$ the subset of married persons in E. Then **joan** $= \pi_{\text{Joan}} : \mathcal{P}(E) \longrightarrow \mathbf{2}$ is a projection quantifier, and

$$\begin{aligned}\mathbf{joan}(\mathbf{married}) &= \pi_{\text{Joan}}(\mathbf{married}) \\ &= \begin{cases} 1 & : \quad \text{Joan} \in \mathbf{married} \\ 0 & : \quad \text{Joan} \notin \mathbf{married} \end{cases}\end{aligned}$$

  The example illustrates that π_{Joan} can indeed be used to evaluate statements of the type "Is Joan X?", where X is a crisp predicate. In particular, it is suitable for modelling the interpretation of the proper name "Joan" as a two-valued quantifier (provided that the individual Joan is contained in the considered domain). See [7, p. 164+] for a linguistic case that proper names be treated as a special type of NL quantifiers.

A corresponding definition of fuzzy projection quantifiers is straightforward.

Definition 3.4.
Let a base set E be given and $e \in E$. The fuzzy projection quantifier $\widetilde{\pi}_e : \widetilde{\mathcal{P}}(E) \longrightarrow \mathbf{I}$ *is defined by*

$$\widetilde{\pi}_e(X) = \mu_X(e)$$

for all $X \in \widetilde{\mathcal{P}}(E)$.

Notes 3.5.

- The term 'fuzzy projection quantifier' has been chosen because the mapping $\widetilde{\pi}_e : \widetilde{\mathcal{P}}(E) \longrightarrow \mathbf{I}$ can be viewed as the projection of the 'E-tuple' $\mu_X : E \longrightarrow \mathbf{I}$ onto its component $\mu_X(e)$, which is indexed by e.
- Fuzzy projection quantifiers provide the missing class of operators which are suited to model gradual membership assessments like "To which grade is Joan X?" through fuzzy quantification. For example, we simply need to evaluate $\widetilde{\pi}_{\text{Joan}}(\mathbf{tall})$ in order to assess the grade to which Joan is tall, and we can compute $\widetilde{\pi}_{\text{Joan}}(\mathbf{rich})$ to determine $\mu_{\mathbf{rich}}(\text{Joan})$, the degree to which Joan is rich. In particular, the fuzzy projection quantifier $\widetilde{\pi}_{\text{Joan}} : \widetilde{\mathcal{P}}(E) \longrightarrow \mathbf{I}$ is suitable for interpreting the proper name "Joan" as a fuzzy quantifier (again assuming that Joan be present in the given base set).

It is apparent from the above reasoning that crisp and fuzzy projection quantifiers play the same basic role of crisp/fuzzy membership assessment. Intuitively, a plausible model of fuzzy quantification should be compatible with the fundamental operation of membership assessment, and thus assign to each crisp projection quantifier π_e its obvious fuzzy counterpart, the corresponding fuzzy projection quantifier $\widetilde{\pi}_e$. In other words, we expect that

$$\mathcal{F}(\pi_e) = \widetilde{\pi}_e\,,$$

regardless of the chosen base set $E \neq \varnothing$ and considered element $e \in E$.
This makes the second requirement to be imposed on all models of fuzzy quantification.

3.4 The Induced Propositional Logic

The linguistic theory of quantification knows various constructions on natural language quantifiers which involve the use of a Boolean connective (like negation) or corresponding set-theoretic operation (e.g. complementation). Examples comprise the negation of quantifiers, antonyms, duals, intersections and unions in the arguments of a quantifier, conjunctions and disjunctions of quantifiers, and others.[1] In order to generalize these constructions on two-valued quantifiers to semi-fuzzy and fuzzy quantifiers, the Boolean connectives must be replaced with suitable continuous-valued counterparts. Similarly, the move from two-valued quantifiers to fuzzy quantifiers necessitates the use of set-theoretic operations on fuzzy sets, and thus forces us to select a particular fuzzy complement, intersection and union, in terms of which the generalized constructions will then be expressed. Of course, one could simply resort to the standard choices, like standard negation $1 - x$, conjunction min and disjunction max, but this would make too strong a commitment and foreclose the theoretical analysis of broader classes of models, which rely on general fuzzy negations, conjunctions and disjunctions.

The intent to cover such general models forces us to admit for general fuzzy set operations (rather than the standard choices), which will be used to define the constructions on quantifiers. We will permit the choice of connectives to depend on the considered QFM because these constructions must match the the model. In addition to associating this canonical choice of fuzzy operations to the QFM of interest, we should show that this choice is well-motivated.

Again, it would be tedious and obstructive to formal analysis if the required decisions were made on an individual case basis. As in the case of QFMs, we need a general solution which controls the transfer from propositional functions to corresponding fuzzy truth functions under a given QFM. This construction must be fully general and hence applicable to arbitrary QFMs.

In order to devise a construction parametrized by the QFM which assigns canonical choices of 'induced' fuzzy truth functions and induced fuzzy set operations to the corresponding crisp concepts, we need a natural embedding of propositional functions into semi-fuzzy quantifiers to which the QFM of interest can then be applied. In turn, we also need an inverse construction which determines the target fuzzy truth function from the resulting fuzzy quantifier. The required operations on fuzzy sets can then be defined from the resulting fuzzy connectives in the apparent way, i.e. by applying the fuzzy connectives to the observed membership grades.

We have investigated two alternative schemes motivated by independent considerations which rest on distinct embeddings and corresponding inverse constructions. We shall see later in theorem Th. 4.12 that in all intended models, both constructions establish the same canonical choice of induced truth functions. This indicates that the construction of induced fuzzy truth functions,

[1] All mentioned constructions will be formally defined and discussed later on.

to be introduced now, indeed results in the appropriate choice of connectives and fuzzy set operations for the given QFM (more details on the alternative construction can be found in Sect. 4.4).
In order to establish the link between logical connectives and quantifiers, we first observe that $\mathbf{2}^n \cong \mathcal{P}(\{1, \ldots, n\})$, using the bijection $\eta : \mathbf{2}^n \longrightarrow \mathcal{P}(\{1, \ldots, n\})$ defined by

$$\eta(x_1, \ldots, x_n) = \{k \in \{1, \ldots, n\} : x_k = 1\}, \tag{3.2}$$

for all $x_1, \ldots, x_n \in \mathbf{2}$. We can use an analogous construction in the fuzzy case. We then have $\mathbf{I}^n \cong \widetilde{\mathcal{P}}(\{1, \ldots, n\})$, based on the bijection $\widetilde{\eta} : \mathbf{I}^n \longrightarrow \widetilde{\mathcal{P}}(\{1, \ldots, n\})$ defined by

$$\mu_{\widetilde{\eta}(x_1,\ldots,x_n)}(k) = x_k \tag{3.3}$$

for all $x_1, \ldots, x_n \in \mathbf{I}$ and $k \in \{1, \ldots, n\}$. These bijections can be utilized for a translation between semi-fuzzy truth functions (i.e. mappings $f : \mathbf{2}^n \longrightarrow \mathbf{I}$) and corresponding semi-fuzzy quantifiers $Q_f : \mathcal{P}(\{1, \ldots, n\}) \longrightarrow \mathbf{I}$, which embeds propositional functions into semi-fuzzy quantifiers, and similarly the translation from fuzzy quantifiers $\widetilde{Q} : \widetilde{\mathcal{P}}(\{1, \ldots, n\}) \longrightarrow \mathbf{I}$ into fuzzy truth functions $\widetilde{f} : \mathbf{I}^n \longrightarrow \mathbf{I}$, which implements the required inverse embedding.

Definition 3.6.
Suppose $\mathcal{F}$ is a QFM and $f : \mathbf{2}^n \longrightarrow \mathbf{I}$ is a mapping (i.e. a 'semi-fuzzy truth function') for some $n \in \mathbb{N}$. The semi-fuzzy quantifier $Q_f : \mathcal{P}(\{1, \ldots, n\}) \longrightarrow \mathbf{I}$ is defined by

$$Q_f(Y) = f(\eta^{-1}(Y))$$

for all $Y \in \mathcal{P}(\{1, \ldots, n\})$. In terms of Q_f, the induced fuzzy truth function *$\widetilde{\mathcal{F}}(f) : \mathbf{I}^n \longrightarrow \mathbf{I}$ is then defined by*

$$\widetilde{\mathcal{F}}(f)(x_1, \ldots, x_n) = \mathcal{F}(Q_f)(\widetilde{\eta}(x_1, \ldots, x_n)),$$

for all $x_1, \ldots, x_n \in \mathbf{I}$.

Notes 3.7.

- If $f : \mathbf{2}^0 \longrightarrow \mathbf{I}$ is a nullary semi-fuzzy truth function (i.e., a constant), then $Q_f : \mathcal{P}(\varnothing) \longrightarrow \mathbf{I}$ turns into a nullary semi-fuzzy quantifier on the universe $E = \{\varnothing\}$, noticing that $\mathcal{P}(\varnothing) = \mathcal{P}(\{\varnothing\})^0 = \{\varnothing\}$. Hence in this case, the fuzzy truth function $\widetilde{\mathcal{F}}(f) : \mathbf{I}^0 \longrightarrow \mathbf{I}$ can be expressed as $\widetilde{\mathcal{F}}(f)(\varnothing) = \mathcal{F}(c)(\varnothing)$, where $c : \mathcal{P}(\{\varnothing\})^0 \longrightarrow \mathbf{I}$ is the constant $c(\varnothing) = f(\varnothing)$.
- We shall not impose restrictions on the induced connectives directly; these will be entailed by the remaining axioms.
- Whenever $\mathcal{F}$ is clear from the context, we shall abbreviate $\widetilde{\mathcal{F}}(f)$ as $\widetilde{f}$. For example, the induced disjunction will be written $\widetilde{\vee}$.

men" is interpreted by inserting the interpretation **men** $\in \mathcal{P}(E)$ of the verbal phrase "are men" into the second argument slot. Compatibility with internal complementation should not be artificially restricted to the n-th argument, though, and the particular indexing of the arguments should be inessential to the outcome of fuzzy quantification. Plausible approaches to fuzzy quantification should therefore respect complementation in all arguments. By means of permutations of arguments, to be discussed in Sect. 4.5, we will be able to reduce the general compatibility condition to the base condition on complementation in the last argument.

- Zadeh [197, p. 165] has proposed a notion of antonymy for proportional quantifiers, which is defined on their representations in terms of membership functions $\mu_Q : \mathbf{I} \longrightarrow \mathbf{I}$.

TGQ also knows the concept of the *dual* of a two-valued quantifier (written as $Q\widetilde{\Box}$ in our notation), which is the antonym of the negation of a quantifier (or equivalently, the negation of the antonym) [7, p. 197], [50, p. 238]:

Definition 3.11.
The dual $Q\widetilde{\Box} : \mathcal{P}(E)^n \longrightarrow \mathbf{I}$ *of a semi-fuzzy quantifier* $Q : \mathcal{P}(E)^n \longrightarrow \mathbf{I}$, $n > 0$ *is defined by*

$$Q\widetilde{\Box}(Y_1, \ldots, Y_n) = \widetilde{\neg}\, Q(Y_1, \ldots, Y_{n-1}, \neg Y_n)\,,$$

for all $Y_1, \ldots, Y_n \in \mathcal{P}(E)$*. The dual* $\widetilde{Q}\widetilde{\Box} = \widetilde{\neg}\,\widetilde{Q}\widetilde{\neg}$ *of a fuzzy quantifier* $\widetilde{Q}$ *is defined analogously.*

For example, "some" is the dual of "all". Again resorting to crisp concepts like "men" and "mammals", we may therefore assert that

$$\mathbf{all}(\mathbf{men}, \mathbf{mammals}) = \widetilde{\neg}\, \mathbf{some}(\mathbf{men}, \neg\mathbf{mammals})\,.$$

The latter relationship ensures that "All men are mammals" can be paraphrased as "It is not the case that some men are not mammals".

Acknowledging the significance of these constructions to the description of natural language, we expect plausible models of fuzzy quantification to be homomorphic with respect to these constructions on quantifiers. Hence $\mathcal{F}(\mathbf{no})$ should be the negation of $\mathcal{F}(\mathbf{some})$, $\mathcal{F}(\mathbf{no})$ should be the antonym of $\mathcal{F}(\mathbf{all})$, and $\mathcal{F}(\mathbf{some})$ should be the dual of $\mathcal{F}(\mathbf{all})$. Concerning external negation, this will ensure, for example, that "No rich is young" can be paraphrased as "It is not the case that some rich is young", which is justified because $\mathcal{F}(\mathbf{no})(\mathbf{rich}, \mathbf{young}) = \widetilde{\neg}\, \mathcal{F}(\mathbf{some})(\mathbf{rich}, \mathbf{young})$. As to internal complementation, we obtain that $\mathcal{F}(\mathbf{no})(\mathbf{rich}, \mathbf{young}) = \mathcal{F}(\mathbf{all})(\mathbf{rich}, \widetilde{\neg}\, \mathbf{young})$, and thus "No rich are young" can also be phrased as "All rich are not young". Finally in the case of the dual, we conclude from $\mathcal{F}(\mathbf{all})(\mathbf{rich}, \mathbf{old}) = \widetilde{\neg}\, \mathcal{F}(\mathbf{some})(\mathbf{rich}, \widetilde{\neg}\, \mathbf{old})$ that "All rich are old" means the same as "It is not the case that some rich are not old".

The interdependencies of external negation, internal complementation (formation of antonyms) and dualisation are summarized in the *Aristotelian square*.[3] The Aristotelian square of the quantifier "all" is displayed in Fig. 3.1.

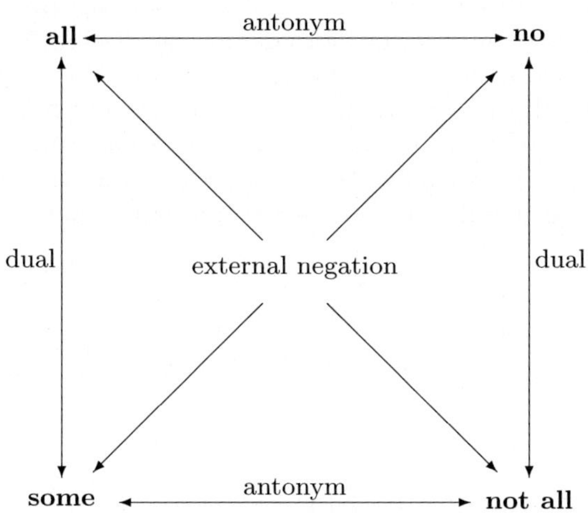

Fig. 3.1. The Aristotelian square of "all"

The Aristotelian square expresses in graphical form that the operators $(\bullet)$ (identity), $\widetilde{\neg}(\bullet)$ (external negation), $(\bullet)\neg$ (antonym) and $(\bullet)\widetilde{\square}$ (dualisation) constitute a Klein group structure on semi-fuzzy quantifiers under the operation $\circ$ of composition:

$\circ$	$(\bullet)$	$\widetilde{\neg}(\bullet)$	$(\bullet)\neg$	$(\bullet)\widetilde{\square}$
$(\bullet)$	$(\bullet)$	$\widetilde{\neg}(\bullet)$	$(\bullet)\neg$	$(\bullet)\widetilde{\square}$
$\widetilde{\neg}(\bullet)$	$\widetilde{\neg}(\bullet)$	$(\bullet)$	$(\bullet)\widetilde{\square}$	$(\bullet)\neg$
$(\bullet)\neg$	$(\bullet)\neg$	$(\bullet)\widetilde{\square}$	$(\bullet)$	$\widetilde{\neg}(\bullet)$
$(\bullet)\widetilde{\square}$	$(\bullet)\widetilde{\square}$	$(\bullet)\neg$	$\widetilde{\neg}(\bullet)$	$(\bullet)$

Or in the case of fuzzy quantifiers,

$\circ$	$(\bullet)$	$\widetilde{\neg}(\bullet)$	$(\bullet)\widetilde{\neg}$	$(\bullet)\widetilde{\square}$
$(\bullet)$	$(\bullet)$	$\widetilde{\neg}(\bullet)$	$(\bullet)\widetilde{\neg}$	$(\bullet)\widetilde{\square}$
$\widetilde{\neg}(\bullet)$	$\widetilde{\neg}(\bullet)$	$(\bullet)$	$(\bullet)\widetilde{\square}$	$(\bullet)\widetilde{\neg}$
$(\bullet)\widetilde{\neg}$	$(\bullet)\widetilde{\neg}$	$(\bullet)\widetilde{\square}$	$(\bullet)$	$\widetilde{\neg}(\bullet)$
$(\bullet)\widetilde{\square}$	$(\bullet)\widetilde{\square}$	$(\bullet)\widetilde{\neg}$	$\widetilde{\neg}(\bullet)$	$(\bullet)$

The square can be properly set up for arbitrary semi-fuzzy quantifiers $Q : \mathcal{P}(E)^n \longrightarrow \mathbf{I}$ and fuzzy quantifiers $\widetilde{Q} : \widetilde{\mathcal{P}}(E)^n \longrightarrow \mathbf{I}$ provided that $\widetilde{\neg} : \mathbf{I} \longrightarrow \mathbf{I}$

[3] Sometimes also referred to as the *square of opposition*, see e.g. Gamut [50, p.238].

is involutive and satisfies $\widetilde{\neg} 1 = 0$, $\widetilde{\neg} 0 = 1$. These conditions will of course be entailed by the DFS axioms stated below.

The requirement that $\mathcal{F}$ be compatible with external negation, internal complementation (formation of antonyms) and dualisation, can then be summarized into the condition that the considered model of fuzzy quantification preserve Aristotelian squares (for more details, see Sect. 4.7).

In order to warrant this, there is no need to explicitly require the compliance of $\mathcal{F}$ with all three constructions involved in the square. As will later be shown in Th. 4.19 and Th. 4.20, the targeted conformance to these constructions can be distilled into a single representative requirement, that of preserving duals of quantifiers. It is hence sufficient to demand that

$$\mathcal{F}(Q\widetilde{\square}) = \mathcal{F}(Q)\widetilde{\square}$$

for all semi-fuzzy quantifiers $Q : \mathcal{P}(E)^n \longrightarrow \mathbf{I}$ of arity $n > 0$.
The latter condition comprises the third requirement, which governs plausible choices of models.

3.6 Unions of Argument Sets

Apart from complementation, we can also perform other set-theoretic operations on the arguments of a quantifier. Due to De Morgan's law, and the known compatiblity of the considered models to complementation, which has been ensured by the previous requirement, it is sufficient to consider unions of arguments, in order to ensure that the given model of fuzzy quantification fully preserves the Boolean argument structure that can be expressed in NL. Let us hence introduce the construction which builds new quantifiers from given ones by means of forming the union of arguments:

Definition 3.12.
Let a semi-fuzzy quantifier $Q : \mathcal{P}(E)^n \longrightarrow \mathbf{I}$ of arity $n > 0$ be given. We define the semi-fuzzy quantifier $Q\cup : \mathcal{P}(E)^{n+1} \longrightarrow \mathbf{I}$ by

$$Q\cup(Y_1, \ldots, Y_{n+1}) = Q(Y_1, \ldots, Y_{n-1}, Y_n \cup Y_{n+1})$$

for all $Y_1, \ldots, Y_{n+1} \in \mathcal{P}(E)$. In the case of fuzzy quantifiers, $\widetilde{Q}\widetilde{\cup}$ is defined analogously, based on the given fuzzy set operation $\widetilde{\cup}$.

Notes 3.13.

- To see how this construction expresses on the level of NL surface, consider the example that "Most men drink or smoke", and its model as a quantifying expression, $\mathbf{most}(\mathbf{men}, \mathbf{drink}\cup\mathbf{smoke})$, where we have assumed for the sake of argument that the extensions $\mathbf{smoke}, \mathbf{drink} \in \mathcal{P}(E)$ of "smoke" and "drink" be crisp. The above quantifying expression can be decomposed into the constructed quantifier "Most Y_1 are Y_2 or Y_3", i.e. $Q' = \mathbf{most}\cup$, which is applied to the argument triple $(\mathbf{men}, \mathbf{drink}, \mathbf{smoke})$.

- The construction is also underlying the definition of the two-place NL quantifier "all" in terms of the unary universal quantifier $\forall$ known from logics, because $\mathbf{all}(Y_1, Y_2) = \forall((\neg Y_1) \cup Y_2)$.
- The dual construction of intersections in arguments can be defined along the same lines, see Sect. 4.9.

In order to ensure the full preservation of Boolean argument structure, conforming models of fuzzy quantification can be expected to comply with the above construction of unions in arguments. We therefore require that $\mathcal{F}$ preserves the union of arguments, which is formally captured by the equality

$$\mathcal{F}(Q\cup) = \mathcal{F}(Q)\widetilde{\cup} \tag{3.4}$$

which must be valid for arbitrary semi-fuzzy quantifiers $Q : \mathcal{P}(E)^n \longrightarrow \mathbf{I}$ of arity $n > 0$.
Hence consider the example that "Most rich are old or lucky", which is modelled by the fuzzy quantifying expression $\mathcal{F}(\mathbf{most})(\mathbf{rich}, \mathbf{old} \mathbin{\widetilde{\cup}} \mathbf{lucky})$ built from the fuzzy extensions $\mathbf{rich}, \mathbf{old}, \mathbf{lucky} \in \widetilde{\mathcal{P}}(E)$ of "rich", "old", and "lucky" persons, respectively. The compliance of $\mathcal{F}$ with equation (3.4) then ensures that the involved composite quantifier $\mathcal{F}(\mathbf{most})\widetilde{\cup}$ can be properly represented by its underlying semi-fuzzy quantifier $Q' = \mathbf{most}\cup$. Hence it does not matter whether "Many rich are old or lucky" is computed by evaluating $\mathcal{F}(\mathbf{many})(\mathbf{rich}, \mathbf{old} \mathbin{\widetilde{\cup}} \mathbf{lucky})$ or by resorting to the decomposition $\mathcal{F}(Q')(\mathbf{rich}, \mathbf{old}, \mathbf{lucky})$ based on $Q'(Y_1, Y_2, Y_3) = \mathbf{many}(Y_1, Y_2 \cup Y_3)$.
As mentioned above, the requirement (3.4) imposed on unions of arguments can be combined with other conditions like compatibility with internal complementation, in order to achieve full preservation of Boolean argument structure.[4] In particular, this will ensure that $\mathcal{F}(\mathbf{all})(X_1, X_2) = \mathcal{F}(\forall)((\neg X_1) \cup X_2)$, i.e. the two-place NL quantifier "all" can be determined from the unary universal quantifier $\forall$ even when there is fuzziness in the arguments.

This completes the discussion of Boolean argument structure, and the corresponding condition of preserving unions in arguments, which formalizes the fourth requirement on models of fuzzy quantification.

3.7 Monotonicity in the Arguments

It is essential for a model of fuzzy quantification to to comply with the expected entailment relationships. The fifth requirement on plausible models, to be introduced now, captures those entailments that stem from monotonicity properties of the involved quantifiers. In order to express the relevant monotonicity properties of semi-fuzzy and fuzzy quantifiers, let us first recall the definition of the fuzzy inclusion relation.

[4] Assuming that none of the Boolean variables X_i occurs more than once.

Definition 3.14 (Fuzzy inclusion relation).
Suppose E is some set and $X_1, X_2 \in \widetilde{\mathcal{P}}(E)$ are fuzzy subsets of E. We say that X_1 is contained in X_2 (in symbols, $X_1 \subseteq X_2$) if

$$\mu_{X_1}(e) \leq \mu_{X_2}(e)$$

for all $e \in E$.

Based on this concept, we can now state precisely what it means for a (semi-)fuzzy quantifier to be monotonic in one of its arguments.

Definition 3.15 (Monotonicity in the i-th argument).
A semi-fuzzy quantifier $Q : \mathcal{P}(E)^n \longrightarrow \mathbf{I}$ is said to be nondecreasing in its i-th argument, $i \in \{1, \ldots, n\}$, *if*

$$Q(Y_1, \ldots, Y_n) \leq Q(Y_1, \ldots, Y_{i-1}, Y_i', Y_{i+1}, \ldots, Y_n)$$

whenever the involved arguments $Y_1, \ldots, Y_n, Y_i' \in \mathcal{P}(E)$ satisfy $Y_i \subseteq Y_i'$. Q is said to be nonincreasing in the i-th argument *if under the same conditions, it always holds that*

$$Q(Y_1, \ldots, Y_n) \geq Q(Y_1, \ldots, Y_{i-1}, Y_i', Y_{i+1}, \ldots, Y_n)\,.$$

The corresponding definitions for fuzzy quantifiers $\widetilde{Q} : \widetilde{\mathcal{P}}(E)^n \longrightarrow \mathbf{I}$ are entirely analogous. In this case, the arguments range over $\widetilde{\mathcal{P}}(E)$, and '$\subseteq$' is the fuzzy inclusion relation.

Notes 3.16.

- To present an example, "all" is nonincreasing in the first argument and nondecreasing in the second argument. From the former property, we then know that

$$\mathbf{all}(\mathbf{married} \cap \mathbf{men}, \mathbf{have\_children}) \geq \mathbf{all}(\mathbf{men}, \mathbf{have\_children}),$$

  which expresses that "All married men have children" makes a weaker requirement than "All men have children". As to the latter property, we may conclude from the nondecreasing monotonicity of "all" in its second argument that

$$\mathbf{all}(\mathbf{men}, \mathbf{have\_daughters}) \leq \mathbf{all}(\mathbf{men}, \mathbf{have\_children})\,.$$

  Consequently "All men have daughters" poses a stronger condition than "All men have children".
- In TGQ, nondecreasing monotonicity in the scope argument (i.e. in the argument slot which accepts the interpretation of the verb phrase) is usually termed *upward monotonicity*, while a quantifier which is nonincreasing in

this position is called *downward monotonic*, see e.g. [50, p. 232+].[5] Recalling the convention of reserving the last argument for denotation of the verb phrase, this means that positive monotonic quantifiers correspond to those quantifiers which are nondecreasing in the n-th argument, and downward monotonicity captures those quantifiers that are nonincreasing in their n-th argument. For example, "all" is upward monotonic and "no" is downward monotonic. The property of upward/downward monotonicity is rather typical of NL quantifiers.

- In the common case of two-place quantification, TGQ has also coined special terms for monotonicity in the first argument. In TGQ, those two-place quantifiers that are nondecreasing in their restriction are dubbed *persistent*, while those that are nonincreasing in the first argument are called *antipersistent*, see e.g. [7, p. 193] and [50, p. 242+]. Simple examples are "some" (persistent) and "all" (antipersistent). Compared to upward/downward monotonicity, there are fewer instances of NL quantifiers which are persistent or antipersistent [50, p. 243]. For example, proportional quantifiers like "most" typically lack both persistence and antipersistence.
- Zadeh [197, p. 164] has coined the similar notions of *monotone nondecreasing* and *monotone nonincreasing* quantifiers, which refer to the membership functions $\mu_Q : \mathbf{I} \longrightarrow \mathbf{I}$ used to define the proportional type.

Let us now discusss the close relationship between monotonicity properties of a quantifier and valid patterns of reasoning, and show how the monotonicity type of a quantifier constrains the inventory of admissible syllogisms. Hence consider the quantifier "more than ten", which is nondecreasing in its first argument. It then holds that

$$\begin{aligned} &\mathbf{more\ than\ ten}(\mathbf{married} \cap \mathbf{men}, \mathbf{have\_children}) \\ &\leq \mathbf{more\ than\ ten}(\mathbf{men}, \mathbf{have\_children})\,. \end{aligned}$$

It is this inequality which justifies the syllogism

More than ten married men have children
All married men are men
____
More than ten men have children

Other quantifiers like "some" can be substituted for "more than ten" here, which are also nondecreasing in the first argument. In the case of a quantifier which is *nonincreasing* in the first argument, like "no" or "all", the above pattern is no longer valid, and must be replaced with a pattern which fits the quantifier type. In this case, we obtain the converse inequality

$$\mathbf{no}(\mathbf{men}, \mathbf{have\_children}) \leq \mathbf{no}(\mathbf{married} \cap \mathbf{men}, \mathbf{have\_children})$$

and a corresponding pattern of reasoning,

[5] Barwise and Cooper [7, p. 184+], however, refer to positive and monotonic kinds as monotone increasing and monotone decreasing, respectively.

No men have children
All married men are men

No married men have children

As concerns monotonicity in the second argument, let us consider the quantifier "most" which is nondecreasing in its second argument. We then obtain that

$$\mathbf{most}(\mathbf{men}, \mathbf{married}) \leq \mathbf{most}(\mathbf{men}, \mathbf{married} \cup \mathbf{divorced}),$$

i.e. "Most men are married" expresses a stronger condition than "Most men are married or divorced". Again, the above inequality justifies the pattern of reasoning,

Most men are married
All married are married or divorced

Most men are married or divorced

Let us now turn to the fuzzy case. Even when there is fuzziness in the arguments, we would certainly expect the above patterns of reasoning to remain applicable. Because the above syllogisms are justified by the underlying inequality that express the monotonicity properties of the considered quantifiers, this means that we must enforce the *preservation of monotonicity properties* in order to maintain the applicable patterns of reasoning. For example, it should hold in a plausible model of fuzzy quantification that

$$\begin{aligned}&\mathcal{F}(\mathbf{more\ than\ ten})(\mathbf{young} \mathbin{\widetilde{\cap}} \mathbf{men}, \mathbf{rich})\\ &\leq \mathcal{F}(\mathbf{more\ than\ ten})(\mathbf{young} \mathbin{\widetilde{\cap}} \mathbf{persons}, \mathbf{rich})\end{aligned}$$

assuming that $\mathbf{men} \subseteq \mathbf{persons}$. It is this inequality which permits us to conclude from "More than ten young men are rich" to the entailed "More than ten young persons are rich", regardless of the fuzziness in $\mathbf{young}, \mathbf{rich} \in \widetilde{\mathcal{P}}(E)$. Similarly, it should hold that

$$\mathcal{F}(\mathbf{most})(\mathbf{young}, \mathbf{very\_poor}) \leq \mathcal{F}(\mathbf{most})(\mathbf{young}, \mathbf{poor}),$$

assuming that the fuzzy subsets $\mathbf{very\_poor}, \mathbf{poor} \in \widetilde{\mathcal{P}}(E)$ satisfy $\mathbf{very\_poor} \subseteq \mathbf{poor}$. We can then conclude from "Most young are very poor" that indeed "Most young are poor". These examples illustrate that the considered entailment relationships will transfer to the general case of fuzzy arguments, provided that the chosen model of quantification $\mathcal{F}$ preserve the underlying monotonicity properties. The relevant adequacy condition on QFMs can then be phrased as follows.

Definition 3.17 (Preservation of monotonicity in the arguments). *A QFM $\mathcal{F}$ is said to* preserve monotonicity in the arguments *if semi-fuzzy quantifiers $Q : \mathcal{P}(E)^n \longrightarrow \mathbf{I}$ which are nondecreasing (nonincreasing) in their i-th argument, $i \in \{1, \ldots, n\}$, are mapped to fuzzy quantifiers $\mathcal{F}(Q)$ which are also nondecreasing (nonincreasing) in their i-th argument.*

Notes 3.18.

- For example, $\mathcal{F}(\mathbf{all})$ should be nonincreasing in the first and nondecreasing in the second argument. The above motivating cases based on $\mathcal{F}(\textbf{more than ten})$ and $\mathcal{F}(\mathbf{most})$ are also instances of the general preservation property, of course.
- When combined with the other requirements on models of fuzzy quantification, the preservation condition can be restricted to the case that Q is *nonincreasing in its last argument.* Again, this restriction will serve the purpose of simplifying the axiom system and ensuring its independence. The fifth requirement on plausible models will be stated in terms of *nonincreasing* (rather than nondecreasing) monotonicity for technical purposes; when presented in this way, the criterion facilitates the proof that $\widetilde{\mathcal{F}}(\neg)$ is a strong negation operator. To give an example where the restricted property applies directly, consider the two-valued quantifier "no", which is nonincreasing in its last argument. By the restricted preservation condition, then, $\mathcal{F}(\mathbf{no})$ is nonincreasing in its last argument also. For instance, this ensures that

$$\mathcal{F}(\mathbf{no})(\mathbf{young}, \mathbf{rich}) \leq \mathcal{F}(\mathbf{no})(\mathbf{young}, \mathbf{very\_rich})\,,$$

  i.e. *"No young are very rich"* is entailed by *"No young are rich"*.

To sum up, every plausible model of fuzzy quantification should preserve the relevant entailment relationships, which are often tied to the monotonicity type of the involved quantifier. It is thus crucial for the theory of fuzzy quantification to consider the underlying monotonicity properties of quantifiers which show up in these entailments. The fifth requirement on models of fuzzy quantification therefore enforces a criterion which is essential to the preservation of general monotonicity properties, and in turn ensures that all corresponding patterns of reasoning remain valid in the presence of fuzziness.

3.8 The Induced Extension Principle

The final requirement on plausible models of fuzzy quantification is concerned with the problem of establishing a systematic relationship between the interpretations of quantifiers on different base sets. To this end, a homomorphism condition will be introduced based on powerset mappings which connect the behaviour of the fuzzification mechanism across domains:

Definition 3.19 (Powerset mapping).
To each mapping $f : E \longrightarrow E'$, we associate a mapping $\widehat{f} : \mathcal{P}(E) \longrightarrow \mathcal{P}(E')$ (the powerset mapping *of f) which is defined by*

$$\widehat{f}(Y) = \{f(e) : e \in Y\}\,,$$

for all $Y \in \mathcal{P}(E)$.

Notes 3.20.

- Often the same symbol f is used to denote both the original mapping and its extension to powersets. In the present context, however, it is important to discern the base mapping f from its associated powerset mapping, and we will therefore use the $\widehat{f}$-notation throughout.
- There is a closely related concept, namely that of the inverse image mapping $f^{-1} : \mathcal{P}(E') \longrightarrow \mathcal{P}(E)$ of a given $f : E \longrightarrow E'$, which is (as usual) defined by

$$f^{-1}(V) = \{e \in E : f(e) \in V\}, \tag{3.5}$$

for all $V \in \mathcal{P}(E')$. Often if V is a singleton, i.e. $V = \{v\}$ for some $v \in E'$, we will simply write $f^{-1}(v)$.

The underlying mechanism which transports f to $\widehat{f}$ can be generalized to the case of fuzzy sets; such a mechanism is then called an *extension principle.* Formally, we define (a pretty general class of) extension principles as follows.

Definition 3.21 (Extension principle).
An extension principle $\mathcal{E}$ *assigns to each mapping* $f : E \longrightarrow E'$ *a corresponding mapping* $\mathcal{E}(f) : \widetilde{\mathcal{P}}(E) \longrightarrow \widetilde{\mathcal{P}}(E')$. *For convenience, we shall assume that* $E, E' \neq \varnothing$.

Notes 3.22.

- Extension principles thus provide the desired mechanism which associates fuzzy powerset mappings $\mathcal{E}(f) : \widetilde{\mathcal{P}}(E) \longrightarrow \widetilde{\mathcal{P}}(E')$ with given base mappings $f : E \longrightarrow E'$. We do not impose any a priori restrictions on the well-behavedness of extension principles. These will later result from the axioms on plausible models of fuzzy quantification. A survey on the intuitive expectations on reasonable choices, along with the theorems that these semantical conditions are fulfilled in the models, is given below in Sect. 4.12. These results and the analysis of standard quantifiers in Sect. 4.16 will further reveal that plausible extension principles can be expressed in terms of existential quantification, i.e. they can be represented by a possibly infinitary formula built from an underlying s-norm.
- The case that $E = \varnothing$ or $E' = \varnothing$ has been excluded in order to allow a simpler definition of induced extension principles that will be associated with QFMs (to be defined below in Def. 3.25). After all, the case of base mappings f with an empty domain or range is irrelevant to our present purposes anyway. However, the issue is likely to be judged differently when discussing extension principles in a broader context. The restriction to nonempty sets is just a matter of convenience because it will simplify the later definition of the induced extension principle.
Every extension principle (in the sense of the above definition) can easily

be completed into a 'full' extension principle, which is also defined in the case of $E = \varnothing$ or $E' = \varnothing$. Hence consider $f : \varnothing \longrightarrow E'$.
In this case, we stipulate that

$$\mathcal{E}(f) = c_\varnothing\,, \tag{3.6}$$

where $c_\varnothing : \widetilde{\mathcal{P}}(\varnothing) \longrightarrow \widetilde{\mathcal{P}}(E')$ is the constant which assigns the set $c_\varnothing(\varnothing) = \varnothing \in \widetilde{\mathcal{P}}(E')$ to $\varnothing \in \widetilde{\mathcal{P}}(\varnothing) = \{\varnothing\}$. In the remaining case that $E' = \varnothing$, we simply observe from the definition of a function that the considered $f : E \longrightarrow \varnothing$ only qualifies as a mapping if $E = \varnothing$ as well. However, the case that $E' = \varnothing$ is already covered by equation (3.6).

The prototypical example of an extension principle has been suggested by Zadeh [189]. Recast in our notation, this 'standard extension principle' is defined as follows.

Definition 3.23 (Standard extension principle).
Let $f : E \longrightarrow E'$ a mapping. The standard extension principle *assigns to f the fuzzy powerset mapping $\hat{\hat{f}} : \widetilde{\mathcal{P}}(E) \longrightarrow \widetilde{\mathcal{P}}(E')$ defined by*

$$\mu_{\hat{\hat{f}}(X)}(y) = \sup\{\mu_X(e) : e \in f^{-1}(y)\}$$

for all $y \in E'$, where $f^{-1}(y) = \{e \in E : y = f(e)\}$.

Note 3.24. The extension principle can also be generalized to n-ary mappings $f : E_1 \times \cdots \times E_n \longrightarrow E'$ which it takes to n-ary fuzzy powerset mappings $\hat{\hat{f}} : \widetilde{\mathcal{P}}(E_1) \times \cdots \times \widetilde{\mathcal{P}}(E_n) \longrightarrow \widetilde{\mathcal{P}}(E')$, defined by

$$\mu_{\hat{\hat{f}}(X_1,\ldots,X_n)}(y) = \sup\{\min_{i=1}^{n} \mu_{X_i}(e_i) : (e_1, \ldots, e_n) \in f^{-1}(y)\}\,,$$

for all $X_1 \in \widetilde{\mathcal{P}}(E_1), \ldots, X_n \in \widetilde{\mathcal{P}}(E_n)$ and $y \in E'$; see Yager [178] for details. This generalized version of the extension principle will be of no importance for the following. However, it is apparent that the general extension principles $\mathcal{E}$ considered here can be generalized to n-place mappings along the same lines.

With each QFM, we can associate a corresponding extension principle through a canonical construction.

Definition 3.25.
Every QFM $\mathcal{F}$ induces an extension principle $\widehat{\mathcal{F}}$ which to each $f : E \longrightarrow E'$ (where $E, E' \neq \varnothing$) assigns the mapping $\widehat{\mathcal{F}}(f) : \widetilde{\mathcal{P}}(E) \longrightarrow \widetilde{\mathcal{P}}(E')$ defined by

$$\mu_{\widehat{\mathcal{F}}(f)(X)}(e') = \mathcal{F}(\chi_{\widehat{f}(\bullet)}(e'))(X)\,,$$

for all $X \in \widetilde{\mathcal{P}}(E)$, $e' \in E'$; or more succinctly:

$$\mu_{\widehat{\mathcal{F}}(f)(X)}(e') = \mathcal{F}(\pi_{e'} \circ \widehat{f})(X)$$

The equivalence of the first and second form is obvious from the definition of projection quantifiers, see Def. 3.2. The powerset functions $\widehat{f} : \mathcal{P}(E) \longrightarrow \mathcal{P}(E')$ and the corresponding fuzzy powerset functions $\widehat{\mathcal{F}}(f) : \widetilde{\mathcal{P}}(E) \longrightarrow \widetilde{\mathcal{P}}(E')$ obtained from the induced extension principle are important in our context because they can be applied to the argument sets of semi-fuzzy quantifiers (crisp case, $\widehat{f}$) and fuzzy quantifiers (fuzzy case, using $\widehat{\mathcal{F}}(f)$).

We require that every 'reasonable' choice of $\mathcal{F}$ be compatible with its induced extension principle in the following sense.
Suppose that $Q : \mathcal{P}(E)^n \longrightarrow \mathbf{I}$ is a semi-fuzzy quantifier and $f_1, \ldots, f_n : E' \longrightarrow E$ are given mappings, $E' \neq \varnothing$. We can construct a semi-fuzzy quantifier $Q' : \mathcal{P}(E')^n \longrightarrow \mathbf{I}$ by composing Q with the powerset mappings $\widehat{f_i}, \ldots, \widehat{f_n}$, i.e.

$$Q'(Y_1, \ldots, Y_n) = Q(\widehat{f_1}(Y_1), \ \ldots, \ \widehat{f_n}(Y_n)),$$

for all $Y_1, \ldots, Y_n \in \mathcal{P}(E)$. This can be expressed more compactly if we recall the concept of *product mapping*. If $g_1, \ \ldots, \ g_m : A \longrightarrow B$ are any mappings, then $\overset{n}{\underset{i=1}{\times}} g_i : A^n \longrightarrow B^n$ is defined by

$$(\overset{n}{\underset{i=1}{\times}} g_i)(x_1, \ \ldots, \ x_n) = (g_1(x_1), \ \ldots, \ g_n(x_n)).$$

By using the product, the above definition of Q' then becomes $Q' = Q \circ \overset{n}{\underset{i=1}{\times}} \widehat{f_i}$, because

$$(Q \circ \overset{n}{\underset{i=1}{\times}} \widehat{f_i})(Y_1, \ldots, Y_n) = Q(\widehat{f_1}(Y_1), \ldots, \widehat{f_n}(Y_n)) \tag{3.7}$$

for all $Y_1, \ldots, Y_n \in \mathcal{P}(E')$; '$\circ$' denotes functional composition. By utilizing the induced extension principle $\widehat{\mathcal{F}}$ of a QFM, we can perform a similar construction on fuzzy quantifiers, thus composing $\widetilde{Q} : \widetilde{\mathcal{P}}(E)^n \longrightarrow \mathbf{I}$ with $\widehat{\mathcal{F}}(f_1), \ldots, \widehat{\mathcal{F}}(f_n)$ to form the fuzzy quantifier $\widetilde{Q} \circ \overset{n}{\underset{i=1}{\times}} \widehat{\mathcal{F}}(f_i) : \widetilde{\mathcal{P}}(E')^n \longrightarrow \mathbf{I}$ defined by

$$(\widetilde{Q} \circ \overset{n}{\underset{i=1}{\times}} \widehat{\mathcal{F}}(f_i))(X_1, \ldots, X_n) = \widetilde{Q}(\widehat{\mathcal{F}}(f_1)(X_1), \ldots, \widehat{\mathcal{F}}(f_n)(X_n)),$$

for all $X_1, \ldots, X_n \in \widetilde{\mathcal{P}}(E')$. We require that a plausible choice of QFM $\mathcal{F}$ comply with this construction, and hence impose the following homomorphism condition with respect to the application of (crisp or fuzzy) powerset functions:

Definition 3.26 (Compatibility with functional application).
Let $\mathcal{F}$ be a given QFM. We say that $\mathcal{F}$ is compatible with functional application *if the equality*

$$\mathcal{F}(Q \circ \mathop{\times}_{i=1}^{n} \widehat{f}_i) = \mathcal{F}(Q) \circ \mathop{\times}_{i=1}^{n} \widehat{\mathcal{F}}(f_i) \tag{3.8}$$

is valid for all semi-fuzzy quantifiers $Q : \mathcal{P}(E)^n \longrightarrow \mathbf{I}$ *and mappings* $f_1, \ldots, f_n : E' \longrightarrow E$ *with domain* $E' \neq \varnothing$.

Notes 3.27.

- The term 'functional application' has been chosen because the constructed quantifier is obtained from applying the extended functions $f_1, \ldots, f_n$ to the corresponding arguments.
- Let us notice in advance that it is not possible to span the intended class of models if the induced powerset mappings $\widehat{\mathcal{F}}(f_i)$ are replaced with the standard choice of fuzzy powerset mappings $\hat{\hat{f}}_i$ obtained from the standard extension principle. As will be shown below in theorem Th. 4.62, the use of the standard extension principle would restrict the admissible models of fuzzy quantification to those that induce the standard disjunction $\widetilde{\mathcal{F}}(\vee) = \max$. This is too restrictive because we would like to have models for arbitrary t- and s-norms. It was the intent to cover such general models which motivated the development of the induced extension principle because any reference to the standard extension principle would result in a restriction to the limited class of 'standard models'. (Of course, the standard models are the preferred choice in most applications, but it deepens the knowledge of fuzzy quantification if other models can also be studied).

As mentioned above, the induced extension principle $\widehat{\mathcal{F}}$ enables us to construct fuzzy quantifiers $Q' = Q \circ \mathop{\times}_{i=1}^{n} \widehat{\mathcal{F}}(f_i)$ on a base set E from a given fuzzy quantifier Q on *another* base set E'. It is this cross-domain characteristic which makes the compliance condition expressed by (3.8) especially important to the theory of fuzzy quantification. In fact, the required compatibility with functional application will constitute the *only* criterion in the proposed axiom system which controls the behaviour of $\mathcal{F}$ on different base sets E, E'. For example, if $\beta : E \longrightarrow E'$ is a bijection, $Q : \mathcal{P}(E')^n \longrightarrow \mathbf{I}$ is a semi-fuzzy quantifier, and $Q' : \mathcal{P}(E)^n \longrightarrow \mathbf{I}$ is defined by

$$Q'(Y_1, \ldots, Y_n) = Q(\widehat{\beta}(Y_1), \ \ldots, \ \widehat{\beta}(Y_n))$$

for all $Y_1, \ldots, Y_n \in \mathcal{P}(E)$, then

$$\mathcal{F}(Q')(X_1, \ldots, X_n) = \mathcal{F}(Q)(\widehat{\mathcal{F}}(\beta)(X_1), \ \ldots, \ \widehat{\mathcal{F}}(\beta)(X_n))$$

for all $X_1, \ldots, X_n \in \widetilde{\mathcal{P}}(E)$, and

$$\mathcal{F}(Q)(X_1, \ldots, X_n) = \mathcal{F}(Q')(\widehat{\mathcal{F}}(\beta^{-1})(X_1), \ \ldots, \ \widehat{\mathcal{F}}(\beta^{-1})(X_n))$$

for all $X_1, \ldots, X_n \in \widetilde{\mathcal{P}}(E')$, which shows that $\mathcal{F}$ may not depend on any particular properties of elements of a base set E.

The sixth and last requirement that will be imposed on models of fuzzy quantification, i.e. the compatibility with functional application, thus enforces a coherent behaviour of $\mathcal{F}$ across domains.

3.9 Determiner Fuzzification Schemes: The DFS Axioms

The requirements on plausible models of fuzzy quantification are summarized in the following axiom system.

Definition 3.28 (Determiner fuzzification schemes).
A QFM $\mathcal{F}$ is called a determiner fuzzification scheme *(DFS) if the following conditions are satisfied for all semi-fuzzy quantifiers $Q : \mathcal{P}(E)^n \longrightarrow \mathbf{I}$.*

Correct generalisation	$\mathcal{U}(\mathcal{F}(Q)) = Q \quad \textit{if } n \leq 1$	(Z-1)
Projection quantifiers	$\mathcal{F}(Q) = \widetilde{\pi}_e \quad \textit{if } Q = \pi_e \textit{ for some } e \in E$	(Z-2)
Dualisation	$\mathcal{F}(Q\widetilde{\square}) = \mathcal{F}(Q)\widetilde{\square} \quad n > 0$	(Z-3)
Internal joins	$\mathcal{F}(Q\cup) = \mathcal{F}(Q)\widetilde{\cup} \quad n > 0$	(Z-4)
Preservation of monotonicity	*If Q is nonincreasing in the n-th arg, then $\mathcal{F}(Q)$ is nonincreasing the in n-th arg, $n > 0$*	(Z-5)
Functional application	$\mathcal{F}(Q \circ \underset{i=1}{\overset{n}{\times}} \widehat{f_i}) = \mathcal{F}(Q) \circ \underset{i=1}{\overset{n}{\times}} \widehat{\mathcal{F}}(f_i)$ *where* $f_1, \ldots, f_n : E' \longrightarrow E$, $E' \neq \varnothing$.	(Z-6)

Notes 3.29.

- The term 'determiner fuzzification scheme' (DFS) rather than 'quantifier fuzzification scheme' has been chosen to reduce the risk of confusion with quantifier fuzzification mechanisms (QFMs). However, a DFS is nothing but a special kind of QFM which satisfies the above conditions.
- These conditions simply combine the six requirements introduced above which capture important expectations on models of fuzzy quantification. Some of the requirements have been restricted to special cases. This will reduce the effort of proving that a given candidate QFM $\mathcal{F}$ is indeed a model of the theory. Most importantly, the restriction to special cases avoids any overlap of the criteria. This is necessary to obtain a system of independent axioms. This consideration motivated the restriction of 'correct generalisation' (Z-1) to quantifiers of arities $n \leq 1$, and the restriction of the monotonicity requirement to quantifiers which are nonincreasing in their last argument. The original, unrestricted requirements are of course entailed by the resulting axiom system; see Th. 4.1 and Th. 4.28 in the next chapter.

- We have taken pains to avoid any reference to the standard connectives and the standard extension principle in the DFS axioms (in favor of the *induced* connectives and induced extension principle of a DFS) because the use of the standard choices would have excluded all models which induce $\widetilde{\mathcal{F}}(\vee) \neq \max$ and $\widetilde{\mathcal{F}}(\wedge) \neq \min$ (see Th. 5.25 below). In order to maintain the possibility of 'non-standard models' based on fuzzy conjunctions different from min, we associated a canonical choice of fuzzy truth functions and general extension principle to each QFM of interest. The constructions on quantifiers which depend on a fuzzy truth function or the extension principle have then been formulated in terms of these induced constructions. This strategy targets at a coherent or 'self-consistent' system compatible with its own induced constructions.
- The original presentation of determiner fuzzification schemes in [53] was based on a set of nine DFS axioms. These axioms contained some interdependencies, though, which motivated the refinement of the original axioms into the equivalent axiom system presented above. See [55] for details on both axiom systems and the proof of their equivalence.

The proposed axioms might appear rather abstract at first sight. This is because these axioms have resulted from a process of simplification and compression. The development of the axiom system started with the formalization of a large number of linguistic plausibility criteria (the full list will be discussed in the next three chapters). By eliminating redundancy and interdependencies, a generating system of basic criteria was then identified. This reduction was necessary to obtain a minimal (i.e. irreducible) axiom system of elementary criteria which can be easily verified for every QFM of interest. The remaining plausibility criteria which are not part of the axiom system must then be *entailed* by these core axioms. Let us now state that the final axiom system is indeed independent, i.e. the axioms (Z-1)–(Z-6) achieve a minimal characterization of the intended class of models:

Theorem 3.30.
The DFS axioms (Z-1)–(Z-6) *are independent, i.e. none of the conditions is entailed by the remaining conditions.*

Thus each criterion identifies a dimension in the space of possible models which is significant with respect to their semantic plausibility.

3.10 Chapter Summary

The novel quantification framework proposed in this book has introduced a rich class of potential models. In this chapter, we made an effort to identify the plausible models of fuzzy quantification within the proposed framework. In order to develop a general solution to the problem of systematic and coherent interpretation, we pursued a strategy which is essentially algebraic, i.e. the plausible models were characterized in terms of observable properties of the considered QFM. It is thus assumed that all important aspects of plausible interpretations can be expressed by formal criteria which constrain the admissible choices of QFMs. The catalogue of formal semantic criteria will then guarantee that the interpretations of quantifiers be restricted to the plausible choices.

Embarking on this general strategy, we considered various intuitive requirements on the models, which were then compiled into 'DFS axioms', the axiomatic framework for plausible models of fuzzy quantification. The relevance of the six basic postulates (Z-1)–(Z-6) can be sketched as follows.

- The first condition of 'correct generalisation' (Z-1) is mandatory to ensure the internal coherence of the quantification framework. Generally speaking, one will always require that the fuzzifications results determined by a fuzzification mechanism properly generalize the original constructs for crisp sets to which the fuzzification mechanism was applied.
- The semantical postulate (Z-2) which demands the proper interpretation of projection quantifiers makes explicit the relationship of crisp/fuzzy membership assessments and quantification. This requirement is essential from a linguistic perspective because it warrants the correct modelling of proper names like "Joan", which are generally viewed as a special type of quantifiers in TGQ.
- The requirement of compatibility with dualisation (Z-3) is also important from a logical and linguistic viewpoint: When combined with the other axioms, it enforces the full compliance of the models with the important constructions of the dual, external negation and antonym of a quantifier.
- The condition on 'internal joins' (Z-4), which demands the conformance of the models to unions of arguments, can be exemplified by NL sentences like *"Most men drink or smoke"*. Apart from its linguistic motivation, the condition is of key relevance to the DFS axioms, because it is the only requirement which mediates the behaviour of a considered QFM for quantifiers of different arities. It is this condition which will achieve the desired *cross-arity coherence* of the model, thus contributing to the proper modelling of fuzzy multi-place quantification.
- The requirement of 'monotonicity in arguments' (Z-5) makes sure that the monotonicity type of a base quantifier translate to the fuzzy case. This is especially important in order to preserve the valid entailments which pertain to these monotonicity properties.
- The final criterion (Z-6) which demands the compatibility with functional application is not directly inspired from linguistics. Due to the fact that all other axioms refer to quantifiers on a single base set E only, it was necessary to add a requirement which links the behaviour of $\mathcal{F}$ across domains. The criterion of functional application is solely concerned with this issue of *cross-domain coherence*, which it traces back to the obvious desideratum of compositionality with respect to powerset mappings.

It should be apparent from this summary that the DFS axioms are well-motivated and linguistically significant. As stated in Th. 3.30, the axioms are also mutually independent, i.e. irreducible to a smaller axiom set. In addition, the proposed axioms are *consistent*. It will be shown in the later Chaps. 7 to 10 that the axioms admit rich and interesting classes of models.

However, an axiom set cannot be appropriately judged when looking at the axioms in isolation. The axiom system had to be rather compressed in order to achieve the desired minimality. This means that most semantical properties of interest are not *included* in the list of axioms, but rather *entailed* by the proposed axiom system. In the following chapters, we will therefore consider a larger catalogue of linguistic requirements and check the validity of each criterion in the models. In this way, we can validate the completeness of the axioms, i.e. their coverage of linguistic expectations on the interpretation of quantifiers. As opposed to the bona fide assumption of linguistic plausibility made by existing approaches, the proposal will then be subjected to a systematic linguistic evaluation, which substantiates that it indeed captures the plausible models of fuzzy quantification.

4

Semantic Properties of the Models

4.1 Motivation and Chapter Overview

In the last chapter, an axiom system for models of fuzzy quantification has been proposed. The system comprises six basic criteria for plausible interpretations of fuzzy quantifiers. The description of these models in terms of the postulates (Z-1)–(Z-6) is rather condensed, though. On the one hand, this offers the advantage of a minimal, non-redundant definition. However, the succinct description of the models makes it difficult to judge whether the requirements capture all important aspects of systematic and coherent interpretations.

The following three chapters will provide evidence that this is indeed the case. The present chapter will be concerned with semantic desiderata which must be considered absolutely basic, i.e. mandatory for all models. Most of these criteria originate from the Theory of Generalized Quantifiers and thus reflect linguistic considerations on the models. Some other postulates will originate from logic. For example, we will discuss the proper interpretation of the universal and existential quantifiers. We will also be interested in constraints on the expected effects of fuzziness on the quantification results, which have something to say about about our understanding of fuzziness in natural language.

The present chapter will formalize all of these criteria and prove that they are valid in arbitrary models of the theory. The other two chapters in the series on semantical properties will develop concepts which apply to models in a given subclass (Chap. 5) and discuss advanced topics including more specialized properties of models and some problematic cases (Chap. 6).

4.2 Correct Generalisation

Let us start with the most basic requirement on any type of fuzzification mechanism, that of *correct generalization*. It is essential to the use of the

I. Glöckner: *Fuzzy Quantifiers*, StudFuzz **193**, 111–148 (2006)
www.springerlink.com © Springer-Verlag Berlin Heidelberg 2006

fuzzification pattern that the generalized model which results from the fuzzification mechanism consistently extend the crisp base model from which it is built [59, p. 419+]. In the case of quantifier fuzzification mechanisms, we would thus like the fuzzy quantifiers $\mathcal{F}(Q)$ that result from the mechanism to coincide with the original semi-fuzzy quantifiers Q on arbitrary crisp arguments. It was already shown on p. 87 how this requirement can be succinctly expressed in terms of the 'underlying semi-fuzzy quantifiers' $\mathcal{U}(\widetilde{Q})$, and it then becomes $\mathcal{U}(\mathcal{F}(Q)) = Q$. Acknowledging its importance, correct generalisation has been made an integral part of the DFS axioms, and We have required in (Z-1) that the equality $\mathcal{U}(\mathcal{F}(Q)) = Q$ be valid for all nullary or unary quantifiers The restriction to quantifiers of arities $n \leq 1$ helped to simplify the axiom system, thus shortening the required proofs that certain QFMs of interest are indeed models of the theory. Let us now overcome the restriction to $n \leq 1$ and establish that $\mathcal{F}(Q)$ consistently extends Q, regardless of the arity $n \in \mathbb{N}$ of the considered quantifier $Q : \mathcal{P}(E)^n \longrightarrow \mathbf{I}$.

Theorem 4.1.
Suppose that $\mathcal{F}$ is a DFS and $Q : \mathcal{P}(E)^n \longrightarrow \mathbf{I}$ is an n-ary semi-fuzzy quantifier. Then $\mathcal{U}(\mathcal{F}(Q)) = Q$, i.e. for all crisp *subsets $Y_1, \ldots, Y_n \in \mathcal{P}(E)$,*

$$\mathcal{F}(Q)(Y_1, \ldots, Y_n) = Q(Y_1, \ldots, Y_n) .$$

For example, if E is a set of persons, and **women**, **married** $\in \mathcal{P}(E)$ are the crisp sets of "women" and "married persons" in E, then

$$\mathcal{F}(\mathbf{some})(\mathbf{women}, \mathbf{married}) = \mathbf{some}(\mathbf{women}, \mathbf{married}) ,$$

i.e. the 'fuzzy some' obtained by applying $\mathcal{F}$ coincides with the (original) 'crisp some' whenever the latter is defined, which is of course highly desirable.

4.3 Properties of the Induced Truth Functions

Let us now turn to the fuzzy truth functions induced by a DFS, which constitute the propositional part of its induced logic. The first theorem is concerned with the identity truth function $\mathrm{id}_\mathbf{2} : \mathbf{2} \longrightarrow \mathbf{2}$ defined by $\mathrm{id}_\mathbf{2}(0) = 0$, $\mathrm{id}_\mathbf{2}(1) = 1$.

Theorem 4.2 (Identity truth function).
In every DFS $\mathcal{F}$, $\widetilde{\mathcal{F}}(\mathrm{id}_\mathbf{2}) = \mathrm{id}_\mathbf{I}$.

The identity truth function is hence generalised to the fuzzy identity truth function $\mathrm{id}_\mathbf{I} : \mathbf{I} \longrightarrow \mathbf{I}$ which maps every gradual truth value $x \in \mathbf{I}$ to itself, which is quite satisfactory.

As for negation, the standard choice in fuzzy logic is certainly $\neg : \mathbf{I} \longrightarrow \mathbf{I}$, defined by $\neg x = 1 - x$ for all $x \in \mathbf{I}$. The essential properties of this prototypical choice of negation are captured by the following definition, which guides reasonable choices of negation operators.

Definition 4.3 (Strong negation).
$\widetilde{\neg} : \mathbf{I} \longrightarrow \mathbf{I}$ *is called a* strong negation operator *if it satisfies*

a. $\widetilde{\neg} 0 = 1$ *(boundary condition)*
b. $\widetilde{\neg} x_1 \geq \widetilde{\neg} x_2$ *for all* $x_1, x_2 \in \mathbf{I}$ *such that* $x_1 \leq x_2$ *(i.e.* $\widetilde{\neg}$ *is monotonically nonincreasing)*
c. $\widetilde{\neg} \circ \widetilde{\neg} = \mathrm{id}_{\mathbf{I}}$ *(i.e.* $\widetilde{\neg}$ *is involutive).*

Notes 4.4.

- Whenever the standard negation $\neg x = 1 - x$ is being assumed, we will drop the 'tilde'-notation. Hence the standard fuzzy complement is denoted $\neg X$, where $\mu_{\neg X}(e) = 1 - \mu_X(e)$. Similarly, the external negation of a (semi-) fuzzy quantifier with respect to the standard negation is written $\neg Q$, and the antonym of a fuzzy quantifier with respect to the standard fuzzy complement is written as $\underset{\sim}{Q}\neg$.
- Apart from the standard choice, further examples of these negation operators are e.g. Sugeno's λ-complements [154].

As witnessed by the following theorem, the fuzzy negation induced by a DFS $\mathcal{F}$ is plausible in the sense of belonging to the class of strong negation operators.

Theorem 4.5 (Negation).
In every DFS $\mathcal{F}$, $\widetilde{\neg} = \widetilde{\mathcal{F}}(\neg)$ *is a strong negation operator.*

With conjunction, there are several common choices in fuzzy logic (although the standard is certainly $\wedge = \min$). All of these belong to the class of t-norms, which capture the expectations on reasonable conjunction operators, cf. Schweizer/Sklar [148], Klir/Yuan [99, p. 61+].

Definition 4.6 (t-norms).
$\widetilde{\wedge} : \mathbf{I} \times \mathbf{I} \longrightarrow \mathbf{I}$ *is called a t-norm if it satisfies the following conditions.*

a. $x \mathbin{\widetilde{\wedge}} 0 = 0$, *for all* $x \in \mathbf{I}$
b. $x \mathbin{\widetilde{\wedge}} 1 = x$, *for all* $x \in \mathbf{I}$
c. $x_1 \mathbin{\widetilde{\wedge}} x_2 = x_2 \mathbin{\widetilde{\wedge}} x_1$ *for all* $x_1, x_2 \in \mathbf{I}$ *(commutativity)*
d. *If* $x_1 \leq x_1'$, *then* $x_1 \mathbin{\widetilde{\wedge}} x_2 \leq x_1' \mathbin{\widetilde{\wedge}} x_2$, *for all* $x_1, x_1', x_2 \in \mathbf{I}$ *(i.e.* $\widetilde{\wedge}$ *is monotonically nondecreasing)*
e. $(x_1 \mathbin{\widetilde{\wedge}} x_2) \mathbin{\widetilde{\wedge}} x_3 = x_1 \mathbin{\widetilde{\wedge}} (x_2 \mathbin{\widetilde{\wedge}} x_3)$, *for all* $x_1, x_2, x_3 \in \mathbf{I}$ *(associativity).*

The prototypical t-norm is the standard fuzzy conjunction $\wedge = \min$ (it is the largest t-norm). Other well-known examples of t-norms are the *algebraic product* $\widetilde{\wedge}_a$ and the *bounded product* $\widetilde{\wedge}_b$, which are defined by

$$x_1 \mathbin{\widetilde{\wedge}_a} x_2 = x_1 \cdot x_2$$
$$x_1 \mathbin{\widetilde{\wedge}_b} x_2 = \max(0, x_1 + x_2 - 1)$$

for all $x_1, x_2 \in \mathbf{I}$. For a less well-known examplar, consider $\widetilde{\wedge}_m$, defined by

$$x_1 \mathbin{\widetilde{\wedge}_m} x_2 = \begin{cases} 2x_1x_2 & : \max(x_1, x_2) < \frac{1}{2} \\ 1 + 2x_1x_2 - x_1 - x_2 & : \min(x_1, x_2) \geq \frac{1}{2} \\ \min(x_1, x_2) & : \text{else} \end{cases} \tag{4.1}$$

for all $x_1, x_2 \in \mathbf{I}$, see [53, p. 30]. Of course, there are many other t-norms; for a recent discussion of advanced topics and pointers to other publications, see e.g. [96, 113].
Let us now state that the fuzzy conjunction induced by a DFS is 'reasonable', in the sense of belonging to the class of t-norms.

Theorem 4.7 (Conjunction).
In every DFS $\mathcal{F}$, $\widetilde{\wedge} = \widetilde{\mathcal{F}}(\wedge)$ is a t-norm.

The dual concept of t-norm is that of an s-norm or t-co-norm, which expresses the essential properties of fuzzy disjunction operators.

Definition 4.8 (s-norms).
$\widetilde{\vee} : \mathbf{I} \times \mathbf{I} \longrightarrow \mathbf{I}$ is called an s-norm if it satisfies the following conditions.

a. *$x \mathbin{\widetilde{\vee}} 1 = 1$, for all $x \in \mathbf{I}$*
b. *$x \mathbin{\widetilde{\vee}} 0 = x$, for all $x \in \mathbf{I}$*
c. *$x_1 \mathbin{\widetilde{\vee}} x_2 = x_2 \mathbin{\widetilde{\vee}} x_1$ for all $x_1, x_2 \in \mathbf{I}$ (commutativity)*
d. *If $x_1 \leq x_1'$, then $x_1 \mathbin{\widetilde{\vee}} x_2 \leq x_1' \mathbin{\widetilde{\vee}} x_2$, for all $x_1, x_1', x_2 \in \mathbf{I}$ (i.e. $\widetilde{\vee}$ is monotonically nondecreasing)*
e. *$(x_1 \mathbin{\widetilde{\vee}} x_2) \mathbin{\widetilde{\vee}} x_3 = x_1 \mathbin{\widetilde{\vee}} (x_2 \mathbin{\widetilde{\vee}} x_3)$, for all $x_1, x_2, x_3 \in \mathbf{I}$ (associativity).*

Examples of s-norms are the standard choice $\vee = \max$ (it is the smallest s-norm), the algebraic sum $\widetilde{\vee}_a$ and the bounded sum $\widetilde{\vee}_b$ defined by

$$x_1 \mathbin{\widetilde{\vee}_a} x_2 = x_1 + x_2 - x_1 \cdot x_2$$
$$x_1 \mathbin{\widetilde{\vee}_b} x_2 = \min(1, x_1 + x_2)$$

for all $x_1, x_2 \in \mathbf{I}$.

Theorem 4.9 (Disjunction).
In every DFS, $x_1 \mathbin{\widetilde{\vee}} x_2 = \widetilde{\neg}(\widetilde{\neg} x_1 \mathbin{\widetilde{\wedge}} \widetilde{\neg} x_2)$, i.e. $\widetilde{\vee}$ is the dual s-norm of $\widetilde{\wedge}$ under $\widetilde{\neg}$.

Hence the fuzzy disjunction induced by a DFS is also plausible, and it is definable in terms of $\widetilde{\wedge}$ and $\widetilde{\neg}$. A similar point can be made about the other two-place logical connectives, see [53, p. 29 and p. 32]. For example, the induced implication of a DFS can be expressed in terms of $\widetilde{\vee}$ and $\widetilde{\neg}$ (and hence also in terms of $\widetilde{\wedge}$ and $\widetilde{\neg}$):

Theorem 4.10 (Implication).
In every DFS, $x_1 \mathbin{\widetilde{\rightarrow}} x_2 = \widetilde{\neg} x_1 \mathbin{\widetilde{\vee}} x_2$.

The only connectives which are not apparently reducible to $\widetilde{\wedge}$ and $\widetilde{\neg}$, because their construction involves a subtle dependency between variables, are the antivalence xor and the equivalence $\leftrightarrow$. We will consider these connectives later because they require more effort (see remarks on pp. 155 and 166).

4.4 A Different View of the Induced Propositional Logic

The definition of induced fuzzy truth functions presented in Def. 3.6 is not the only possible choice, and it has already been mentioned that an alternative construction exists which is equally straightforward. In fact, the first publication on DFS theory [53] relied on the construction which has now become 'alternative'. It was only the desire to obtain an independent axioms system which later guided the decision in [55] to introduce a new construction. This new construction has now become standard, because it helped to eliminate some subtle interdependencies in the initial definition of the DFS axioms. In the following, we will review the principles that underly the 'old' construction of induced connectives. It will then be shown that the 'new' and 'old' constructions coincide in every DFS, which makes an additional point that these models are well-motivated. Hence both constructions provide different views of the induced propositional logic from their specific perspective.

The alternative construction of induced truth function, which we define now, rests on the observation that (a) the set of crisp truth values $\mathbf{2} = \{0, 1\}$ and the powerset $\mathcal{P}(\{*\})$ are isomorphic; and (b) the set of continuous truth values $\mathbf{I} = [0, 1]$ and the fuzzy powerset $\widetilde{\mathcal{P}}(\{*\})$ are also isomorphic, where $\{*\}$ is an arbitrary singleton set, e.g. $\{*\} = \{\varnothing\}$. Suitable mappings which establish these isomorphisms are the apparent bijections (a) $\pi_* : \mathcal{P}(\{*\}) \longrightarrow \mathbf{2}$; and (b) $\widetilde{\pi}_* : \widetilde{\mathcal{P}}(\{*\}) \longrightarrow \mathbf{I}$. The basic idea is to utilize the former bijection for a transfer of the original truth function to a semi-fuzzy quantifier, to which the considered QFM $\mathcal{F}$ can be applied. Now utilizing the latter bijection, the resulting fuzzy quantifier can then be translated into the desired fuzzy truth function. The alternative construction can thus be described as follows.

Definition 4.11.
Let a mapping $f : \mathbf{2}^n \longrightarrow \mathbf{I}$ be given. We can view f as a semi-fuzzy quantifier $f^* : \mathcal{P}(\{*\})^n \longrightarrow \mathbf{I}$ by defining

$$f^*(Y_1, \dots, Y_n) = f(\pi_*(Y_1), \ \dots, \ \pi_*(Y_n)) \, .$$

for all $Y_1, \dots, Y_n \in \mathcal{P}(\{*\})$. By applying the considered QFM $\mathcal{F}$, f^* is generalized to a fuzzy quantifier $\mathcal{F}(f^*) : \widetilde{\mathcal{P}}(\{*\})^n \longrightarrow \mathbf{I}$, from which we obtain a fuzzy truth function $\widetilde{\widetilde{\mathcal{F}}}(f) : \mathbf{I}^n \longrightarrow \mathbf{I}$,

$$\widetilde{\widetilde{\mathcal{F}}}(f)(x_1, \ \dots x_n) = \mathcal{F}(f^*)(\widetilde{\pi}_*^{-1}(x_1), \ \dots, \ \widetilde{\pi}_*^{-1}(x_n))$$

for all $x_1, \ \dots, \ x_n \in \mathbf{I}$.

More details on this construction can be found in [53]. As already remarked above, it results in the same canonical choice of fuzzy truth functions, if the considered $\mathcal{F}$ is a DFS.

Theorem 4.12.
In every DFS, $\widetilde{\mathcal{F}} = \widetilde{\widetilde{\mathcal{F}}}$.

$\widetilde{\mathcal{F}}$ can be distinct from $\widetilde{\widetilde{\mathcal{F}}}$ if $\mathcal{F}$ is not a DFS, though, and indeed the construction used in $\widetilde{\mathcal{F}}$ provided better support for developing DFS theory further and extracting an independent system of axioms. The alternative definition $\widetilde{\widetilde{\mathcal{F}}}$ was less suited to accomplish this task because the construction of $\widetilde{\widetilde{\mathcal{F}}}(\wedge)$ and $\widetilde{\widetilde{\mathcal{F}}}(\vee)$ using the 'old' construction involves multi-place quantification ($n = 2$), while the computation of $\widetilde{\mathcal{F}}(\wedge)$ and $\widetilde{\mathcal{F}}(\vee)$ using the 'new' method is only based on one-place quantification. It is this simplification which made it possible to formalize DFS theory in terms of independent conditions and develop the current set of DFS axioms (Z-1)–(Z-6).

Let us now discuss homomorphism properties of every DFS with respect to operations on the argument sets.

4.5 Argument Permutations

Definition 4.13 (Argument permutations).
Let $Q : \mathcal{P}(E)^n \longrightarrow \mathbf{I}$ be a semi-fuzzy quantifier and $\beta : \{1, \ldots, n\} \longrightarrow \{1, \ldots, n\}$ a permutation. By $Q\beta : \mathcal{P}(E)^n \longrightarrow \mathbf{I}$ we denote the semi-fuzzy quantifier defined by

$$Q\beta(Y_1, \ldots, Y_n) = Q(Y_{\beta(1)}, \ldots, Y_{\beta(n)}),$$

for all $Y_1, \ldots, Y_n \in \mathcal{P}(E)$. In the case of fuzzy quantifiers $\widetilde{Q} : \widetilde{\mathcal{P}}(E)^n \longrightarrow \mathbf{I}$, the quantifier $\widetilde{Q}\beta : \widetilde{\mathcal{P}}(E)^n \longrightarrow \mathbf{I}$ is defined analogously.

It is well-known that every permutation can be decomposed into a composition of transpositions. These express a single permutation step which simply swaps two selected elements in a given finite sequence:

Definition 4.14 (Transpositions).
For all $n \in \mathbb{N}$ ($n > 0$) and $i, j \in \{1, \ldots, n\}$, the transposition $\tau_{i,j}$ *: $\{1, \ldots, n\} \longrightarrow \{1, \ldots, n\}$ is defined by*

$$\tau_{i,j}(k) = \begin{cases} i & : \quad k = j \\ j & : \quad k = i \\ k & : \quad else \end{cases}$$

for all $k \in \{1, \ldots, n\}$. Let us further stipulate a succinct notation for $\tau_{i,n}$, which is abbreviated by $\tau_i = \tau_{i,n}$. The transposition τ_i has the effect of exchanging positions i and n.

Notes 4.15.

- The restricted class of transpositions τ_i, $i \in \{1, \ldots, n\}$, is still capable of composing arbitrary permutations, because every transposition $\tau_{i,j}$ can be expressed as $\tau_{i,j} = \tau_i \circ \tau_j \circ \tau_i$.

- In the case of these simple transpositions, $Q\tau_i(X_1, \ldots, X_n)$ becomes

$$Q\tau_i(X_1, \ldots, X_n) = Q(X_1, \ldots, X_{i-1}, X_n, X_{i+1}, \ldots, X_{n-1}, X_i)\,.$$

Due to the conceptual and notational simplicity of the transpositions τ_i, these offer the preferred representation of permutations that will be used throughout the monograph.
The importance of argument permutations/transpositions is witnessed by the following examples.

1. There is a meaning of **only** : $\mathcal{P}(E)^2 \longrightarrow \mathbf{2}$ where **only** = **all**τ_1, i.e.

$$\mathbf{only}(Y_1, Y_2) = \mathbf{all}(Y_2, Y_1)$$

   for all $Y_1, Y_2 \in \mathcal{P}(E)$. For example, if E is a set of persons, **men** $\in \mathcal{P}(E)$ the set of those $v \in E$ which are men, and **smokers** $\in \mathcal{P}(E)$ is the set[1] of persons who are smokers, then

$$\mathbf{only}(\mathbf{men}, \mathbf{smokers}) = \mathbf{all}\tau_1(\mathbf{men}, \mathbf{smokers}) = \mathbf{all}(\mathbf{smokers}, \mathbf{men})\,,$$

   i.e. the meaning of "Only men are smokers" coincides with that of "All smokers are men".
2. Argument transpositions render it possible to express symmetry properties of quantifiers. For example, the quantifier **some** : $\mathcal{P}(E)^2 \longrightarrow \mathbf{2}$ is symmetrical in its arguments, which can be stated as

$$\mathbf{some} = \mathbf{some}\tau_1\,.$$

   This warrants that in the above domain of persons/smokers,

$$\mathbf{some}(\mathbf{men}, \mathbf{smokers}) = \mathbf{some}\tau_1(\mathbf{men}, \mathbf{smokers}) = \mathbf{some}(\mathbf{smokers}, \mathbf{men})\,,$$

   which expresses our intuition that the meanings of "Some men are smokers" and "Some smokers are men" coincide.

As stated by the following theorem, the DFS axioms ensure that all considered models of fuzzy quantification commute with argument transpositions, i.e. the corresponding constructions based on fuzzy quantifiers and fuzzy arguments will carry the desired semantics.

Theorem 4.16.
Every DFS $\mathcal{F}$ is compatible with argument transpositions, i.e. for every a semi-fuzzy quantifier $Q : \mathcal{P}(E)^n \longrightarrow \mathbf{I}$ $i \in \{1, \ldots, n\}$,

$$\mathcal{F}(Q\tau_i) = \mathcal{F}(Q)\tau_i\,.$$

[1] we shall assume that this set be crisp for the sake of argument

Recalling that permutations can be expressed as a sequence of transpositions, Th. 4.16 actually ensures that $\mathcal{F}$ commutes with arbitrary permutations of the arguments of a quantifier. In particular, symmetry properties of a quantifier Q carry over to its fuzzified analogue $\mathcal{F}(Q)$. For example, it holds in every DFS that $\mathcal{F}(\textbf{some}) = \mathcal{F}(\textbf{some})\tau_1$ and thus

$$\mathcal{F}(\textbf{some})(\textbf{rich}, \textbf{young}) = \mathcal{F}(\textbf{some})(\textbf{young}, \textbf{rich}),$$

i.e. the meaning of "Some rich people are young" and "Some young people are rich" coincide. Of course, we also obtain that

$$\mathcal{F}(\textbf{only})(\textbf{old}, \textbf{rich}) = \mathcal{F}(\textbf{all})\tau_1(\textbf{old}, \textbf{rich}) = \mathcal{F}(\textbf{all})(\textbf{rich}, \textbf{old}),$$

i.e. "Only old people are rich" means that "All rich people are old", where the extensions $\textbf{old}, \textbf{rich} \in \widetilde{\mathcal{P}}(E)$ of old and rich people are now assumed to be fuzzy.

4.6 Cylindrical Extensions

Let us now consider another property related to the argument structure of quantifiers. In the case that an argument is 'redundant' or 'vacuous', by having no effect on the actual outcome of a semi-fuzzy quantifier Q, we would certainly expect that the argument will also have no effect on the quantification results obtained from the associated fuzzy quantifier $\mathcal{F}(Q)$. This intuitive criterion can be formalized as follows.

Definition 4.17.
Let $\mathcal{F}$ be a QFM. We say that $\mathcal{F}$ is compatible with cylindrical extensions *if the following condition holds for every semi-fuzzy quantifier $Q : \mathcal{P}(E)^n \longrightarrow \mathbf{I}$. If $Q' : \mathcal{P}(E)^{n'} \longrightarrow \mathbf{I}$ is defined by*

$$Q'(Y_1, \ldots, Y_{n'}) = Q(Y_{i_1}, \ldots, Y_{i_{n'}})$$

for all $Y_1, \ldots, Y_{n'} \in \mathcal{P}(E)$, where $n' \in \mathbb{N}$, $n' \geq n$ and $i_1, \ldots, i_n \in \{1, \ldots, n'\}$ such that $1 \leq i_1 < i_2 < \cdots < i_n \leq n'$, then

$$\mathcal{F}(Q')(X_1, \ldots, X_{n'}) = \mathcal{F}(Q)(X_{i_1}, \ldots, X_{i_n}),$$

for all $X_1, \ldots, X_{n'} \in \widetilde{\mathcal{P}}(E)$.

This property of being compatible with cylindrical extensions is very fundamental. It simply states that vacuous argument positions of a quantifier can be eliminated. For example, if $Q' : \mathcal{P}(E)^4 \longrightarrow \mathbf{I}$ is a semi-fuzzy quantifier and if there exists a semi-fuzzy quantifier $Q : \mathcal{P}(E) \longrightarrow \mathbf{I}$ such that $Q'(Y_1, Y_2, Y_3, Y_4) = Q(Y_3)$ for all $Y_1, \ldots, Y_4 \in \mathcal{P}(E)$, then we know that Q' does not really depend on all arguments; it is apparent that the choice of

Y_1, Y_2 and Y_4 has no effect on the quantification result. In this case, it is straightforward to require that $\mathcal{F}(Q')(X_1, X_2, X_3, X_4) = \mathcal{F}(Q)(X_3)$ for all $X_1, \ldots, X_4 \in \widetilde{\mathcal{P}}(E)$, i.e. $\mathcal{F}(Q')$ is also independent of X_1, X_2, X_4, and it can be computed from $\mathcal{F}(Q)$.

Theorem 4.18.

Every DFS $\mathcal{F}$ is compatible with cylindrical extensions.

Hence every model of DFS theory fulfills a property which is vital to the plausible analysis of multi-place quantification.

4.7 Negation and Antonyms

So far we have considered the elementary adequacy properties of QFMs under the semantical constructions of permuting and vacuously extending arguments. Despite the rather abstract appearence of these criteria, they are essential for the approach to be internally consistent. We will now turn to properties that also express on the linguistic surface. We first consider those constructions that involve a negation step, which can either be applied inside the quantifying expression (internal complementation of arguments, antonyms), or applied from 'outside' to the quantifying expression as a whole (external negation). The compatibility of the models of DFS theory to these semantical constructions is in each case ensured by their known compatibility with dualisation (Z-3) and by the elementary properties underlying the framework that were discussed above.

Theorem 4.19.
Every DFS $\mathcal{F}$ is compatible with the formation of antonyms. Thus if $Q : \mathcal{P}(E)^n \longrightarrow \mathbf{I}$ is a semi-fuzzy quantifier of arity $n > 0$, then $\mathcal{F}(Q\neg) = \mathcal{F}(Q)\widetilde{\neg}$.

The theorem guarantees e.g. that $\mathcal{F}(\mathbf{all})(\mathbf{rich}, \widetilde{\neg}\,\mathbf{lucky}) = \mathcal{F}(\mathbf{no})(\mathbf{rich}, \mathbf{lucky})$. Let us notice that by Th. 4.16, the theorem generalises to arbitrary argument positions. Hence every DFS is fully compatible to the complementation of arguments.

Theorem 4.20.
Every DFS $\mathcal{F}$ is compatible with the negation of quantifiers. Hence if $Q : \mathcal{P}(E)^n \longrightarrow \mathbf{I}$ is a semi-fuzzy quantifier, then $\mathcal{F}(\widetilde{\neg}\,Q) = \widetilde{\neg}\,\mathcal{F}(Q)$.

For example, the theorem ensures that

$$\mathcal{F}(\mathbf{at\ most\ 10})(\mathbf{young}, \mathbf{rich}) = \widetilde{\neg}\,\mathcal{F}(\mathbf{more\ than\ 10})(\mathbf{young}, \mathbf{rich})$$

because $\mathbf{at\ most\ 10} = \neg\mathbf{more\ than\ 10}$, i.e. "at most 10" is the negation of "more than 10".

Definition 4.21.
Let $Q : \mathcal{P}(E)^n \longrightarrow \mathbf{I}$ be a semi-fuzzy quantifier ($n > 0$) and $A \in \mathcal{P}(E)$ a crisp subset of E. By $Q\triangle A : \mathcal{P}(E)^n \longrightarrow \mathbf{I}$ we denote the semi-fuzzy quantifier defined by

$$Q\triangle A(Y_1, \dots, Y_n) = Q(Y_1, \dots, Y_{n-1}, Y_n \triangle A)$$

for all $Y_1, \dots, Y_n \in \mathcal{P}(E)$, where $\triangle$ denotes the symmetrical set difference. For fuzzy quantifiers $\widetilde{Q}$, we define $\widetilde{Q}\,\widetilde{\triangle}\,A$ analogously, where the fuzzy symmetrical difference $X_1 \,\widetilde{\triangle}\, X_2 \in \widetilde{\mathcal{P}}(E)$ is defined by $\mu_{X_1 \widetilde{\triangle} X_2}(e) = \mu_{X_1}(e) \,\widetilde{\mathrm{xor}}\, \mu_{X_2}(e)$ for all $e \in E$.

By using the symmetrical difference, we can negate the membership values of only *part* of the elements of an argument set (namely of those which are contained in A). Hence if $Q : \mathcal{P}(E)^n \longrightarrow \mathbf{I}$ is a semi-fuzzy quantifier, then

$$Q\triangle A(Y_1, \dots, Y_n) = Q(Y_1, \dots, Y_{n-1}, Z)$$

where

$$\chi_Z(e) = \begin{cases} \neg\chi_{Y_n}(e) & : \quad e \in A \\ \chi_{Y_n}(e) & : \quad e \notin A \end{cases}$$

for all $Y_1, \dots, Y_n, A \in \mathcal{P}(E)$ and $e \in E$. Likewise if $Q : \widetilde{\mathcal{P}}(E)^n \longrightarrow \mathbf{I}$ is a fuzzy quantifier and $A \in \mathcal{P}(E)$ is crisp, then

$$Q\widetilde{\triangle}A(X_1, \dots, X_n) = Q(X_1, \dots, X_{n-1}, Z)$$

where

$$\mu_Z(e) = \begin{cases} \widetilde{\neg}\,\mu_{X_n}(e) & : \quad e \in A \\ \mu_{X_n}(e) & : \quad e \notin A \end{cases}$$

for all $X_1, \dots, X_n \in \widetilde{\mathcal{P}}(E)$, $e \in E$. Let us now state that every DFS is compatible even with this more fine-structured type of negation.

Theorem 4.22.
Suppose $\mathcal{F}$ is a DFS. Then for every semi-fuzzy quantifier $Q : \mathcal{P}(E)^n \longrightarrow \mathbf{I}$ ($n > 0$) and every crisp subset $A \in \mathcal{P}(E)$, $\mathcal{F}(Q\triangle A) = \mathcal{F}(Q)\,\widetilde{\triangle}\,A$.

4.9 Intersections of Arguments

The DFS axiom (Z-4) explicitly requires that a plausible model of fuzzy quantification be compatible with unions of arguments. In addition, Th. 4.19 has shown that the models of DFS theory preserve antonymy, and are thus compatible with the complementation of arguments as well. Now we will turn to the construction of intersecting arguments which has not been considered so far. It is again convenient to introduce an operator-based notation.

Definition 4.23 (Internal meets).
Suppose $Q : \mathcal{P}(E)^n \longrightarrow \mathbf{I}$ is a semi-fuzzy quantifier, $n > 0$. The semi-fuzzy quantifier $Q\cap : \mathcal{P}(E)^{n+1} \longrightarrow \mathbf{I}$ is defined by

$$Q\cap(Y_1, \ldots, Y_{n+1}) = Q(Y_1, \ldots, Y_{n-1}, Y_n \cap Y_{n+1}),$$

for all $Y_1, \ldots, Y_{n+1} \in \mathcal{P}(E)$. In the case of a fuzzy quantifiers $\widetilde{Q} : \widetilde{\mathcal{P}}(E)^n \longrightarrow \mathbf{I}$, $\widetilde{Q}\widetilde{\cap} : \widetilde{\mathcal{P}}(E)^{n+1} \longrightarrow \mathbf{I}$ is defined analogously, based on the induced intersection $\widetilde{\cap}$.

The compatibility of a QFM with intersections in the arguments of a quantifier is then expressed in the apparent way, i.e. $\mathcal{F}$ should satisfy $\mathcal{F}(Q\cap) = \mathcal{F}(Q)\widetilde{\cap}$ for all $Q : \mathcal{P}(E)^n \longrightarrow \mathbf{I}$ of arity $n > 0$. Knowing that every DFS commutes with unions of arguments, it is not surprising that we also obtain a positive result in the dual case of intersections:

Theorem 4.24.
Every DFS $\mathcal{F}$ is compatible with the intersection of arguments.

For example, $\mathcal{F}(\mathbf{some}) = \mathcal{F}(\exists)\widetilde{\cap}$, because the two-place quantifier **some** can be expressed as $\mathbf{some} = \exists\cap$.

By Th. 4.16, then, every DFS is also known to commute with intersections in arbitrary argument positions. For example, consider the NL statement "Most of the young and rich are tall". In order to interpret this statement and construct its meaning from the fuzzy extensions **young**, **rich** and $\mathbf{tall} \in \widetilde{\mathcal{P}}(E)$, we use a decomposition into the quantifier Q' defined by $Q'(Y_1, Y_2, Y_3) = \mathbf{most}(Y_1 \cap Y_2, Y_3)$, which expresses "Most Y_1's and Y_2's are Y_3's". It is then reasonable to expect that the above statement can be interpreted by applying the resulting fuzzy quantifier $\mathcal{F}(Q')$ to the arguments **young**, **rich** and **tall**. This is exactly where the considered property applies, and we conclude that indeed $\mathcal{F}(Q')(\mathbf{young}, \mathbf{rich}, \mathbf{tall}) = \mathcal{F}(\mathbf{many})(\mathbf{young} \,\widetilde{\cap}\, \mathbf{rich}, \mathbf{tall})$, as desired.

4.10 Argument Insertion

Let us now consider the operation of inserting an argument into a quantifier. By isolating the insertion of a single argument, we obtain the following construction on semi-fuzzy quantifiers:

Definition 4.25 (Argument insertion).
Let $Q : \mathcal{P}(E)^n \longrightarrow \mathbf{I}$ be a semi-fuzzy quantifier of arity $n > 0$ and $A \in \mathcal{P}(E)$. By $Q \triangleleft A : \mathcal{P}(E)^{n-1} \longrightarrow \mathbf{I}$ we denote the semi-fuzzy quantifier defined by

$$Q \triangleleft A(Y_1, \ldots, Y_{n-1}) = Q(Y_1, \ldots, Y_{n-1}, A)$$

for all $Y_1, \ldots, Y_{n-1} \in \mathcal{P}(E)$. (An analogous definition of $\widetilde{Q} \triangleleft A$ is assumed for fuzzy quantifiers).

Note 4.26. As usual, attention is placed on the last argument. The insertion of other arguments can be modelled by a preceeding argument transposition. For example, we can insert A into the $(n-1)$-th argument slot by constructing $Q\tau_{n-1}\triangleleft A$. The insertion of multiple arguments can then be decomposed into a sequence of single argument insertions.

The significance of argument insertion mainly stems from its ability to model an important natural language construction known as *adjectival restriction*, see e.g. [9, p. 448] and [50, p. 247]. To present an example, an NL sentence like "Many married men have children" can be interpreted by inserting the arguments **married**, **have_children** $\in \mathcal{P}(E)$, which represent the extensions of the NL concepts "married" and "have children", respectively, into the composite quantifier "Many married Y_1's are Y_2's". The involved composite quantifier is then said to be constructed from the base quantifier "many" by 'adjectival restriction', in this case based on the crisp adjective "married". To see how the above construction of argument insertion supports the modelling of adjectival restriction, let us simply notice that the construction of the composite quantifier Q' corresponds to an intersection with the denotation of the adjective, in this case $Q'(Y_1, Y_2) = \mathbf{many}(Y_1 \cap \mathbf{married}, Y_2)$. Adjectival restriction can hence be decomposed into the constructions of intersecting argument sets and the insertion of constant arguments like "married". In terms of the relevant operators on quantifiers, we can then express the composite quantifier Q' as

$$Q' = Q\tau_i \cap \triangleleft A \tau_i \,.$$

This has the desired effect of restricting the i-th argument of the semi-fuzzy quantifier Q to a considered subset $A \in \mathcal{P}(E)$, which usually results from the interpretation of a crisp extensional adjective like "married".

The concept of argument insertion is also important because it allows an incremental interpretation of complex expressions in the sense of Frege's *principle of compositionality* (see e.g. Gamut [50, p.140]), which states that the meaning of a complex expression is a function of the meanings of its subexpressions. To provide an example (which again illustrates the mechanism of restricting an argument), let us consider the sentence "Most male persons are married". Suppose that E is a base set, and that **person**, **male**, **married** $\in \mathcal{P}(E)$ are the extensions of "person", "male" and "married" in E, respectively. Informally, we can think of this sentence as being interpreted in the following steps.[4]

$$\begin{aligned} Q &= \text{Most } X_1\text{'s are } X_2\text{'s} \\ Q' &= \text{Most male } Y_1\text{'s are } Y_2\text{'s} \\ Q'' &= \text{Most male persons are } Z\text{'s} \\ Q''' &= \text{Most male persons are married} \end{aligned}$$

[4] This decomposition only serves illustrative purposes. A formal linguist is likely to prefer a different decomposition of the above sentence.

Ever since R. Montague published his influential work on 'the proper treatment of quantification in ordinary English' [115] (reprinted in [163]), formal linguists would typically model the incremental interpretation process in some variant of the typed λ-calculus. This might look roughly[5] as follows:

$$\begin{aligned} Q &= \lambda X_1 X_2.\mathbf{most}(X_1,\, X_2) = \mathbf{most} \\ Q' &= \lambda Y_1 Y_2.\mathbf{most}(Y_1 \cap \mathbf{male},\, Y_2) \\ Q'' &= \lambda Z.Q'(\mathbf{person},\, Z) \\ Q''' &= Q''(\mathbf{married})\,. \end{aligned}$$

By means of the operators defined on semi-fuzzy quantifiers we can recast this,

$$\begin{aligned} Q &= \mathbf{most} \\ Q' &= \mathbf{most}\tau_1 \cap\!\triangleleft \mathbf{male}\tau_1 \\ Q'' &= Q'\tau_1 \triangleleft \mathbf{person} \\ Q''' &= Q'' \triangleleft \mathbf{married}\,. \end{aligned}$$

The following theorem establishes the compatibility of a DFS to the insertion of arguments like "married".

Theorem 4.27.
Every $\mathcal{F}$ is compatible with argument insertions, i.e. $\mathcal{F}(Q\triangleleft A) = \mathcal{F}(Q)\triangleleft A$ for all semi-fuzzy quantifiers $Q : \mathcal{P}(E)^n \longrightarrow \mathbf{I}$ of arity $n > 0$ and all crisp subsets $A \in \mathcal{P}(E)$.

The theorem thus states that every DFS commutes with the insertion of *crisp* arguments. It is limited in scope to the case of crisp arguments because, given a semi-fuzzy quantifier $Q : \mathcal{P}(E)^n \longrightarrow \mathbf{I}$ and a *fuzzy* subset $A \in \widetilde{\mathcal{P}}(E)$, $Q\triangleleft A$ would be undefined. (Q is a *semi*-fuzzy quantifier and hence defined on crisp arguments only!) Nevertheless, the axiom ensures that

$$\mathcal{F}(\mathbf{all}\tau_1 \cap\!\triangleleft \mathbf{married}\tau_1)(\mathbf{men},\, \mathbf{lucky}) = \mathcal{F}(\mathbf{all})(\mathbf{married} \mathbin{\widetilde{\cap}} \mathbf{men},\, \mathbf{lucky})$$

provided that $\mathbf{married} \in \mathcal{P}(E)$ is crisp, i.e. we obtain the same result when applying $\mathcal{F}$ to the composite quantifier

$$Q = \text{"All married } Y_1\text{'s are } Y_2\text{"}$$

and then evaluating $\mathcal{F}(Q)(\mathbf{men},\, \mathbf{lucky})$, or when first applying $\mathcal{F}$ to **all** and then inserting the extensions of "married men" and "lucky". The theorem is also useful because it ensures that *boundary conditions* (with respect to the crisp case) are valid. For example, it is apparent from the theorem that

[5] For convenience, we will use a variant of the typed λ-calculus which offers product types. In monadic type theory, the expressions might look somewhat different.

$$x_1 \mathbin{\widetilde{\wedge}} 1 = x_1$$

for all $x_1 \in \mathbf{I}$, which is one of the defining conditions of t-norms.

A direct modelling of adjectival restriction with a fuzzy adjective is not possible because fuzzy arguments cannot be inserted into a semi-fuzzy quantifier. Therefore a construction different from $Q \triangleleft A$ is needed to handle fuzzy adjectival restriction, which will depend on the chosen model $\mathcal{F}$. It will be shown later in Sect. 6.8 how the insertion of fuzzy arguments into semi-fuzzy quantifiers can be modelled in the framework of DFS theory. The latter construction will then permit a compositional interpretation of fuzzy adjectival restriction in dedicated models as well.

4.11 Monotonicity Properties

In this section, we will discuss monotonicity properties of quantifiers, i.e. semantical characteristics that can be expressed through a comparison of gradual quantification results under the natural order '$\leq$'. Such comparisons are of obvious relevance to logic because small results (close to 'false') generally reflect stronger conditions, while a weakening of conditions expresses in larger membership grades (close to 'true').

Theorem 4.28.
Suppose $\mathcal{F}$ is a DFS and $Q : \mathcal{P}(E)^n \longrightarrow \mathbf{I}$. Then Q is monotonically nondecreasing (nonincreasing) in its i-th argument ($i \leq n$) if and only if $\mathcal{F}(Q)$ is monotonically nondecreasing (nonincreasing) in its i-th argument.

Compared to the original postulate (Z-5), which is restricted to nonincreasing monotonicity in the last argument, the theorem now covers both types of monotonicity in arbitrary argument positions. For example, **some** : $\mathcal{P}(E)^2 \longrightarrow \mathbf{2}$ is monotonically nondecreasing in both arguments. By the theorem, then, $\mathcal{F}(\mathbf{some}) : \widetilde{\mathcal{P}}(E)^2 \longrightarrow \mathbf{I}$ is nondecreasing in both arguments also. In particular,

$$\mathcal{F}(\mathbf{some})(\mathbf{young\_men},\, \mathbf{very\_tall}) \leq \mathcal{F}(\mathbf{some})(\mathbf{men},\, \mathbf{tall})\,,$$

i.e. "Some young men are very tall" entails "Some men are tall", if $\mathbf{young\_men} \subseteq \mathbf{men}$ and $\mathbf{very\_tall} \subseteq \mathbf{tall}$.

So far, we have only considered global monotonicity properties, i.e. monotonicity properties which hold unconditionally and for arbitrary choices of argument sets. In some cases, it can also be instructive to consider monotonicity properties which hold only locally, in a specified range of argument sets.

Definition 4.29.
Let $Q : \mathcal{P}(E)^n \longrightarrow \mathbf{I}$ and $U, V \in \mathcal{P}(E)^n$ be given. We say that Q is locally nondecreasing *in the range (U, V) if for all $Y_1, \ldots, Y_n, Y'_1, \ldots, Y'_n \in \mathcal{P}(E)$ such that $U_i \subseteq X_i \subseteq X'_i \subseteq V_i$ $(i = 1, \ldots, n)$, we have $Q(Y_1, \ldots, Y_n) \leq$*

$Q(Y_1', \ldots, Y_n')$. We will say that Q is locally nonincreasing in the range (U, V) if under the same conditions, $Q(Y_1, \ldots, Y_n) \leq Q(Y_1', \ldots, Y_n')$. On fuzzy quantifiers, local monotonicity is defined analogously, but the arguments are taken from $\widetilde{\mathcal{P}}(E)$, and '$\subseteq$' is the fuzzy inclusion relation.

To present an example, consider the proportional quantifier

$$\textbf{more than 10 percent} = [\textbf{rate} > 0.1]\,,$$

which is neither nonincreasing nor nondecreasing in its first argument. Nevertheless, some characteristics of the quantifier express themselves in its local monotonicity properties. For example, suppose that $A, B \in \mathcal{P}(E)$ are subsets of E, where A is assumed to be nonempty. Then **more than 10 percent** is locally nonincreasing in the range $((A, B), (A \cup \neg B, B))$, and it is locally nondecreasing in the range $((A, B), (A \cup B, B))$.
A frequently observed case of local monotonicity properties is that of a quantifier which is locally constant in some range (U, V), i.e. both nondecreasing and nonincreasing. Hence let again $E \neq \varnothing$ be some finite base set, $e \in E$ an arbitrary element of E, and $r \in (0, 1]$. Then, $[\textbf{rate} \geq r]$ is locally constant in the range (U, V), where $U = (\{e\}, \varnothing)$ and $V = (E, \varnothing)$. If $\{e\} \subseteq Y_1 \subseteq E$ and $\varnothing \subseteq Y_2 \subseteq \varnothing$, i.e. $Y_2 = \varnothing$, we always have

$$\frac{|Y_1 \cap \varnothing|}{|Y_1|} = \frac{|\varnothing|}{|Y_1|} = \frac{0}{|Y_1|} = 0\,,$$

and hence

$$[\textbf{rate} \geq r](Y_1, \varnothing) = 0\,.$$

It is natural to require that $\mathcal{F}$ preserves such local monotonicity properties, i.e. if $Q : \mathcal{P}(E)^n \longrightarrow \mathbf{I}$ is locally nondecreasing (nonincreasing) in some range (U, V), then we expect $\mathcal{F}(Q) : \widetilde{\mathcal{P}}(E)^n \longrightarrow \mathbf{I}$ to be nondecreasing (nonincreasing) in that range as well.
Let us now state that every DFS preserves monotonicity properties of semi-fuzzy quantifiers even if these hold only locally, i.e. all considered models of fuzzy quantification comply with this requirement:

Theorem 4.30.
Suppose $\mathcal{F}$ is a DFS, $Q : \mathcal{P}(E)^n \longrightarrow \mathbf{I}$ a semi-fuzzy quantifier and $U, V \in \mathcal{P}(E)^n$. Then Q is locally nondecreasing (nonincreasing) in the range (U, V) if and only if $\mathcal{F}(Q)$ is locally nondecreasing (nonincreasing) in the range (U, V).

The theorem therefore ensures that those characteristics of a quantifier which become visible through its local monotonicity properties be preserved when applying a DFS. Hence in the second example above, we obtain that $\mathcal{F}([\textbf{rate} \geq r])(X_1, \varnothing) = 0$ for all fuzzy subsets with nonempty core, which is quite satisfying.

The models can also be shown to be *monotonic* in the sense of preserving inequalities between quantifiers. Let us firstly define a partial order $\leq$ on (semi-)fuzzy quantifiers.

Definition 4.31.
Suppose $Q, Q' : \mathcal{P}(E)^n \longrightarrow \mathbf{I}$ are semi-fuzzy quantifiers. Let us write $Q \leq Q'$ if for all $Y_1, \ldots, Y_n \in \mathcal{P}(E)$, $Q(Y_1, \ldots, Y_n) \leq Q'(Y_1, \ldots, Y_n)$. On fuzzy quantifiers, we define $\leq$ analogously, based on arguments in $\widetilde{\mathcal{P}}(E)$.

For example, $[\mathbf{rate} > 0.5] \leq [\mathbf{rate} > 0.2]$, which reflects our intuition that "More than 50 percent of the Y_1's are Y_2's" is a stronger condition than "More than 20 percent of the Y_1's are Y_2's".

Theorem 4.32.
Suppose that $\mathcal{F}$ is a DFS and let $Q, Q' : \mathcal{P}(E)^n \longrightarrow \mathbf{I}$ be semi-fuzzy quantifiers. Then $Q \leq Q'$ if and only if $\mathcal{F}(Q) \leq \mathcal{F}(Q')$.

The theorem ensures that inequalities between quantifiers carry over to the corresponding fuzzy quantifiers. Hence

$$\begin{aligned}&\mathcal{F}(\textbf{more than 50 percent})(\textbf{blonde}, \textbf{tall})\\ &\leq \mathcal{F}(\textbf{more than 20 percent})(\textbf{blonde}, \textbf{tall})\,,\end{aligned}$$

as desired.
Let us also consider another example, which illustrates the utility of cylindrical extensions. We firstly observe that by Def. 4.31, $Q \leq Q'$ is defined only if Q and Q' have the same arity n. But it may be useful to express inequalities also in the case of quantifiers involving a different number of arguments. For example, $\mathbf{all}\cap : \mathcal{P}(E)^3 \longrightarrow \mathbf{2}$ can be said to be smaller than $\mathbf{all} : \mathcal{P}(E)^2 \longrightarrow \mathbf{2}$ in the sense that

$$\mathbf{all}\cap(Y_1, Y_2, Y_3) = \mathbf{all}(Y_1, Y_2 \cap Y_3) \leq \mathbf{all}(Y_1, Y_2)\,,$$

for all $Y_1, Y_2, Y_3 \in \mathcal{P}(E)$. Let us now establish that this kind of inequality is preserved by every DFS $\mathcal{F}$. We define the following cylindrical extension $Q : \mathcal{P}(E)^3 \longrightarrow \mathbf{2}$ of $\mathbf{all} : \mathcal{P}(E)^2 \longrightarrow \mathbf{2}$, viz

$$Q(Y_1, Y_2, Y_3) = \mathbf{all}(Y_1, Y_2)\,,$$

for all $Y_1, Y_2, Y_3 \in \mathcal{P}(E)$. This derived semi-fuzzy quantifier Q has the same arity as $\mathbf{all}\cap$, which means that the above theorem Th. 4.32 is now applicable. We can further utilize the earlier theorem Th. 4.18 and conclude that

$$\begin{aligned}\mathcal{F}(\mathbf{all})(X_1, X_2) &= \mathcal{F}(Q)(X_1, X_2, X_3) && \text{by Th. 4.18}\\ &\leq \mathcal{F}(\mathbf{all}\cap)(X_1, X_2, X_3) && \text{by Th. 4.32}\\ &= \mathcal{F}(\mathbf{all})(X_1, X_2 \mathbin{\widetilde{\cap}} X_3)\end{aligned}$$

for all $X_1, X_2, X_3 \in \widetilde{\mathcal{P}}(E)$.

In particular, if $\mathbf{old}, \mathbf{bald}, \mathbf{rich} \in \widetilde{\mathcal{P}}(E)$ are the (fuzzy) extensions of "old", "bald" and "rich", respectively, in our example universe E, then

$$\mathcal{F}(\mathbf{all})(\mathbf{old}, \mathbf{bald} \mathbin{\widetilde{\cap}} \mathbf{rich}) \leq \mathcal{F}(\mathbf{all})(\mathbf{old}, \mathbf{bald})\,,$$

i.e. "All old are bald and rich" expresses a stronger condition than "All old are bald".

Definition 4.33.
Suppose $Q_1, Q_2 : \mathcal{P}(E)^n \longrightarrow \mathbf{I}$ *are semi-fuzzy quantifiers and* $U, V \in \mathcal{P}(E)^n$. *We say that* Q_1 *is (not necessarily strictly) smaller than* Q_2 *in the range* (U, V), *in symbols:* $Q_1 \leq_{(U,V)} Q_2$, *if for all* $Y_1, \ldots, Y_n \in \mathcal{P}(E)$ *such that* $U_1 \subseteq Y_1 \subseteq V_1, \ldots, U_n \subseteq Y_n \subseteq V_n$,

$$Q_1(X_1, \ldots, X_n) \leq Q_2(X_1, \ldots, X_n) .$$

On fuzzy quantifiers, $\widetilde{Q}_1 \leq_{(U,V)} \widetilde{Q}_2$ *is defined analogously, based on arguments in* $\widetilde{\mathcal{P}}(E)$ *and the fuzzy inclusion relation '*$\subseteq$*'.*

For example, the two-place quantifier "all" is smaller than "some" whenever the first argument is nonempty, i.e.

$$\mathbf{all} \leq_{((\{e\},E\}),(E,E))} \mathbf{some} ,$$

for all $e \in E$.
As we will now state, every model preserves inequalities between quantifiers even if these hold only locally.

Theorem 4.34.
Let $\mathcal{F}$ *be a DFS,* $Q_1, Q_2 : \mathcal{P}(E)^n \longrightarrow \mathbf{I}$ *and* $U, V \in \mathcal{P}(E)^n$. *Then*

$$Q_1 \leq_{(U,V)} Q_2 \Leftrightarrow \mathcal{F}(Q_1) \leq_{(U,V)} \mathcal{F}(Q_2) .$$

The theorem thus ensures that local inequalities, like those observed in the case of "some" and "all", are preserved when applying a DFS. In particular, if $\mathbf{tall}, \mathbf{lucky} \in \widetilde{\mathcal{P}}(E)$ are fuzzy subsets of E and **tall** has nonempty support, then

$$\mathcal{F}(\mathbf{all})(\mathbf{tall}, \mathbf{lucky}) \leq \mathcal{F}(\mathbf{some})(\mathbf{tall}, \mathbf{lucky}) ,$$

as desired.

4.12 Properties of the Induced Extension Principle

Let us now investigate the formal properties of the induced extension principle of $\mathcal{F}$. This analysis will substantiate that the induced extension principle determines a plausible assignment of fuzzy powerset mappings to given base mappings. Building on this foundation, it becomes possible to show that all considered models of fuzzy quantification preserve the important semantical properties of quantitativity and extensionality, which are closely tied to the induced extension principle. In addition, every DFS has the property of being 'contextual', i.e. it captures an intuitive requirement on the intended effects of fuzziness in a quantifier's arguments.
Let us now state the formal results that underly the later theorems of linguistic relevance.

Theorem 4.35.
Suppose $\mathcal{F}$ is a DFS and $\widehat{\mathcal{F}}$ the extension principle induced by $\mathcal{F}$. Then for all $f : E \longrightarrow E'$, $g : E' \longrightarrow E''$ (where $E \neq \varnothing$, $E' \neq \varnothing$, $E'' \neq \varnothing$),

a. $\widehat{\mathcal{F}}(g \circ f) = \widehat{\mathcal{F}}(g) \circ \widehat{\mathcal{F}}(f)$
b. $\widehat{\mathcal{F}}(\mathrm{id}_E) = \mathrm{id}_{\widetilde{\mathcal{P}}(E)}$

Note 4.36. A reader familiar with category theory will recognize this as the statement that $\widehat{\mathcal{F}}$ is a covariant functor from the category of non-empty sets to the category of fuzzy power sets, provided that on objects E, we define $\widehat{\mathcal{F}}(E) = \widetilde{\mathcal{P}}(E)$.[6] It should be apparent from the remarks on p. 102 how to dispense with the restriction to non-empty sets if so desired. $\widehat{\mathcal{F}}$ is a faithful functor because $f \neq g$ implies $\widehat{\mathcal{F}}(f)|_{\mathcal{P}(E)} = \widehat{f} \neq \widehat{g} = \widehat{\mathcal{F}}(g)|_{\mathcal{P}(E)}$, i.e. $\widehat{\mathcal{F}}(f) \neq \widehat{\mathcal{F}}(g)$. $\widehat{\mathcal{F}}$ is also injective on objects; $E \neq E'$ implies $\widetilde{\mathcal{P}}(E) \neq \widetilde{\mathcal{P}}(E')$.

The induced extension principles of all DFSes coincide on injective mappings with the apparent plausible definition:

Theorem 4.37.
Suppose that $\mathcal{F}$ is a DFS and $f : E \longrightarrow E'$ is an injection. Then for all $X \in \widetilde{\mathcal{P}}(E)$, $e' \in E'$,

$$\mu_{\widehat{\mathcal{F}}(f)(X)}(e') = \begin{cases} \mu_X(f^{-1}(e')) & : \quad e' \in \operatorname{Im} f \\ 0 & : \quad e' \notin \operatorname{Im} f \end{cases}$$

where $f^{-1}(e') = \{e \in E : f(e) = e'\}$.

This result on the extension of injective mappings turned out to be rather useful. For example, it can be used to show that a DFS is compatible with exactly one extension principle. Knowing this might facilitate the proof that a given DFS induces a certain extension principle.

Theorem 4.38.
Suppose $\mathcal{F}$ is a DFS and $\mathcal{E}$ an extension principle such that for every semi-fuzzy quantifier $Q : \mathcal{P}(E')^n \longrightarrow \mathbf{I}$ and all $f_1 : E \longrightarrow E'$, ..., $f_n : E \longrightarrow E'$, $\mathcal{F}(Q \circ \overset{n}{\underset{i=1}{\times}} \widehat{f_i}) = \mathcal{F}(Q) \circ \overset{n}{\underset{i=1}{\times}} \mathcal{E}(f_i)$. Then $\widehat{\mathcal{F}} = \mathcal{E}$.

The extension principle $\widehat{\mathcal{F}}$ of a DFS $\mathcal{F}$ is uniquely determined by the fuzzy existential quantifiers $\mathcal{F}(\exists) = \mathcal{F}(\exists_E) : \widetilde{\mathcal{P}}(E) \longrightarrow \mathbf{I}$ induced by $\mathcal{F}$.

Theorem 4.39.
Suppose $\mathcal{F}$ is a given DFS. For every mapping $f : E \longrightarrow E'$ and all $e' \in E'$, $\mu_{\widehat{\mathcal{F}}(f)(\bullet)}(e') = \mathcal{F}(\exists)\widetilde{\cap}\triangleleft f^{-1}(e')$.

---

[6] By the category of fuzzy power sets we mean the category in which the objects are fuzzy power sets $\widetilde{\mathcal{P}}(E)$, the morphisms are mappings $f : \widetilde{\mathcal{P}}(E) \longrightarrow \widetilde{\mathcal{P}}(E')$ which to each fuzzy subset $X \in \widetilde{\mathcal{P}}(E)$ assign a fuzzy subset $f(X) \in \widetilde{\mathcal{P}}(E')$, and $\circ$ is the usual composition of functions.

The converse can also be shown: the fuzzy existential quantifiers obtained from a DFS $\mathcal{F}$ are uniquely determined by its extension principle $\widehat{\mathcal{F}}$.

Theorem 4.40.
Let a DFS $\mathcal{F}$ be given. If $E \neq \varnothing$ and $\exists = \exists_E : \mathcal{P}(E) \longrightarrow \mathbf{2}$, then $\mathcal{F}(\exists) = \widetilde{\pi}_\varnothing \circ \widehat{\mathcal{F}}(!)$, where $! : E \longrightarrow \{\varnothing\}$ is the mapping defined by $!(e) = \varnothing$ for all $e \in E$.

This completes the analysis of the formal characteristics underlying the extension principles that are induced by the models of DFS theory. Some applications in which this analysis supported the proof of semantical properties of the models are stated in Sects. 4.13 (quantitativity), 4.14 (extensionality) and 4.15 (contextuality).

The analysis of the extension principle presented here is complemented with an analysis of the interpretation of the standard quantifiers, which is presented below in Sect. 4.16. The subsequent investigation of $\forall$ and $\exists$ is instructive because the induced extension principle is closely related to the interpretation of existential quantifiers. Its precise structure becomes apparent once we combine the above Th. 4.39 with the later explicit formula for existential quantification in the models, see Th. 4.61.

4.13 Quantitativity

Many quantifiers of interest, like "almost all", "most" etc., do not depend on any particular characteristics of the elements in the base set. It is not the specific choice of elements that determines the quantification result, but only the quantitative aspects. In the finite case, such quantifiers can be defined from the cardinalities of the arguments (or cardinalities of Boolean combinations). Other quantifiers, like "John" (proper names) or "Most married X_1's are X_2's" (adjectival restriction), though, are closer tied to the domain and its particular elements. Cardinality information about combinations of the arguments is not sufficient to determine the semantical interpretation of these quantifiers in any given situation, and these quantifiers are therefore called 'non-quantitative'.[7] In TGQ, the notion of quantitativity is commonly given an elegant and convincing definition in terms of automorphism invariance, which traces back to Mostowski [116]. This formalisation is easily adapted to the case of semi-fuzzy quantifiers and even fuzzy quantifiers:

Definition 4.41 (Quantitative semi-fuzzy quantifier).
A semi-fuzzy quantifier $Q : \mathcal{P}(E)^n \longrightarrow \mathbf{I}$ is called quantitative *if for all automorphisms*[8] *$\beta : E \longrightarrow E$ and all $Y_1, \dots, Y_n \in \mathcal{P}(E)$,*

$$Q(Y_1, \dots, Y_n) = Q(\widehat{\beta}(Y_1), \; \dots, \; \widehat{\beta}(Y_n)) \, .$$

---

[7] Non-quantitative quantifiers are sometimes also dubbed 'qualitative'.
[8] i.e. bijections of E into itself

It is apparent that for each $E \neq \varnothing$, $n \in \mathbb{N}$ and $X_1, \ldots, X_n \in \widetilde{\mathcal{P}}(E)$, the resulting relation of contextual equality $\sim_{(X_1,\ldots,X_n)}$ is an equivalence relation on the set of all semi-fuzzy quantifiers $Q : \mathcal{P}(E)^n \longrightarrow \mathbf{I}$.

Definition 4.47.
A QFM $\mathcal{F}$ is said to be contextual *if for all $Q, Q' : \mathcal{P}(E)^n \longrightarrow \mathbf{I}$ and every choice of fuzzy argument sets $X_1, \ldots, X_n \in \widetilde{\mathcal{P}}(E)$,*

$$Q \sim_{(X_1,\ldots,X_n)} Q'$$

entails that

$$\mathcal{F}(Q)(X_1, \ldots, X_n) = \mathcal{F}(Q')(X_1, \ldots, X_n)\,.$$

As illustrated by the motivating example, it is highly desirable that a QFM satisfies this very elementary but fundamental adequacy condition. And indeed, every DFS will exhibit the desired property:

Theorem 4.48.
Every DFS $\mathcal{F}$ is contextual.

To give an example of how contextuality is useful in establishing properties of QFMs, consider the following theorem which generalizes Th. 4.27.

Theorem 4.49.
Suppose that $\mathcal{F}$ is a contextual QFM compatible with cylindrical extensions. Then $\mathcal{F}$ is compatible with argument insertions, i.e. $\mathcal{F}(Q \triangleleft A) = \mathcal{F}(Q) \triangleleft A$ for all semi-fuzzy quantifiers $Q : \mathcal{P}(E)^n \longrightarrow \mathbf{I}$ of arity $n > 0$ and all crisp subsets $A \in \mathcal{P}(E)$.

Other applications of contextuality are presented below in Chap. 6, which substantiate the significance of the novel concept to the modelling of fuzzy quantification.

4.16 Semantics of the Standard Quantifiers

We have already seen in Th. 4.39 and Th. 4.40 that the existential quantifier can be expressed in terms of the induced extension principle and vice versa. This section will present some further results on the interpretation of the universal and existential quantifiers in the considered models of fuzzy quantification.

Let us first recall Thiele's analysis of fuzzy universal and existential quantification [156, 157, 158]. Thiele has developed an axiomatic characterisation of these dual types of quantifiers and their expected semantics in the presence of fuzziness, which culminates in his definitions of T-quantifiers and S-quantifiers. Thiele also develops representation theorems for T- and

S-quantifiers which show how these quantifiers can be decomposed into possibly infinitary formulas involving t- or s-norms. These theorems, which will be presented below, have proven invaluable for establishing the results on the interpretation of the standard quantifiers in DFS theory.

Thiele's definition of T-quantifiers captures the essential requirements on fuzzy universal quantifiers. Making use of the concepts developed here, the proposed definition can be expressed as follows.

Definition 4.50 (T-quantifiers).
A fuzzy quantifier $\widetilde{Q} : \widetilde{\mathcal{P}}(E) \longrightarrow \mathbf{I}$ *is called a* T-quantifier *if* $\widetilde{Q}$ *satisfies the following axioms:*

a. *For all* $X \in \widetilde{\mathcal{P}}(E)$ *and* $e \in E$, $\widetilde{Q}(X \cup \neg\{e\}) = \mu_X(e)$;
b. *For all* $X \in \widetilde{\mathcal{P}}(E)$ *and* $e \in E$, $\widetilde{Q}(X \cap \neg\{e\}) = 0$;
c. $\widetilde{Q}$ *is nondecreasing, i.e. for all* $X, X' \in \widetilde{\mathcal{P}}(E)$ *such that* $X \subseteq X'$, *it holds that* $\widetilde{Q}(X) \leq \widetilde{Q}(X')$;
d. $\widetilde{Q}$ *is quantitative, i.e. for every automorphism (permutation)* $\beta : E \longrightarrow E$, $\widetilde{Q} \circ \hat{\hat{\beta}} = \widetilde{Q}$.

Note 4.51. In the above definition, $\cap$ is the standard fuzzy intersection based on min, and $\cup$ is the standard fuzzy union based on max. However, all fuzzy intersections based on t-norms and all fuzzy unions based on s-norms will give the same results, because one of the arguments is a crisp subset of E.

There is a close relationship between T-quantifiers and t-norms. Following Thiele, we define the connective $\widetilde{\wedge}_{\widetilde{Q}}$ which corresponds to the T-quantifier.

Definition 4.52.
Suppose $\widetilde{Q} : \widetilde{\mathcal{P}}(E) \longrightarrow \mathbf{I}$ *is a T-quantifier and* $|E| > 1$. $\widetilde{\wedge}_{\widetilde{Q}} : \mathbf{I} \times \mathbf{I} \longrightarrow \mathbf{I}$ *is defined by*

$$x_1 \mathbin{\widetilde{\wedge}_{\widetilde{Q}}} x_2 = \widetilde{Q}(X)$$

for all $x_1, x_2 \in \mathbf{I}$, *where* $X \in \widetilde{\mathcal{P}}(E)$ *is defined by*

$$\mu_X(e) = \begin{cases} x_1 & : \quad e = e_1 \\ x_2 & : \quad e = e_2 \\ 1 & : \quad \mathit{else} \end{cases} \tag{4.7}$$

and $e_1 \neq e_2$, $e_1, e_2 \in E$ *are two arbitrary, distinct elements of* E.

Note 4.53. It is evident from the quantitativity of T-quantifiers that $x_1 \mathbin{\widetilde{\wedge}_{\widetilde{Q}}} x_2$ does not depend on the choice of $e_1, e_2 \in E$.

It is Thiele's merit of having shown that T-quantifiers are exactly those quantifiers that can be decomposed into the following (possibly infinitary) construction based on an underlying t-norm:

Let us notice that if E is finite, i.e. $E = \{e_1, \ldots, e_m\}$ where the e_i are pairwise distinct, then the expressions presented in the above theorems can be simplified into

$$\mathcal{F}(\forall)(X) = \mathop{\widetilde{\wedge}}_{i=1}^{m} \mu_X(e_i)\,,$$
$$\mathcal{F}(\exists)(X) = \mathop{\widetilde{\vee}}_{i=1}^{m} \mu_X(e_i)\,.$$

Hence the fuzzy universal (existential) quantifiers of $\mathcal{F}$ are reasonable in the sense that the important relationship between $\forall$ and $\wedge$ ($\exists$ and $\vee$, resp.), which holds in the finite case, is preserved by the DFS. In particular, the above theorems show that in every DFS, the fuzzy existential and fuzzy universal quantifiers are uniquely determined by the induced fuzzy disjunction and conjunction.

Theorem 4.62.

Suppose $\mathcal{F}$ is a DFS, $\widehat{\mathcal{F}}$ its induced extension principle and $\widetilde{\vee} = \widetilde{\mathcal{F}}(\vee)$.

a. *$\widehat{\mathcal{F}}$ is uniquely determined by $\widetilde{\vee}$, in the way described by Th. 4.39 and Th. 4.61, i.e.*

$$\mu_{\widehat{\mathcal{F}}(f)(X)}(e') = \sup\{\mathop{\widetilde{\vee}}_{i=1}^{m} \mu_X(a_i) : A = \{a_1, \ldots, a_m\} \in f^{-1}(e') \text{ finite},\ a_i \neq a_j \text{ if } i \neq j\}$$

for all $f : E \longrightarrow E'$, $X \in \widetilde{\mathcal{P}}(E)$ and $e' \in E'$, where $E, E' \neq \varnothing$.

b. *$\widetilde{\vee}$ is uniquely determined by $\widehat{\mathcal{F}}$, viz. $x_1 \,\widetilde{\vee}\, x_2 = (\widetilde{\pi}_\varnothing \circ \widehat{\mathcal{F}}(!))(X)$ for all $x_1, x_2 \in \mathbf{I}$, where $X \in \widetilde{\mathcal{P}}(\{1, 2\})$ is defined by $\mu_X(1) = x_1$ and $\mu_X(2) = x_2$, and ! is the unique mapping $! : \{1, 2\} \longrightarrow \{\varnothing\}$.*

In particular, if $\widehat{\mathcal{F}} = \hat{\hat{(\bullet)}}$ is the standard extension principle, then $\widetilde{\vee} = \max$. Because I did not want QFMs in which $\widetilde{\vee} \neq \max$ to be a priori excluded from consideration, it was not possible to state (Z-6) in terms of the standard extension principle. Acknowledging this constraining role, it was rather necessary to introduce general extension principles along with the construction of induced extension principles, which selects an appropriate choice of such general extension principle for each given $\mathcal{F}$. Only by developing the theory of fuzzy quantification in this way it was possible to leave open the chance for models based on different choices of fuzzy disjunction, like bounded sum or algebraic sum, and perhaps even the extreme case of drastic sum.

4.17 Fuzzy Inverse Images

We shall now return to a more theoretically oriented construction which is nevertheless worthwhile investigating, because it contributes to a theorem (in the subsequent section) which elucidates the semantic relationship between unary (i.e. one-place) and general multi-place quantification which underlies all models of DFS theory. The required construction is that of *fuzzy inverse images*, a notion closely related to the extension principle.

In the case of crisp sets, we assume the usual definition of inverse images which has already been stated in the previous chapter, see equation (3.5). Generalising this concept, every QFM $\mathcal{F}$ induces fuzzy inverse images by means of the following construction.

Definition 4.63.
Suppose $\mathcal{F}$ is a QFM and $f : E \longrightarrow E'$ is some mapping. $\mathcal{F}$ induces a fuzzy inverse image mapping $\widehat{\mathcal{F}}^{-1}(f) : \widetilde{\mathcal{P}}(E') \longrightarrow \widetilde{\mathcal{P}}(E)$ *which to each $X \in \widetilde{\mathcal{P}}(E')$ assigns the fuzzy subset $\widehat{\mathcal{F}}^{-1}(f)$ defined by*

$$\mu_{\widehat{\mathcal{F}}^{-1}(f)(X)}(e) = \mathcal{F}(\chi_{f^{-1}(\bullet)}(e))(X)\,,$$

for all $e \in E$.

If $\mathcal{F}$ is a DFS, then its induced fuzzy inverse images coincide with the apparent 'reasonable' definition:

Theorem 4.64.
Suppose $\mathcal{F}$ is a DFS, $f : E \longrightarrow E'$ is a mapping and $X \in \widetilde{\mathcal{P}}(E')$. Then for all $e \in E$, $\mu_{\widehat{\mathcal{F}}^{-1}(f)(X)}(e) = \mu_X(f(e))$.

Hence $\mathcal{F}$ not only induces a meaningful extension principle, but also induces a reasonable choice of the reverse construction.

4.18 The Semantics of Fuzzy Multi-Place Quantification

In this section, we will uncover the internal structure of fuzzy multi-place quantification and elucidate its semantical grounding in one-place quantification. The *reduction* of an n-place quantifier to a corresponding unary quantifier will be the key tool for analysing multi-place quantification.

In order to understand how n-place quantifiers can be reduced to one-place quantification, let us recall that for arbitrary sets A, B, C, it always holds that $A^{B\times C} \cong (A^B)^C$, a relationship commonly known as 'currying'. We then have

$$\mathcal{P}(E)^n \cong (\mathbf{2}^E)^n \cong \mathbf{2}^{E\times n} \cong \mathcal{P}(E \times n) \cong \mathcal{P}(E \times \{1, \ldots, n\})\,,$$

and similarly

$$\widetilde{\mathcal{P}}(E)^n \cong (\mathbf{I}^E)^n \cong \mathbf{I}^{E\times n} \cong \widetilde{\mathcal{P}}(E \times n) \cong \widetilde{\mathcal{P}}(E \times \{1, \ldots, n\})\,,$$

where n abbreviates $\{0, \ldots, n-1\}$ as usual. For convenience, $E \times n$ has been replaced by $E \times \{1, \ldots, n\}$, which better suits the convention of numbering the arguments of an n-place quantifier from 1 to n (rather than from 0 to $n-1$). This suggests that by exploiting the bijection $\mathcal{P}(E)^n \cong \mathcal{P}(E \times \{1,\dots,n\})$, n-place quantification as expressed by some $Q : \mathcal{P}(E)^n \longrightarrow \mathbf{I}$ can be replaced with one-place quantification using a unary quantifier $\langle Q \rangle : \mathcal{P}(E \times \{1,\dots,n\}) \longrightarrow \mathbf{I}$, and that conversely $\mathcal{F}(Q) : \widetilde{\mathcal{P}}(E)^n \longrightarrow \mathbf{I}$ can always be recovered from $\mathcal{F}(\langle Q \rangle) : \widetilde{\mathcal{P}}(E \times \{1,\dots,n\}) \longrightarrow \mathbf{I}$. To establish this result, we need some formal machinery. For a given domain E and $n \in \mathbb{N}$, we will abbreviate $E_n = E \times \{1,\dots,n\}$. This will provide a concise notation for the base sets of the resulting unary quantifiers. For $n = 0$, we obtain the empty product $E_0 = \varnothing$.

Definition 4.65 ($\imath_i^{n,E}$).
Let E be a given set, $n \in \mathbb{N} \setminus \{0\}$ and $i \in \{1,\dots,n\}$. By $\imath_i^{n,E} : E \longrightarrow E_n$ we denote the inclusion defined by

$$\imath_i^{n,E}(e) = (e,i)\,,$$

for all $e \in E$.

Note 4.66. The crisp extension (powerset mapping, see Def. 3.19) of $\imath_i^{n,E} : E \longrightarrow E_n$ will be denoted by $\widehat{\imath}_i^{n,E} : \mathcal{P}(E) \longrightarrow \mathcal{P}(E_n)$. The inverse image mapping of $\imath_i^{E,n}$ will be denoted $(\imath_i^{E,n})^{-1} : \mathcal{P}(E_n) \longrightarrow \mathcal{P}(E)$, see equation (3.5).

We can use these injections to define the unary quantifier $\langle Q \rangle$ of interest.

Definition 4.67.
Let $Q : \mathcal{P}(E)^n \longrightarrow \mathbf{I}$ be an n-place semi-fuzzy quantifier, where $n > 0$. Then $\langle Q \rangle : \mathcal{P}(E_n) \longrightarrow \mathbf{I}$ is defined by

$$\langle Q \rangle(Y) = Q((\imath_1^{n,E})^{-1}(Y), \dots, (\imath_n^{n,E})^{-1}(Y))$$

for all $Y \in \mathcal{P}(E_n)$.

For fuzzy quantifiers, $\langle \widetilde{Q} \rangle$ is defined similarly, using the fuzzy inverse image mapping $(\hat{\imath}_i^{n,E})^{-1} : \widetilde{\mathcal{P}}(E_n) \longrightarrow \widetilde{\mathcal{P}}(E)$ of $\imath_i^{n,E}$.

Definition 4.68.
Let $\widetilde{Q} : \widetilde{\mathcal{P}}(E)^n \longrightarrow \mathbf{I}$ be a fuzzy quantifier, $n > 0$. The fuzzy quantifier $\langle \widetilde{Q} \rangle : \widetilde{\mathcal{P}}(E_n) \longrightarrow \mathbf{I}$ is defined by

$$\langle \widetilde{Q} \rangle(X) = \widetilde{Q}((\hat{\imath}_1^{n,E})^{-1}(X), \dots, (\hat{\imath}_n^{n,E})^{-1}(X))\,,$$

for all $X \in \widetilde{\mathcal{P}}(E_n)$.

Let us now establish the relationship between $\langle Q \rangle$ and Q (semi-fuzzy case) and $\langle \widetilde{Q} \rangle$ and $\widetilde{Q}$ (fuzzy case). Let us first introduce a concise notation for iterated unions of a quantifier's arguments:

Definition 4.69.
Suppose that $Q : \mathcal{P}(E)^n \longrightarrow \mathbf{I}$ *is a semi-fuzzy quantifier,* $n > 0$ *and* $k \in \mathbb{N}\setminus\{0\}$. *The semi-fuzzy quantifier* $Q\cup^k : \mathcal{P}(E)^{n+k-1} \longrightarrow \mathbf{I}$ *is inductively defined as follows:*

a. $Q\cup^1 = Q$*;*
b. $Q\cup^k = Q\cup^{k-1}\cup$ *if* $k > 1$.

For fuzzy quantifiers $\widetilde{Q} : \widetilde{\mathcal{P}}(E)^n \longrightarrow \mathbf{I}$, $\widetilde{Q}\widetilde{\cup}^k : \widetilde{\mathcal{P}}(E)^{n+k-1} \longrightarrow \mathbf{I}$ *is defined analogously.*

Theorem 4.70.
For every semi-fuzzy quantifier $Q : \mathcal{P}(E)^n \longrightarrow \mathbf{I}$ *where* $n > 0$,

$$Q = \langle Q\rangle\cup^n \circ \mathop{\times}_{i=1}^{n} \widehat{\imath}_i^{\,n,E} .$$

Note 4.71. This demonstrates that n-place crisp (or 'semi-fuzzy') quantification, where $n > 0$, can always be reduced to one-place quantification. If we allowed for empty base sets, this would also go through for $n = 0$; we only had to exclude this case because $E_0 = \varnothing$, i.e. in this case we have $\langle Q\rangle : \mathcal{P}(\varnothing) \longrightarrow \mathbf{I}$, which does not qualify as a semi-fuzzy quantifier because the base set is empty.

A theorem analogous to Th. 4.70 can also be proven in the fuzzy case:

Theorem 4.72.
Suppose $\widetilde{\vee} : \mathbf{I}\times\mathbf{I} \longrightarrow \mathbf{I}$ *has* $x \,\widetilde{\vee}\, 0 = 0 \,\widetilde{\vee}\, x = x$ *for all* $x \in \mathbf{I}$, *and* $\widetilde{\cup}$ *is the fuzzy union element-wise defined in terms of* $\widetilde{\vee}$. *For every fuzzy quantifier* $\widetilde{Q} : \widetilde{\mathcal{P}}(E)^n \longrightarrow \mathbf{I}$, $n > 0$,

$$\widetilde{Q} = \langle\widetilde{Q}\rangle\widetilde{\cup}^n \circ \mathop{\times}_{i=1}^{n} \hat{\imath}_i^{\,n,E} .$$

Note 4.73. The theorem shows that it is possible to reduce n-place quantification to one-place quantification in the fuzzy case as well. Nullary quantifiers ($n = 0$) had to be excluded for same reason as in Th. 4.70.

The central fact which links these results to QFMs is the following:

Theorem 4.74.
Suppose $\mathcal{F}$ *is a QFM with the following properties:*

a. $x \,\widetilde{\vee}\, 0 = 0 \,\widetilde{\vee}\, x = x$ *for all* $x \in \mathbf{I}$*;*
b. for all semi-fuzzy quantifiers $Q : \widetilde{\mathcal{P}}(E)^n \longrightarrow \mathbf{I}$ *where* $n > 0$,

$$\mathcal{F}(Q\cup) = \mathcal{F}(Q)\widetilde{\cup};$$

c. $\mathcal{F}$ *satisfies* (Z-6) *(functional application);*

d. If E, E' are nonempty sets and $f : E \longrightarrow E'$ is an injective mapping, then $\widehat{\mathcal{F}}(f) = \hat{\hat{f}}$, i.e. $\mathcal{F}$ coincides with the standard extension principle on injections.

Then for all semi-fuzzy quantifiers $Q : \mathcal{P}(E)^n \longrightarrow \mathbf{I}$ of arity $n > 0$,

$$\mathcal{F}(\langle Q \rangle) = \langle \mathcal{F}(Q) \rangle .$$

In particular, every DFS commutes with $\langle \bullet \rangle$, and hence permits the reduction of multi-place quantification to one-place quantification. In other words, the behaviour of $\mathcal{F}$ on general multi-place quantifiers of arbitrary arities $n > 0$ is already implicit in the definition of $\mathcal{F}$ for one-place quantifiers, because the quantification results of multi-place quantification are completely determined by the behaviour of $\mathcal{F}$ in the unary case. In fact, we can actually *define* $\mathcal{F}(Q)$ in terms of $\mathcal{F}(\langle Q \rangle)$, by utilizing Th. 4.70 and Th. 4.72. The semantical relationship discovered here, and the associated constructions of $\langle Q \rangle$ and $\langle \widetilde{Q} \rangle$, hence allow us to *generate* a full-fledged QFM from its core of an underlying 'unary QFM'. In perspective, this would permit us to restrict attention to the behaviour of $\mathcal{F}$ on one-place quantifiers, and to reformulate the DFS axioms (Z-1)–(Z-6) into axioms imposed on the unary base mechanism which underlies $\mathcal{F}$. However, the resulting axioms are likely to become more abstract than the current DFS axioms, and we shall therefore not pursue this idea further here.

From a practical point of view, the techniques developed in this section have already proven useful for establishing the equivalence of the alternative constructions of induced fuzzy truth functions in all models of the theory.

4.19 Chapter Summary

This chapter has shown that the proposed axiomatization of plausible models in terms of the six basic requirements (Z-1)–(Z-6) accounts for a wide range of expectations on models of fuzzy quantification.

We first considered the property of correct generalization, i.e. the fuzzy quantifier $\mathcal{F}(Q)$ must properly generalize Q in the sense that $\mathcal{F}(Q)(Y_1, \ldots, Y_n) = Q(Y_1, \ldots, Y_n)$ for crisp arguments. The corresponding condition (Z-1) in the axiom system was artificially restricted to quantifiers of arity $n \leq 1$. However, as shown by Th. 4.1, the condition also holds unconditionally for quantifiers of arbitrary arities.

We then considered the propositional fragment of the models, which is determined by the construction of induced truth functions. It was shown that the identity truth function maps to the corresponding identity on the unit interval, which is of course the only acceptable choice for the continuous-valued case. The models were further shown to translate the original crisp negation into a strong negation operator, the usual abstraction of negation operators in fuzzy set theory. Next we investigated the two-place connectives. First of all, the induced conjunctions of the models belong to the familiar class of t-norms. As to disjunction, it is the notion of an s-norm which constitutes the class of 'or'-like connectives, and it was shown that in all considered models, the disjunction translates into the $\widetilde{\neg}$-dual s-norm of the induced conjunction. The induced implication can then be expressed in terms of the induced disjunction and negation.

In addition to discussing individual truth functions, we also reviewed the proposed construction of induced connectives. By comparing the results of two alternative constructions of induced truth functions, we gathered some evidence that the choice of induced connectives is indeed canonical, because it is shared by two independent constructions.

Following this discussion of their propositional structure, we then investigated the compliance of the models with certain operations on the quantifiers' arguments. Some of these operations are motivated by linguistic constructions and familiar from TGQ, while others are concerned with the coherence of interpretations.

First of all, we considered a construction which permutes argument positions. The construction can be used, for example, to express symmetry properties of quantifiers. The proposed axiom system ensures that all conforming models be compatible with argument transpositions (and thus also to argument permutations). In particular, these models will preserve the symmetry pattern of a given quantifier, which reveals an important aspect of its meaning.

Having treated the issue of argument symmetry, we turned to cylindrical extensions, a construction which augments a given quantifier by vacuous argument positions. This construction is mainly required to fit a number of given quantifiers to some construction on these quantifiers which requires that the involved quantifiers refer to a common argument list. After stating the formal definition of cylindrical extensions, it was shown that the considered models of fuzzy quantification comply with this construction. In other words, if an argument of a semi-fuzzy quantifier has no effect on the computed interpretations, then it will also have no effect on the quantification results of the corresponding fuzzy quantifier.

We then investigated the various types of negation in order to assess the degree to which the models preserve Boolean structure. This analysis revealed the full compliance of a DFS with the formation of antonyms (complementation of arguments) and the external negation of quantifiers (where the negation operator is applied to the quantifying expression as a whole). Because every model also complies with dualisation by (Z-3), this means that every DFS preserves Aristotelian squares of quantifiers. These visualize the interrelations between the various types of negation which can be applied to a quantifier. The relevance of the Aristotelian squares stems from their close relationship to the Piaget group of transformations, which catches some findings of developmental psychology. Apart from the block-wise type of negation/complementation considered so far, we also discussed a more fine-grained type of negation, which can be modelled by the symmetrical difference of arguments with given crisp sets. In this way, only part of an argument can be complemented, while the remaining portion is left unchanged. This fine-structured type of negation does not pose problems to the proposed models, though, which can handle it unconditionally, just like the simple negation and complementation.

Having discussed the Boolean structure which expresses in the various types of negation, attention was then shifted to the structure imposed by unions and intersections. As to the formation of unions, it is explicitly required by (Z-4) that all models comply with this construction when it occurs in the last argument. Due to the compatibility with argument permutations, it is also clear that every DFS conforms to unions in arbitrary argument position. Recalling the compliance of the models with complementation, and applying De Morgan's law, we conclude that every model also conforms to intersections in arbitrary argument positions. In this sense, the models are fully compatible with Boolean argument structure.

The models were further shown to comply with another construction of relevance to natural language, the insertion of arguments into a quantifier. The construction is linguistically significant because adjectival restriction can be analysed in terms of intersections and argument insertion. In order to investigate this construction, we focused on an elementary insertion step. The models were then shown to to comply with the insertion operation. Thus every DFS is compatible with adjectival restriction by a crisp adjective (like "married"). Due to the simple fact that a fuzzy argument cannot be inserted into a semi-fuzzy quantifier, the related issues of fuzzy argument insertion and fuzzy adjectival restriction must be handled with a rather different approach. The discussion of fuzzy argument insertion has therefore been delayed to the Sect. 6.8 which discusses more advanced topics.

Monotonicity properties are especially important because they constrain the valid patterns of reasoning. The monotonicity postulate (Z-5) requires that legal models preserve the nonincreasing monotonicity of a quantifier in its last argument. This artificial restriction of the criterion helped to keep the DFS axioms as succinct as possible. However, every reasonable model of fuzzy quantification should certainly preserve monotonicity of a quantifier in other arguments as well, regardless of its monotonicity type (nondecreasing/nonincreasing). Apart from this general monotonicity property, the models were further shown to preserve monotonicity properties that hold only locally, a notion which we defined in terms of closed ranges of crisp sets. The proposed relativization of monotonicity properties to any desired choice of argument ranges, is not critical to the models. In fact, every DFS can be shown to preserve any type of monotonicity properties, even if they hold only locally.

We were also concerned with the monotonicity of the fuzzification mechanism itself (as opposed to the preservation of monotonicity properties of a quantifier). In other words, we would expect that the models preserve inequalities between quantifiers, and indeed every DFS was shown to be monotonic in this sense. Again, the models will also preserve local inequalities between quantifiers which hold only in a given range of arguments.

The subsequent analysis of the induced extension principle also showed some interesting findings. First of all, it revealed that the induced extension principle is functorial, i.e. compatible with the composition of powerset mappings. In addition, it was shown that the induced extension principles will always assign the intended interpretation to injective mappings. Most importantly, we discussed the relationship between the induced extension principle and existential quantification: It was shown that the induced powerset mappings can always be expressed in terms of existential quantification and conversely, existential quantification can always be reduced to an application of the extension principle.

We then considered several properties of linguistic relevance which make use of powerset mappings and thus depend on the induced extension principle. First of all, we generalized the well-known concept of quantitativity to semi-fuzzy and fuzzy quantifiers, adopting the usual modelling of quantitativity in terms of automorphism invariance. The subsequent analysis of the models revealed that quantitative semi-fuzzy quantifiers will be mapped to quantitative counterparts, while non-quantitative examples remain non-quantitative when applying a DFS. This demonstrates the suitability of the proposed models to express both types of quantifiers. Existing approaches, by contrast, only support a limited fragment of the quantitative type.

These results on the interpretation of fuzzy powerset mappings made it possible to investigate another important issue from the linguistic point of view, viz the property of 'having extension'. Natural language quantifiers are typically insensitive to the precise choice of the domain. Intuitively, the domain constitutes a background for quantification but its total extent should be inessential as long as it is large enough to contain the considered choice of arguments. Int the chapter, we introduced the notion of an extensional QFM, i.e. model of fuzzy quantification which preserves the property of having extension. It was then shown that every DFS is indeed extensional. This is an important result from a linguistic perspective because (apart from a few possible exceptions), NL quantifiers are generally extensional.

We also considered the novel concept of contextuality, which is concerned with our intuitive understanding of fuzziness and the relationship between fuzzy sets and corresponding crisp sets. Hence let us consider a fuzzy subset $X \in \widetilde{\mathcal{P}}(E)$ involved in a quantification. It is then only the intermediate cases with gradual membership $\mu_X(e) \in (0, 1)$, but not the clear cases with crisp membership $\mu_X(e) \in \{0, 1\}$, which are ambiguous with respect to the decision if "$e \in X$?". Based on the familiar notions of core and support (envelope) of a fuzzy set, the collection of these unclear cases was resolved into a closed range of crisp sets, the ambiguity range $\mathrm{cxt}(X)$. The requirement of contextuality states that the model should only depend on the behaviour of a quantifier for crisp arguments in the ambiguity ranges. Thus, the fuzzy arguments create an interpretation context, and the behaviour of the quantifier outside the context range of the arguments is considered irrelevant to the quantification result. Every plausible model was then shown to comply with this contextuality requirement. We shall experience later that contextuality conflicts with another desideratum, that of preserving general convexity properties of quantifiers. However, contextuality captures so elementary an aspect of fuzzy quantification that it clearly outweighs other desiderata.

The chapter was also concerned with the interpretation of the universal and existential quantifiers in the models. For that purpose, we reviewed Thiele's analysis of the standard quantifiers in terms of so-called T- and S-quantifiers, and the universal and existential quantifiers induced by a DFS were shown to belong to these classes. By applying Thiele's decomposition theorems an explicit representation of these quantifiers was achieved, which can be expressed directly in terms of the induced t-norm/s-norm of the model. The decomposition reveals in particular that the important relationship between conjunction and universal quantification (disjunction and existential quantification), which holds in the crisp case, is preserved when applying a DFS. We also obtained a new representation of the induced extension principle which reveals the internal structure of the induced fuzzy powerset mappings.

Finally, we considered two more theoretically oriented topics. We first reviewed the notion of inverse images (the converse of powerset mappings). Unlike extension principles, there is only one natural generalisation of inverse images to the fuzzy case. We should then expect that the model of fuzzy quantification induce this natural choice of fuzzy inverse images. The proposed models were shown to be plausible in this respect as well. This result contributed to the solution of the last problem discussed in the chapter, viz the reducibility of multi-place quantification to a special kind of unary quantification. By exploiting an apparent Currying relationship, we reduced multi-place semi-fuzzy and fuzzy quantifiers to 'simple' unary quantifiers. We then discussed the precise conditions that a QFM $\mathcal{F}$ must fulfill in order to be compatible with these reductions. This analysis revealed that in the proposed models, multi-place quantification can always be reduced to a special form of unary quantification.

5

Special Subclasses of Models

5.1 Motivation and Chapter Overview

The last chapter has shed some light on the semantical properties shared by *all* models of the theory. By contrast, we will now be concerned with the structuring of these models into natural subclasses with certain joint properties. Furthermore these subclasses provide a context for defining concepts which require a minumum degree of homogeneity on the models' side. For example, we will only be able to compare models which are sufficiently similar for a comparison to make sense. Finally, the research into subclasses also serves to identify a class of standard models for fuzzy quantification which best meet our intuitive expectations.

In the chapter, the models are first grouped by their induced negation. A model transformations scheme is then proposed which is capable of fitting a given QFM to different choices of negation operators. In this way, we can switch from any choice of induced negation to any other choice of induced negation. Without loss of generality, we will therefore focus on those models which induce the standard negation, called $\neg$-DFSes. These models will be further grouped by their induced disjunction $\widetilde{\vee}$, thus forming the classes of $\widetilde{\vee}$-DFSes. These classes will be sufficiently homogeneous to allow the definition of a model aggregation scheme which combines a given collection of $\widetilde{\vee}$-DFSes into a new model. In addition we can define a specificity order on the $\widetilde{\vee}$-DFSes which judges the degree to which a model commits to crisp quantification results. We will also consider the aggregation of quantifiers by conjunctions and disjunctions and state some results concerning the compatibility of a restricted class of models with these novel constructions. Finally we will identify the class of standard models for fuzzy quantification within the proposed framework. Intuitively, the standard models should conform to the standard concepts of fuzzy set theory, like the standard negation $1 - x$, the standard conjunction and disjunction min and max, the standard extension principle etc. The chapter presents an axiomatization of the standard models which accounts for all of these considerations.

I. Glöckner: *Fuzzy Quantifiers*, StudFuzz **193**, 149–160 (2006)
www.springerlink.com © Springer-Verlag Berlin Heidelberg 2006

5.2 Models Which Induce the Standard Negation

To begin with, let us classify the models according to their induced negation.

Definition 5.1.
Let $\widetilde{\neg} : \mathbf{I} \longrightarrow \mathbf{I}$ *be a strong negation operator. A DFS* $\mathcal{F}$ *is called a* $\widetilde{\neg}$*-DFS if its induced negation coincides with* $\widetilde{\neg}$*, i.e.* $\widetilde{\mathcal{F}}(\neg) = \widetilde{\neg}$*. In particular, we will call* $\mathcal{F}$ *a* $\neg$*-DFS if it induces the standard negation* $\neg x = 1 - x$*.*

In the following it will be shown that without loss of generality, one can restrict attention to $\neg$-DFSes. To achieve this, a mechanism is needed which allows us to transform a given DFS $\mathcal{F}$ into another. A suitable model transformation scheme is defined as follows.

Definition 5.2 (Model transformation scheme).
Suppose $\mathcal{F}$ *is a DFS and* $\sigma : \mathbf{I} \longrightarrow \mathbf{I}$ *a bijection. For every semi-fuzzy quantifier* $Q : \mathcal{P}(E)^n \longrightarrow \mathbf{I}$ *and all* $X_1, \ldots, X_n \in \widetilde{\mathcal{P}}(E)$*, we define*

$$\mathcal{F}^\sigma(Q)(X_1, \ldots, X_n) = \sigma^{-1}\mathcal{F}(\sigma Q)(\sigma X_1, \ \ldots, \sigma X_n)\,,$$

where σQ *abbreviates* $\sigma \circ Q$*, and* $\sigma X_i \in \widetilde{\mathcal{P}}(E)$ *is the fuzzy subset with* $\mu_{\sigma X_i} = \sigma \circ \mu_{X_i}$*.*

Let us now establish that every proper instantiation of the transformation scheme indeed constructs a new model of fuzzy quantification:

Theorem 5.3.
If $\mathcal{F}$ *is a DFS and* $\sigma : \mathbf{I} \longrightarrow \mathbf{I}$ *an increasing bijection, then* $\mathcal{F}^\sigma$ *is a DFS.*

It is well-known [99, Th-3.7] that for every strong negation $\widetilde{\neg} : \mathbf{I} \longrightarrow \mathbf{I}$ there is a monotonically increasing bijection $\sigma : \mathbf{I} \longrightarrow \mathbf{I}$ such that $\widetilde{\neg} x = \sigma^{-1}(1 - \sigma(x))$ for all $x \in \mathbf{I}$. The mapping σ is called the *generator* of $\widetilde{\neg}$.

Theorem 5.4.
Suppose $\mathcal{F}$ *is a* $\widetilde{\neg}$*-DFS and* $\sigma : \mathbf{I} \longrightarrow \mathbf{I}$ *is the generator of* $\widetilde{\neg}$*. Then* $\mathcal{F}' = \mathcal{F}^{\sigma^{-1}}$ *is a* $\neg$*-DFS and* $\mathcal{F} = \mathcal{F}'^{\sigma}$*.*

The theorem states that the model transformation accomplishes a bidirectional translation between the models which is capable of adapting the induced negation.

The transformation scheme thus demonstrates a universal translation property with respect to strong negation operators: Every model can be fitted to any choice of negation operator and vice versa. Put differently, each class of $\widetilde{\neg}$-DFSes is representative of the full class of models, because the remaining classes of $\widetilde{\neg}'$-DFSes (based on different choices of the negation $\widetilde{\neg}'$) can be generated from the given source class. In particular, no models of interest are lost if we focus on $\neg$-DFSes only, i.e. on those models which induce the standard negation. Due to the universal translation property, the definitions of novel concepts and constructions on the models can now be restricted to $\neg$-DFSes whenever convenient.

In the following, the models based on the standard negation are further grouped according to their induced disjunction.

Definition 5.5.
A $\neg$-DFS $\mathcal{F}$ which induces a fuzzy disjunction $\widetilde{\vee}$ is called a $\widetilde{\vee}$-DFS.

The benefit of introducing these classes of models is that $\widetilde{\vee}$-DFSes are sufficiently homogeneous for a straightforward definition of constructions that act on collections of these models. In particular, it now becomes possible to introduce the following model aggregation scheme, which combines a collection of $\widetilde{\vee}$-DFSes into a new $\widetilde{\vee}$-DFS, in accordance with a given aggregation operator Ψ.

Theorem 5.6 (Model aggregation scheme).
Suppose $\mathcal{J}$ is a non-empty index set and $(\mathcal{F}_j)_{j\in\mathcal{J}}$ is a $\mathcal{J}$-indexed collection of $\widetilde{\vee}$-DFSes. Further suppose that $\Psi : \mathbf{I}^{\mathcal{J}} \longrightarrow \mathbf{I}$ satisfies the following conditions:

a. If $f \in \mathbf{I}^{\mathcal{J}}$ is constant, i.e. if there is a $c \in \mathbf{I}$ such that $f(j) = c$ for all $j \in \mathcal{J}$, then $\Psi(f) = c$.
b. $\Psi(1-f) = 1-\Psi(f)$, where $1-f \in \mathbf{I}^{\mathcal{J}}$ is point-wise defined by $(1-f)(j) = 1-f(j)$, for all $j \in \mathcal{J}$.
c. Ψ is monotonically increasing, i.e. if $f(j) \leq g(j)$ for all $j \in \mathcal{J}$, then $\Psi(f) \leq \Psi(g)$.

If we define $\Psi[(\mathcal{F}_j)_{j\in\mathcal{J}}]$ by

$$\Psi[(\mathcal{F}_j)_{j\in\mathcal{J}}](Q)(X_1,\ldots,X_n) = \Psi((\mathcal{F}_j(Q)(X_1,\ldots,X_n))_{j\in\mathcal{J}})$$

for all semi-fuzzy quantifiers $Q : \mathcal{P}(E)^n \longrightarrow \mathbf{I}$ and $X_1,\ldots,X_n \in \widetilde{\mathcal{P}}(E)$, then $\Psi[(\mathcal{F}_j)_{j\in\mathcal{J}}]$ is a $\widetilde{\vee}$-DFS.

In particular, convex combinations (e.g., arithmetic mean) and stable symmetric sums [150] of $\widetilde{\vee}$-DFSes are again $\widetilde{\vee}$-DFSes.

The $\neg$-DFSes can be partially ordered by 'specificity' or 'fuzziness', in the sense of closeness to $\frac{1}{2}$. We define a partial order $\preceq_c \subseteq \mathbf{I} \times \mathbf{I}$ by

$$x \preceq_c y \Leftrightarrow y \leq x \leq \tfrac{1}{2} \text{ or } \tfrac{1}{2} \leq x \leq y\,, \tag{5.1}$$

for all $x, y \in \mathbf{I}$. $\preceq_c$ is Mukaidono's ambiguity relation, see [118]. This basic definition of $\preceq_c$ for scalars can be extended to the case of DFSes in the obvious way:

Definition 5.7.
Let $\mathcal{F}$, $\mathcal{F}'$ be $\neg$-DFSes. We say that $\mathcal{F}$ is consistently less specific *than $\mathcal{F}'$, in symbols: $\mathcal{F} \preceq_c \mathcal{F}'$, if for all semi-fuzzy quantifiers $Q : \mathcal{P}(E)^n \longrightarrow \mathbf{I}$ and all $X_1,\ldots,X_n \in \widetilde{\mathcal{P}}(E)$, $\mathcal{F}(Q)(X_1,\ldots,X_n) \preceq_c \mathcal{F}'(Q)(X_1,\ldots,X_n)$.*

Let us now establish the existence of consistently least specific $\widetilde{\vee}$-DFSes. As it turns out, the greatest lower specificity bound of a collection of $\widetilde{\vee}$-DFSes can be expressed using the fuzzy median, defined as follows.

Definition 5.8.
The fuzzy median $\mathrm{med}_{\frac{1}{2}} : \mathbf{I} \times \mathbf{I} \longrightarrow \mathbf{I}$ *is defined by*

$$\mathrm{med}_{\frac{1}{2}}(u_1, u_2) = \begin{cases} \min(u_1, u_2) & : \quad \min(u_1, u_2) > \frac{1}{2} \\ \max(u_1, u_2) & : \quad \max(u_1, u_2) < \frac{1}{2} \\ \frac{1}{2} & : \quad \text{else} \end{cases}$$

The plot of $\mathrm{med}_{\frac{1}{2}}$ is displayed in Fig. 5.1.

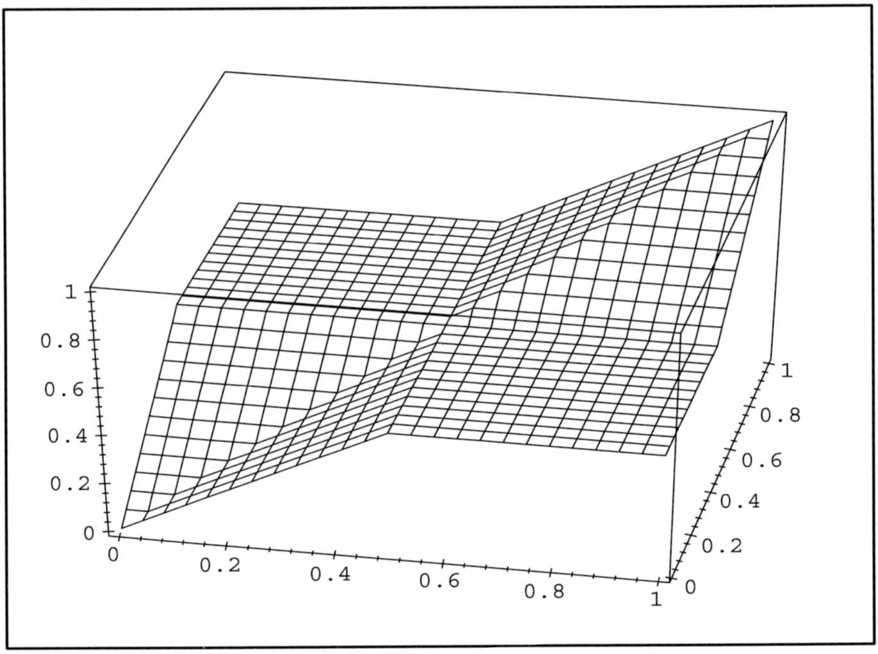

Fig. 5.1. The fuzzy median $\mathrm{med}_{\frac{1}{2}}$

$\mathrm{med}_{\frac{1}{2}}$ is an associative mean operator [16] and the only stable (i.e. idempotent) associative symmetric sum [150].
The fuzzy median can be generalised to an operator $\mathrm{m}_{\frac{1}{2}} : \mathcal{P}(\mathbf{I}) \longrightarrow \mathbf{I}$ which accepts arbitrary subsets of $\mathbf{I}$ as its arguments. Firstly, because it is associative, idempotent and commutative, $\mathrm{med}_{\frac{1}{2}}$ can be generalized to arbitrary *finite* sets of arguments (just apply $\mathrm{med}_{\frac{1}{2}}$ in any order). Noting that for all

finite $X = \{x_1, \ldots, x_n\} \subseteq \mathbf{I}$, $n \geq 2$, it holds that

$$\mathrm{m}_{\frac{1}{2}} X = \mathrm{med}_{\frac{1}{2}}(\min X,\ \max X)\,,$$

the proper definition of $\mathrm{m}_{\frac{1}{2}} X$ in the case $n = 0$, $n = 1$ becomes

$$\mathrm{m}_{\frac{1}{2}} \varnothing = \mathrm{med}_{\frac{1}{2}}(\min \varnothing,\ \max \varnothing) = \mathrm{med}_{\frac{1}{2}}(1,\, 0) = \tfrac{1}{2}\,,$$
$$\mathrm{m}_{\frac{1}{2}} \{u\} = \mathrm{med}_{\frac{1}{2}}(\min\{u\},\ \max\{u\}) = \mathrm{med}_{\frac{1}{2}}(u,\, u) = u$$

We can extend $\mathrm{m}_{\frac{1}{2}}$ to arbitrary subsets $X \subseteq \mathbf{I}$ as follows.

Definition 5.9.
The generalized fuzzy median $\mathrm{m}_{\frac{1}{2}} : \mathcal{P}(\mathbf{I}) \longrightarrow \mathbf{I}$ *is defined by*

$$\mathrm{m}_{\frac{1}{2}} X = \mathrm{med}_{\frac{1}{2}}(\inf X,\ \sup X)\,, \qquad \textit{for all } X \in \mathcal{P}(\mathbf{I}).$$

Note 5.10. This definition is obviously compatible with the above considerations on the proper definition of $\mathrm{m}_{\frac{1}{2}} X$ for finite subsets of $\mathbf{I}$.

Based on the generalized fuzzy median, we can now state the desired theorem on the existence and representation of lower specificity bounds on given collections of $\widetilde{\vee}$-DFSes.

Theorem 5.11.
Let $\widetilde{\vee}$ be an s-norm, and $\mathbb{F}$ a non-empty collection of $\widetilde{\vee}$-DFSes $\mathcal{F} \in \mathbb{F}$. Then there exists a greatest lower specificity bound on $\mathbb{F}$, i.e. a $\widetilde{\vee}$-DFS $\mathcal{F}_{\mathrm{glb}}$ such that $\mathcal{F}_{\mathrm{glb}} \preceq_c \mathcal{F}$ for all $\mathcal{F} \in \mathbb{F}$ (i.e. $\mathcal{F}_{\mathrm{glb}}$ is a lower specificity bound), and for all other lower specificity bounds $\mathcal{F}'$, $\mathcal{F}' \preceq_c \mathcal{F}_{\mathrm{glb}}$.
$\mathcal{F}_{\mathrm{glb}}$ is defined by

$$\mathcal{F}_{\mathrm{glb}}(Q)(X_1, \ldots, X_n) = \mathrm{m}_{\frac{1}{2}}\{\mathcal{F}(Q)(X_1, \ldots, X_n) : \mathcal{F} \in \mathbb{F}\}\,,$$

for all $Q : \mathcal{P}(E)^n \longrightarrow \mathbf{I}$ and $X_1, \ldots, X_n \in \widetilde{\mathcal{P}}(E)$.

In particular, the theorem asserts the existence of a least specific $\widetilde{\vee}$-DFSes, i.e. whenever $\widetilde{\vee}$ is an s-norm such that $\widetilde{\vee}$-DFSes exist at all, then there exists a least specific $\widetilde{\vee}$-DFS (just apply the above theorem to the collection of all $\widetilde{\vee}$-DFSes).
As concerns the converse issue of most specific models, i.e. least upper bounds with respect to $\preceq_c$, the following definition of 'specificity consistency' turns out to provide the key concept:

Definition 5.12.
Suppose $\widetilde{\vee}$ is an s-norm and $\mathbb{F}$ is a non-empty collection of $\widetilde{\vee}$-DFSes $\mathcal{F} \in \mathbb{F}$. $\mathbb{F}$ is called specificity consistent *if for all $Q : \mathcal{P}(E)^n \longrightarrow \mathbf{I}$ and $X_1, \ldots, X_n \in \widetilde{\mathcal{P}}(E)$, either $R_{Q,X_1,\ldots,X_n} \subseteq [0, \frac{1}{2}]$ or $R_{Q,X_1,\ldots,X_n} \subseteq [\frac{1}{2}, 1]$, where*

$$R_{Q,X_1,\ldots,X_n} = \{\mathcal{F}(Q)(X_1, \ldots, X_n) : \mathcal{F} \in \mathbb{F}\}\,.$$

Based on this definition of specificity consistency, we can express the exact conditions under which a collection of $\widetilde{\vee}$-DFSes has a least upper specificity bound, and provide an explicit description of the resulting bound in those cases where it exists.

Theorem 5.13.
Suppose $\widetilde{\vee}$ is an s-norm and $\mathbb{F}$ is a non-empty collection of $\widetilde{\vee}$-DFSes $\mathcal{F} \in \mathbb{F}$.

a. $\mathbb{F}$ has upper specificity bounds exactly if $\mathbb{F}$ is specificity consistent.
b. If $\mathbb{F}$ is specificity consistent, then its least upper specificity bound is the $\widetilde{\vee}$-DFS $\mathcal{F}_{\mathrm{lub}}$ defined by

$$\mathcal{F}_{\mathrm{lub}}(Q)(X_1, \ldots, X_n) = \begin{cases} \sup R_{Q,X_1,\ldots,X_n} & : \quad R_{Q,X_1,\ldots,X_n} \subseteq [\frac{1}{2}, 1] \\ \inf R_{Q,X_1,\ldots,X_n} & : \quad R_{Q,X_1,\ldots,X_n} \subseteq [0, \frac{1}{2}] \end{cases}$$

where $R_{Q,X_1,\ldots,X_n} = \{\mathcal{F}(Q)(X_1, \ldots, X_n) : \mathcal{F} \in \mathbb{F}\}$.

5.3 Models Which Induce the Standard Disjunction

The models of fuzzy quantification can also be grouped according to their induced negation and disjunction:

Definition 5.14.
A DFS $\mathcal{F}$ such that $\widetilde{\neg} = \widetilde{\mathcal{F}}(\neg)$ and $\widetilde{\vee} = \widetilde{\mathcal{F}}(\vee)$ is called a $(\widetilde{\neg}, \widetilde{\vee})$-DFS.

The $(\widetilde{\neg}, \max)$-DFSes in particular, comprise those models which induce the standard disjunction and an arbitrary strong negation operator. These are the models for which we will now state a theorem on the interpretation of conjunctions and disjunctions of quantifiers. To this end, let us first introduce the relevant constructions which build conjunctions and disjunctions from given quantifiers.

In the Theory of Generalized Quantifiers there are constructions $Q \wedge Q'$ and $Q \vee Q'$ of forming the conjunction (disjunction) of two-valued quantifiers $Q, Q' : \mathcal{P}(E)^n \longrightarrow \mathbf{2}$, see e.g. [7, p. 194], [50, p. 234]. These constructions are easily generalized to (semi)-fuzzy quantifiers.

Definition 5.15.
Suppose $\widetilde{\wedge}, \widetilde{\vee} : \mathbf{I} \times \mathbf{I} \longrightarrow \mathbf{I}$ are given (usually the connectives induced by an assumed QFM). For all semi-fuzzy quantifiers Q, $Q' : \mathcal{P}(E)^n \longrightarrow \mathbf{I}$, the conjunction $Q \,\widetilde{\wedge}\, Q' : \mathcal{P}(E)^n \longrightarrow \mathbf{I}$ and the disjunction $Q \,\widetilde{\vee}\, Q' : \mathcal{P}(E)^n \longrightarrow \mathbf{I}$ of Q and Q' are defined by

$$(Q \,\widetilde{\wedge}\, Q')(Y_1, \ldots, Y_n) = Q(Y_1, \ldots, Y_n) \,\widetilde{\wedge}\, Q'(Y_1, \ldots, Y_n)$$
$$(Q \,\widetilde{\vee}\, Q')(Y_1, \ldots, Y_n) = Q(Y_1, \ldots, Y_n) \,\widetilde{\vee}\, Q'(Y_1, \ldots, Y_n)$$

for all $Y_1, \ldots, Y_n \in \mathcal{P}(E)$. For fuzzy quantifiers, $\widetilde{Q} \,\widetilde{\wedge}\, \widetilde{Q}'$ and $\widetilde{Q} \,\widetilde{\vee}\, \widetilde{Q}'$ are defined analogously.

Note 5.16. Conjunctions and disjunctions of (semi-)fuzzy quantifiers of different arities can be formed through cylindrical extensions, i.e. by adding vacuous arguments, see Sect. 4.6.

As to the interpretation of these conjunctions and disjunctions of quantifiers, the following can be shown for $(\widetilde{\neg}, \max)$-DFSes.

Theorem 5.17.
Suppose $\mathcal{F}$ is a $(\widetilde{\neg}, \max)$-DFS. Then for all $Q, Q' : \mathcal{P}(E)^n \longrightarrow \mathbf{I}$,

a. $\mathcal{F}(Q \wedge Q') \leq \mathcal{F}(Q) \wedge \mathcal{F}(Q')$
b. $\mathcal{F}(Q \vee Q') \geq \mathcal{F}(Q) \vee \mathcal{F}(Q')$.

Notes 5.18.

- It should be remarked that the theorem does *not* state the compatibility of $(\widetilde{\neg}, \max)$-DFSes to conjunction and disjunction, it simply establishes bounds on the quantification results. See Sect. 6.4 for a discussion why the inequalities in the above theorem cannot be replaced with equalities on formal grounds.
- Because the theorem refers to the standard fuzzy conjunction and disjunction, the constructions on quantifiers have been written $Q \wedge Q'$ and $Q \vee Q'$, omitting the 'tilde' notation for fuzzy connectives. Similarly, the standard fuzzy intersection and standard fuzzy union will be written $X \cap Y$ and $X \cup Y$, resp., where $\mu_{X \cap Y}(e) = \min(\mu_X(e), \mu_Y(e))$ and $\mu_{X \cup Y}(e) = \max(\mu_X(e), \mu_Y(e))$. The same conventions are stipulated for intersections $\widetilde{Q}\cap$ and unions $\widetilde{Q}\cup$ of the arguments of a fuzzy quantifier, as well as for duals $Q\square$ of semi-fuzzy quantifiers or $\widetilde{Q}\square$ of fuzzy quantifiers, based on the standard negation.

We have not made any claims yet concerning the interpretation of $\widetilde{\leftrightarrow} = \widetilde{\mathcal{F}}(\leftrightarrow)$ (equivalence) and $\widetilde{\text{xor}} = \widetilde{\mathcal{F}}(\text{xor})$ (exclusive or) in a given DFS $\mathcal{F}$, and indeed, there are currently no results available for the full class of models. In the special case of $(\widetilde{\neg}, \max)$-DFSes, though, these connectives are assigned the following interpretation.

Theorem 5.19.
Suppose $\mathcal{F}$ is a $(\widetilde{\neg}, \max)$-DFS. Then for all $x_1, x_2 \in \mathbf{I}$,

a. $x_1 \mathbin{\widetilde{\leftrightarrow}} x_2 = (x_1 \wedge x_2) \vee (\widetilde{\neg} x_1 \wedge \widetilde{\neg} x_2)$
b. $x_1 \mathbin{\widetilde{\text{xor}}} x_2 = (x_1 \wedge \widetilde{\neg} x_2) \vee (\widetilde{\neg} x_1 \wedge x_2)$.

5.4 The Standard Models of Fuzzy Quantification

The most restricted subclass of models we will consider – and the best-behaved – is that of *standard DFSes*, i.e. the class of those models which comply with the standard operations of fuzzy set theory (min, max etc.). Due to the excellent adequacy properties of these models, and their conformance to the established core of fuzzy set theory, it is suggested that this type of DFSes be considered the standard models of fuzzy quantification. Formally, we define standard DFSes as follows.

Definition 5.20.
By a standard DFS *we denote a* $(\neg, \max)$*-DFS.*

Building on the earlier theorems, the fuzzy truth functions induced by a standard DFS can now be summarized as follows.

Theorem 5.21 (Truth functions in standard DFSes).
In every standard DFS $\mathcal{F}$,

$$\begin{aligned}
\widetilde{\neg}\, x_1 &= 1 - x_1 \\
x_1 \mathbin{\widetilde{\vee}} x_2 &= \max(x_1, x_2) \\
x_1 \mathbin{\widetilde{\wedge}} x_2 &= \min(x_1, x_2) \\
x_1 \mathbin{\widetilde{\rightarrow}} x_2 &= \max(1 - x_1, x_2) \\
x_1 \mathbin{\widetilde{\leftrightarrow}} x_2 &= \max(\min(x_1, x_2), \ \min(1 - x_1, 1 - x_2)) \\
x_1 \mathbin{\widetilde{\mathrm{xor}}} x_2 &= \max(\min(x_1, 1 - x_2), \ \min(1 - x_1, x_2))
\end{aligned}$$

The standard DFSes therefore induce the standard connectives of fuzzy logic, i.e. the propositional fragment of a standard model coincides with the well-known K-standard sequence logic of Dienes [37]. In particular, Kleene's three-valued logic [95, p. 344] is obtained when restricting to the three-valued fragment.[1] By the above theorem Th. 4.62, then, standard DFSes are also known to induce the standard extension principle.

As concerns the standard quantifiers, we obtain the familiar choices as well. For example, it is apparent from theorems Th. 4.55 and Th. 4.61 that

$$\begin{aligned}
\mathcal{F}(\exists)(X) &= \sup\{\mu_X(e) : e \in E\} \\
\mathcal{F}(\forall)(X) &= \inf\{\mu_X(e) : e \in E\}
\end{aligned}$$

for all $X \in \widetilde{\mathcal{P}}(E)$. This is quite satisfactory.

The standard models represent a boundary case of DFSes because they induce the smallest fuzzy existential quantifiers, the smallest extension principle, and the largest fuzzy universal quantifiers. This is immediate from Th. 4.61, Th. 4.39 and Th. 4.55, respectively, keeping in mind that min is the largest t-norm and max the smallest s-norm.

[1] Some readers might prefer a different choice of the implication operator, namely $x_1 \mathbin{\widetilde{\rightarrow}} x_2 = \min(1, 1 - x_1 + x_2)$. However, it is clear that every QFM with the highly desirable property of preserving Aristotelian squares will also preserve the interdefinability of the propositional connectives, and therefore differ from Łukasiewicz logic.

Finally, we consider the interpretation of two-valued quantifiers in a standard DFS. Interestingly, there is absolutely no freedom concerning the interpretation assigned to these quantifiers, which is fully determined by the requirements imposed on the standard models:

Theorem 5.22.
All standard DFSes coincide on two-valued quantifiers. Hence if $\mathcal{F}, \mathcal{F}'$ are standard DFSes and $Q : \mathcal{P}(E)^n \longrightarrow \mathbf{2}$ is a two-valued quantifier, then $\mathcal{F}(Q) = \mathcal{F}'(Q)$.

This result establishes a link between the standard models and other fuzzification mechanisms described in the literature. This is because the examples of standard models introduced in [53] are known to coincide with the fuzzification mechanism of Gaines [49] on two-valued quantifiers. We can then conclude from the above theorem that all standard models are indeed compatible with this mechanism. Standard DFSes are more powerful in scope, though, and consistently generalize the Gainesian mechanism to arbitrary semi-fuzzy truth functions and semi-fuzzy quantifiers (see Sect. 10.9 below for a more detailed discussion of the Gainesian fuzzification mechanism).

In terms of the classification of quantification types introduced by Kerre and Liu [108, p. 2], the theorem asserts that all Type II quantifications in the sense of the proposed classification are assigned the very same interpretation in all standard models. In other words, the axioms for standard models of fuzzy quantification are strong enough to identify a unique admissible interpretation for arbitrary Type-II quantifications.

5.5 Axiomatisation of the Standard Models

This section presents a system of conditions on the considered fuzzification mechanism $\mathcal{F}$ which precisely characterise the proposed class of standard models. It should be apparent from the previous remarks which changes to the base axioms (Z-1)–(Z-6) are necessary to effect a reduction to this class. These considerations are summarized in the following adapted system of conditions on $\mathcal{F}$:

Definition 5.23.
Let $\mathcal{F}$ be a QFM. For all semi-fuzzy quantifiers $Q : \mathcal{P}(E)^n \longrightarrow \mathbf{I}$, we stipulate the following conditions.

Correct generalisation	$\mathcal{U}(\mathcal{F}(Q)) = Q \quad \text{if } n \leq 1$	(S-1)
Projection quantifiers	$\mathcal{F}(Q) = \widetilde{\pi}_e \quad \text{if there exists } e \in E \text{ s.th. } Q = \pi_e$	(S-2)
Dualisation	$\mathcal{F}(Q\square) = \mathcal{F}(Q)\square \quad n > 0$	(S-3)
Internal joins	$\mathcal{F}(Q\cup) = \mathcal{F}(Q)\cup \quad n > 0$	(S-4)
Preservation of monotonicity	*If Q is nonincreasing in n-th arg, then* $\mathcal{F}(Q)$ *is nonincreasing in n-th arg,* $n > 0$	(S-5)
Functional application	$\mathcal{F}(Q \circ \overset{n}{\underset{i=1}{\times}} \widehat{f_i}) = \mathcal{F}(Q) \circ \overset{n}{\underset{i=1}{\times}} \hat{\hat{f}}_i$ *where* $f_1, \ldots, f_n : E' \longrightarrow E,\ E' \neq \varnothing$.	(S-6)

Note 5.24. Let us briefly discuss the differences compared to the original DFS axioms (Z-1)–(Z-6). In (S-3), the induced negation/complement has been replaced with the standard choice of fuzzy negation/complement. In (S-4), the induced fuzzy union has been replaced with the standard fuzzy union. Finally, (S-6) is built upon the standard extension principle, rather than requiring the compatibility of $\mathcal{F}$ with its induced extension principle.

As stated by the next theorem, these conditions indeed capture the precise requirements for $\mathcal{F}$ to be a standard DFS.

Theorem 5.25.
The conditions (S-1)–(S-6) *are necessary and sufficient for $\mathcal{F}$ to be a standard DFS.*

5.6 Chapter Summary

Building on the discussion of general properties in the last chapter, the models have now been grouped into natural subclasses and their specific properties and mutual relationships have been investigated. In particular, the goal was to develop those concepts that require a certain homogeneity on part of the models. For example, certain aggregating constructions, but also comparisons between DFSes, are only possible if the considered models show a similar structure, and thus sufficient compatibility. In the chapter, we first grouped the models according to their induced negation into the classes of $\widetilde{\neg}$-DFSes. The proposed model transformation scheme then achieved a bidirectional translation between different classes of $\widetilde{\neg}$-DFSes. By instantiating the transformation scheme, a given model of fuzzy quantification can be adapted to any desired choice of induced negation. In this sense, all negation operators are universal to DFSes. Due to this universal translation property, we can now restrict attention to a single representative choice, $\neg x = 1 - x$, and the corresponding class of models which induce the standard negation. These $\neg$-DFSes form a representative class to which all other classes of models can be reduced. Essentially, this means that any new concepts of fuzzy quantification theory need only be defined for $\neg$-DFSes. By applying the model transformation scheme, these concepts can easily be generalized to other types of models if so desired.

Keeping the standard negation fixed, we further grouped the $\neg$-DFSes by their induced negation. We then introduced a number of concepts for the resulting $\widetilde{\vee}$-DFSes (i.e. models which induce the s-norm $\widetilde{\vee}$ and the standard negation). The relative homogeneity of these models made it possible to define a construction of model aggregation; the precise conditions which constrain the admissible aggregation operators were also stated. This investigation revealed that the $\widetilde{\vee}$-DFSes are closed under important operations like stable symmetric sums or convex combinations. A concrete application has also been described, viz. the construction of the greatest lower specificity bound from a given collection of models.

The partitioning into $\widetilde{\vee}$-DFSes is sufficiently discriminative that the members of a given subclass can now be compared in a meaningful way. Due to the symmetry of DFSes with respect to negation, it is not possible to relate the models by the linear order $\leq$, however. We therefore resorted to a specificity order $\preceq_c$, which is basically Mukaidono's ambiguity relation. Roughly speaking, the models are compared by the degree to which they commit to the two-valued poles. The investigation of this aspect of the models is of particular importance because one is often interested in obtaining as specific results as possible. Unlike $\leq$, the proposed relation $\preceq_c$ is not a total order, i.e. there are choices of models which are comparable under $\preceq_c$. Such models will be considered incompatible, because there are situations in which they tend to opposite crisp outcomes. The chapter has presented the necessary formal machinery for discussing such specificity issues. For example, it was shown that by instantiating the model aggregation scheme with the generalized fuzzy median, we can effectively construct the greatest lower specificity bound for every given collection of $\widetilde{\vee}$-DFSes. As concerns the converse issue of upper specificity bounds, we required the notion of 'specificity consistency' of a given collection of models. Such a collection is considered specificity consistent if all models are mutually consistent in the sense explained above, i.e. comparable under $\preceq_c$. The criterion was then shown to decide upon the existence of least upper specificity bounds. The precise structure of such upper bounds has also been described.

Following this discussion of specificity issues, we focused on two constructions not yet addressed in the previous chapter, viz. conjunctions and disjunctions of quantifiers. Some first results on the interpretation of conjunctions and disjunctions in ($\widetilde{\neg}$, max)-DFSes have been presented which also cover all standard models. Specifically, we have established upper and lower bounds on the possible outcome of the conjunction and disjunction. In addition, the interpretation of the equivalence and antivalence/xor truth functions was investigated. It will become obvious in the next chapter why these truth function pose special difficulties.

Finally, we were concerned with those standard models of fuzzy quantification in which the fuzzy connectives, logical quantifiers, and the associated extension principle are all tied to their standard interpretation in fuzzy logic. The chosen definition of these models only makes the extra requirement that the given $\mathcal{F}$ induce the standard negation and standard disjunction. However, this condition is sufficient to ensure the standard interpretation of all constructions in the resulting standard DFSes. In particular, this means that the propositional part of each standard model coincide with the K-standard sequence logic of Dienes, and that the three-valued fragment coincide with well-known Kleene's logic. The latter observation will be the starting point for a construction of concrete models described in chapter Chap. 7. All standard DFSes coincide for two-valued quantifiers. Among other things, this means that all standard models consistently extend the fuzzification mechanism, proposed by Gaines [49]. Unlike the Gainesian scheme, however, the standard DFSes are not limited to two-valued propositional functions but also cover arbitrary quantifiers.

6

Special Semantical Properties and Theoretical Limits

6.1 Motivation and Chapter Overview

In the previous chapters, the proposed class of models was checked against a catalog of formal requirements in order to elucidate the properties shared by every DFS. In addition, the models were partitioned into natural subgroups which allowed the investigation of specificity issues, for example. In this chapter we shall be concerned with properties like *continuity* (smoothness) which are desirable in many, but not all, cases. Continuity, for example, will certainly be expected of every practical model. From a methodical perspective, however, it is important to retain the discontinuous models because these include many boundary cases of theoretical interest. In the chapter, we will discuss two continuity criteria which capture different aspects of the robustness of a model against small changes in its inputs. Both criteria will be important from an application perspective. In addition, we will pay attention to the way in which a model handles the unspecificity observed in the inputs. Intuitively, one would expect that the model's outputs cannot become more specific when there is less information in the inputs. These concerns regarding the *propagation of fuzziness* in the model will be formalized in terms of the specificity order $\preceq_c$ introduced in the last chapter. Like continuity, the propagation of fuzziness will be considered an important but optional requirement.

We will discuss properties which are natural to require but in conflict with other basic desiderata. This situation is familiar in fuzzy set theory, due to the fact that it is impossible to define a Boolean algebra on the set of continuous truth values in $[0, 1]$. In particular, it is well-known that idempotence/distributivity and the law of contradiction exclude each other in the fuzzy framework. It is not surprising that these conflicts also express in fuzzy quantifications. The (partially) conflicting criteria that we will consider are the following.

First we will be concerned with *conjunctions and disjunctions of quantifiers* and show that no standard models comply with these constructions on formal grounds. Similar results will be obtained for *multiple occurrences of*

I. Glöckner: *Fuzzy Quantifiers*, StudFuzz **193**, 161–177 (2006)
www.springerlink.com © Springer-Verlag Berlin Heidelberg 2006

variables in the argument list of a quantifier. Few models will be compatible with these constructions because every conforming model must fulfill the law of contradiction.

We then turn to three properties of special linguistic relevance. First we discuss a generalized form of monotonicity which shows up in so-called *convex* quantifiers. The preservation of unimodal or bell-shaped monotonicity patterns will turn out pretty difficult for the proposed models. This will force us to restrict preservation of convexity to the most important cases. An important property of NL quantifiers which has attracted much interest in the literature on TGQ is that of *conservativity*. We will discuss two alternative generalizations of conservativity to fuzzy quantifiers, one of which is universally valid in the models, while the other cannot be required at all. In fact, the latter criterion will conflict with assumptions much weaker than the DFS axioms.

Finally we review the construction of argument insertion which is important for modelling adjectival restriction. It will be shown how a compositional treatment of fuzzy argument insertion, and thus of *adjectival restriction by fuzzy adjectives*, can be achieved in the given framework. Acknowledging the importance of Frege's compositionality principle to the analysis of natural language, the models which admit this construction will be optimally plausible from the point of view of NL semantics.

6.2 Continuity Conditions

Let us first consider two adequacy criteria concerned with distinct aspects of the 'smoothness' or 'continuity' of a DFS. These conditions are essential for the models to be *practical* because it is extremely important for applications that the results of a DFS be stable under slight changes in the inputs. These 'changes' can either occur in the fuzzy argument sets (e.g. due to noise), or they can affect the semi-fuzzy quantifier. For example, if a person A has a slightly different interpretation of quantifier Q compared to person B, then we still want them to understand each other, and the quantification results obtained from the two models of the target quantifier should be very similar in such cases.

In order to express the robustness criterion with respect to slight changes in the fuzzy arguments, a metric on fuzzy subsets is needed, which serves as a numerical measure of the similarity of the arguments. For all base sets $E \neq \varnothing$ and all $n \in \mathbb{N}$, we define the metric $d : \widetilde{\mathcal{P}}(E)^n \times \widetilde{\mathcal{P}}(E)^n \longrightarrow \mathbf{I}$ by

$$d((X_1, \ldots, X_n), (X'_1, \ldots, X'_n)) = \max_{i=1}^{n} \sup\{|\mu_{X_i}(e) - \mu_{X'_i}(e)| : e \in E\}, \quad (6.1)$$

for all $X_1, \ldots, X_n, X'_1, \ldots, X'_n \in \widetilde{\mathcal{P}}(E)$. Based on this metric, we can now express the desired criterion for continuity *in arguments*.

Definition 6.1.
We say that a QFM $\mathcal{F}$ is arg-continuous *if and only if $\mathcal{F}$ maps all $Q : \mathcal{P}(E)^n \longrightarrow \mathbf{I}$ to continuous fuzzy quantifiers $\mathcal{F}(Q)$, i.e. for all $X_1, \dots, X_n \in \widetilde{\mathcal{P}}(E)$ and $\varepsilon > 0$ there exists $\delta > 0$ such that*

$$|\mathcal{F}(Q)(X_1, \dots, X_n) - \mathcal{F}(Q)(X_1', \dots, X_n')| < \varepsilon$$

for all $X_1', \dots, X_n' \in \widetilde{\mathcal{P}}(E)$ with $d((X_1, \dots, X_n), (X_1', \dots, X_n')) < \delta$.

Arg-continuity means that a small change in the membership grades $\mu_{X_i}(e)$ of the argument sets does not change $\mathcal{F}(Q)(X_1, \dots, X_n)$ drastically; it thus expresses an important robustness condition with respect to noise.
The second robustness criterion is intended to capture the idea that slight changes in a semi-fuzzy quantifier should not cause the quantification results to change drastically. To introduce this criterion, we must first define suitable distance measures for semi-fuzzy quantifiers and for fuzzy quantifiers. Hence for all semi-fuzzy quantifiers $Q, Q' : \mathcal{P}(E)^n \longrightarrow \mathbf{I}$,

$$d(Q, Q') = \sup\{|Q(Y_1, \dots, Y_n) - Q'(Y_1, \dots, Y_n)| : Y_i \in \mathcal{P}(E)\}, \qquad (6.2)$$

and similarly for all fuzzy quantifiers $\widetilde{Q}, \widetilde{Q}' : \widetilde{\mathcal{P}}(E)^n \longrightarrow \mathbf{I}$,

$$d(\widetilde{Q}, \widetilde{Q}') = \sup\{|\widetilde{Q}(X_1, \dots, X_n) - \widetilde{Q}'(X_1, \dots, X_n)| : X_i \in \widetilde{\mathcal{P}}(E)\}. \qquad (6.3)$$

Definition 6.2.
We say that a QFM $\mathcal{F}$ is Q-continuous *if and only if for each semi-fuzzy quantifier $Q : \mathcal{P}(E)^n \longrightarrow \mathbf{I}$ and all $\varepsilon > 0$, there exists $\delta > 0$ such that $d(\mathcal{F}(Q), \mathcal{F}(Q')) < \varepsilon$ whenever $Q' : \mathcal{P}(E)^n \longrightarrow \mathbf{I}$ satisfies $d(Q, Q') < \delta$.*

Q-continuity captures an important aspect of robustness with respect to imperfect knowledge about the intended choice of a quantifier; i.e. slightly different definitions of Q will produce similar quantification results.
Both conditions are crucial to the utility of a DFS and must be possessed by every practical model. They are not part of the DFS axioms because we would like to have models for general t-norms (including the discontinuous variety).

6.3 Propagation of Fuzziness

The next two criteria are concerned with the 'propagation of fuzziness', i.e. the way in which the amount of imprecision in the model's inputs affects changes of the model's outputs. To this end, let us recall the partial order $\preceq_c \subseteq \mathbf{I} \times \mathbf{I}$ defined by equation (5.1). We can extend $\preceq_c$ to fuzzy sets $X \in \widetilde{\mathcal{P}}(E)$, semi-fuzzy quantifiers $Q : \mathcal{P}(E)^n \longrightarrow \mathbf{I}$ and fuzzy quantifiers $\widetilde{Q} : \widetilde{\mathcal{P}}(E)^n \longrightarrow \mathbf{I}$ as follows.

$$X \preceq_c X' \iff \mu_X(e) \preceq_c \mu_{X'}(e) \qquad \text{for all } e \in E;$$
$$Q \preceq_c Q' \iff Q(Y_1, \ldots, Y_n) \preceq_c Q'(Y_1, \ldots, Y_n) \qquad \text{for all } Y_1, \ldots, Y_n \in \mathcal{P}(E);$$
$$\widetilde{Q} \preceq_c \widetilde{Q}' \iff \widetilde{Q}(X_1, \ldots, X_n) \preceq_c \widetilde{Q}'(X_1, \ldots, X_n) \qquad \text{for all } X_1, \ldots, X_n \in \widetilde{\mathcal{P}}(E).$$

Intuitively, we expect that the quantification results become less specific whenever the quantifier or the argument sets become less specific: the fuzzier the input, the fuzzier the output. For example, consider a base set $E = \{\text{Joan, Lucas, Mary}\}$ and fuzzy subsets **lucky**, **lucky**$' \in \widetilde{\mathcal{P}}(E)$ defined by

$$\begin{aligned}\textbf{lucky} &= 1/\text{Joan} + 0.8/\text{Lucas} + 0.2/\text{Mary},\\ \textbf{lucky}' &= 0.6/\text{Joan} + 0.5/\text{Lucas} + 0.4/\text{Mary}.\end{aligned}$$

Then $\textbf{lucky}' \preceq_c \textbf{lucky}$, i.e. the interpretation of "lucky" as **lucky**$'$ is less committed to the possible crisp decisions. We should therefore expect that $\mathcal{F}(\textbf{most})(\textbf{rich}, \textbf{lucky}') \preceq_c \mathcal{F}(\textbf{most})(\textbf{rich}, \textbf{lucky})$ as well, i.e. the quantification results based on **lucky**$'$ should be less decided than those computed from the fuzzy subset **lucky**, which bears more specific information.

Definition 6.3.
Let a QFM $\mathcal{F}$ be given.

a. We say that $\mathcal{F}$ propagates fuzziness in arguments *if the following property is satisfied for all $Q : \mathcal{P}(E)^n \longrightarrow \mathbf{I}$ and $X_1, \ldots, X_n, X'_1, \ldots, X'_n \in \widetilde{\mathcal{P}}(E)$: If $X_i \preceq_c X'_i$ for all $i = 1, \ldots, n$, then $\mathcal{F}(Q)(X_1, \ldots, X_n) \preceq_c \mathcal{F}(Q)(X'_1, \ldots, X'_n)$.*
b. We say that $\mathcal{F}$ propagates fuzziness in quantifiers *if $\mathcal{F}(Q) \preceq_c \mathcal{F}(Q')$ whenever $Q \preceq_c Q'$.*

Notes 6.4.

- Both conditions are certainly natural to require, and will therefore be considered desirable but optional. A more thorough discussion of propagation of fuzziness and its tradeoffs can be found in Chap. 8, p. 234.
- The intuitive expectation that the output cannot get more detailed when the input gets fuzzier, is satisfied by the standard connectives $\neg x = 1 - x$, $\wedge = \min$ and $\vee = \max$. Thus the standard models of the DFS axioms make good candidates for propagation of fuzziness. We shall see below in Th. 8.38 that both conditions are possible but in fact optional for standard models, and also that the conditions are mutually independent. This indicates that propagation of fuzziness in quantifiers and propagation of fuzziness in arguments open distinct semantical dimensions in the full space of models for fuzzy quantification.
- It should be remarked that the standard conjunction and disjunction are not the only choice of t- and s-norms which propagate fuzziness in their arguments. In fact, the t-norm $\widetilde{\wedge}_m$ defined by equation (4.1) apparently also propagates fuzziness in its arguments, a property which transfers to the dual s-norm $\widetilde{\vee}_m$ of $\widetilde{\wedge}_m$ under the standard negation. The corresponding class of $\widetilde{\vee}_m$-DFSes might then witness the existence of nonstandard models

which propagate fuzziness. However, it is currently not known whether models exist at all in this formal class.

6.4 Conjunctions and Disjunctions of Quantifiers

Let us now review the construction introduced in Def. 5.15 which builds a new quantifier from a conjunction or disjunction of given ones. In the last chapter, we presented a result on the interpretation of such conjunctions and disjunctions in $(\widetilde{\neg}, \max)$-DFSes. However, we only established upper and lower bounds rather than an equality. Let us now show that an improved result stating the full compatibility of $(\widetilde{\neg}, \max)$-DFSes with this construction cannot be achieved because the t-norm min induced by these models violates the law of contradiction.

Theorem 6.5.
Suppose a DFS $\mathcal{F}$ has the property that

$$\mathcal{F}(Q \mathbin{\widetilde{\wedge}} Q') = \mathcal{F}(Q) \mathbin{\widetilde{\wedge}} \mathcal{F}(Q') \tag{6.4}$$

for all semi-fuzzy quantifiers $Q, Q' : \mathcal{P}(E)^n \longrightarrow \mathbf{I}$, where $\widetilde{\wedge} = \widetilde{\mathcal{F}}(\wedge)$. Then

$$x \mathbin{\widetilde{\wedge}} \widetilde{\neg} x = 0$$

for all $x \in \mathbf{I}$.

A DFS $\mathcal{F}$ which is homomorphic with respect to conjunctions (or equivalently, disjunctions) of quantifiers therefore induces a t-norm which respects the law of contradiction. This requirement would exclude many interesting t-norms; in particular, the standard choice $\widetilde{\mathcal{F}}(\wedge) = \min$. Consequently we have *not* required in general that a DFS be compatible with conjunctions/disjunctions of quantifiers.

6.5 Multiple Occurrences of Variables

The discussion of the propositional connectives $\leftrightarrow$ and xor has been separated from the discussion of the simple cases like conjunction and disjunction in Sect. 4.3. The difficulties in proving properties of the models for these connectives are due to the fact that the definition of $\leftrightarrow$ and xor involves *multiple occurrences* of propositional variables, viz.

$$\begin{aligned} x_1 \leftrightarrow x_2 &= (x_1 \wedge x_2) \vee (\neg x_1 \wedge \neg x_2) \\ &= (\neg x_1 \vee x_2) \wedge (x_1 \vee \neg x_2) \\ x_1 \operatorname{xor} x_2 &= (x_1 \wedge \neg x_2) \vee (\neg x_1 \wedge x_2) \\ &= (x_1 \vee x_2) \wedge (\neg x_1 \vee \neg x_2)\,, \end{aligned}$$

for all $x_1, x_2 \in \{0, 1\}$. A glance at the DFS axioms reveals that there is *no* axiom which describes this case of multiple occurrences of variables. In order to be able to discuss the issue on the formal level, let us introduce the following construction:

Definition 6.6.
Suppose $Q : \mathcal{P}(E)^m \longrightarrow \mathbf{I}$ *is a semi-fuzzy quantifier and* $\xi : \{1, \ldots, n\} \longrightarrow \{1, \ldots, m\}$ *is a mapping. By* $Q\xi : \mathcal{P}(E)^n \longrightarrow \mathbf{I}$ *we denote the semi-fuzzy quantifier defined by* $Q\xi(Y_1, \ldots, Y_n) = Q(Y_{\xi(1)}, \ldots, Y_{\xi(n)})$*, for all* $Y_1, \ldots, Y_n \in \mathcal{P}(E)$. *We use an analogous definition for fuzzy quantifiers.*

The interesting case is that of a non-injective ξ, which inserts the same variable in two (or more) argument positions of the original quantifier Q.

Theorem 6.7.
Let $\mathcal{F}$ *be a DFS compatible with the duplication of variables, i.e. for* $Q : \mathcal{P}(E)^m \longrightarrow \mathbf{I}$ *and* $\xi : \{1, \ldots, n\} \longrightarrow \{1, \ldots, m\}$ *for some* $n \in \mathbb{N}$, *it always holds that* $\mathcal{F}(Q\xi) = \mathcal{F}(Q)\xi$. *Then the induced conjunction satisfies*

$$x \,\widetilde{\wedge}\, \widetilde{\neg} x = 0$$

for all $x \in \mathbf{I}$.

Again, this will be considered too restrictive because this type of conjunction excludes the standard models. Consequently, it was not required in general that $\mathcal{F}$ be homomorphic with respect to the duplication of variables.

6.6 Convex Quantifiers

Apart from the monotonicity type, TGQ has also developed more sophisticated concepts which describe the characteristic shape of a given quantifier.

The following definition of convex quantifiers covers unimodal, bell-shaped, trapezoidal and similar cases.

Definition 6.8.
Let $Q : \mathcal{P}(E)^n \longrightarrow \mathbf{I}$ *be an* n*-ary semi-fuzzy quantifier,* $n > 0$. Q *is said to be* convex in its i-th argument, *where* $i \in \{1, \ldots, n\}$, *if*

$$\begin{aligned} Q(Y_1, \ldots, Y_n) \geq \min(\, & Q(Y_1, \ldots, Y_{i-1}, Y_i', Y_{i+1}, \ldots, Y_n), \\ & Q(Y_1, \ldots, Y_{i-1}, Y_i'', Y_{i+1}, \ldots, Y_n)) \end{aligned}$$

whenever $Y_1, \ldots, Y_n, Y_i', Y_i'' \in \mathcal{P}(E)$ *and* $Y_i' \subseteq Y_i \subseteq Y_i''$.

Convexity of a fuzzy quantifier $\widetilde{Q} : \widetilde{\mathcal{P}}(E)^n \longrightarrow \mathbf{I}$ *in the* i*-th argument is defined analogously based on arguments in* $\widetilde{\mathcal{P}}(E)$ *and the fuzzy inclusion relation.*

Notes 6.9.

- To present an example, the absolute quantifier "between 10 and 20" is convex in both arguments, and the proportional quantifier "about 30 percent" is convex in the second argument.
- In TGQ, those quantifiers which we call 'convex' are usually dubbed 'continuous', see e.g. [160] and [50, Def. 16, p. 250]. The terminology has been changed here in order to avoid the possible confusion with 'continuous' in the sense of 'smooth'.

Some well-known properties of two-valued convex quantifiers also carry over to semi-fuzzy and fuzzy quantifiers.

Theorem 6.10 (Conjunctions of convex semi-fuzzy quantifiers). *Let $Q, Q' : \mathcal{P}(E)^n \longrightarrow \mathbf{I}$ be semi-fuzzy quantifiers of arity $n > 0$ which are convex in the i-th argument, where $i \in \{1, \ldots, n\}$. Then the semi-fuzzy quantifier $Q \wedge Q' : \mathcal{P}(E)^n \longrightarrow \mathbf{I}$, defined by*

$$(Q \wedge Q')(Y_1, \ldots, Y_n) = \min(Q(Y_1, \ldots, Y_n), Q'(Y_1, \ldots, Y_n))$$

for all $Y_1, \ldots, Y_n \in \mathcal{P}(E)$, is also convex in the i-th argument.

Note 6.11. The theorem states that conjunctions of convex semi-fuzzy quantifiers are convex (provided the standard fuzzy conjunction $\wedge = \min$ is chosen).

A similar point can be made about fuzzy quantifiers.

Theorem 6.12 (Conjunctions of convex fuzzy quantifiers). *Let $\widetilde{Q}, \widetilde{Q}' : \widetilde{\mathcal{P}}(E)^n \longrightarrow \mathbf{I}$ be fuzzy quantifiers of arity $n > 0$ which are convex in the i-th argument, where $i \in \{1, \ldots, n\}$. Then the fuzzy quantifier $\widetilde{Q} \wedge \widetilde{Q}' : \widetilde{\mathcal{P}}(E)^n \longrightarrow \mathbf{I}$, defined by*

$$(\widetilde{Q} \wedge \widetilde{Q}')(X_1, \ldots, X_n) = \min(\widetilde{Q}(X_1, \ldots, X_n), \widetilde{Q}'(X_1, \ldots, X_n))$$

for all $X_1, \ldots, X_n \in \widetilde{\mathcal{P}}(E)$, is also convex in the i-th argument.

Let us further observe that every convex semi-fuzzy quantifier can be decomposed into a conjunction of a nonincreasing and a nondecreasing semi-fuzzy quantifier:

Theorem 6.13 (Decomposition of convex semi-fuzzy quantifiers). *A semi-fuzzy quantifier $Q : \mathcal{P}(E)^n \longrightarrow \mathbf{I}$ is convex in its i-th argument, $i \in \{1, \ldots, n\}$, if and only if Q is the conjunction of a nondecreasing and a nonincreasing semi-fuzzy quantifier, i.e. if there exist $Q^+, Q^- : \mathcal{P}(E)^n \longrightarrow \mathbf{I}$ such that Q^+ is nondecreasing in its i-th argument; Q^- is nonincreasing in its i-th argument, and $Q = Q^+ \wedge Q^-$.*

Again, a similar point can be made about fuzzy quantifiers.

Theorem 6.14 (Decomposition of convex fuzzy quantifiers).
A fuzzy quantifier $\widetilde{Q} : \widetilde{\mathcal{P}}(E)^n \longrightarrow \mathbf{I}$ *is convex in its i-th argument,* $i \in \{1, \ldots, n\}$*, if and only if* $\widetilde{Q}$ *is the conjunction of a nondecreasing and a nonincreasing fuzzy quantifier, i.e. if there exist* $\widetilde{Q}^+, \widetilde{Q}^- : \widetilde{\mathcal{P}}(E)^n \longrightarrow \mathbf{I}$ *such that* $\widetilde{Q}^+$ *is nondecreasing in its i-th argument;* $\widetilde{Q}^-$ *is nonincreasing in its i-th argument, and* $\widetilde{Q} = \widetilde{Q}^+ \wedge \widetilde{Q}^-$.

We say that a QFM $\mathcal{F}$ preserves convexity if the convexity of a quantifier in its arguments is preserved when applying $\mathcal{F}$.

Definition 6.15.
A QFM $\mathcal{F}$ *is said to* preserve convexity of n-ary quantifiers, *where* $n \in \mathbb{N} \setminus \{0\}$, *if and only if every n-ary semi-fuzzy quantifier* $Q : \mathcal{P}(E)^n \longrightarrow \mathbf{I}$ *which is convex in its i-th argument is mapped to a fuzzy quantifier* $\mathcal{F}(Q)$ *which is also convex in its i-th argument.* $\mathcal{F}$ *is said to* preserve convexity *if* $\mathcal{F}$ *preserves the convexity of n-ary quantifiers for all* $n > 0$.

As we shall now see, preservation of convexity expresses a semantical criterion which in its strong form conflicts with other desirable properties. Let us first notice that contextuality of a QFM excludes preservation of convexity.

Theorem 6.16.

Suppose that $\mathcal{F}$ *is a contextual QFM with the following properties: for every base set* $E \neq \varnothing$,

a. *the quantifier* $\mathbb{O} : \mathcal{P}(E) \longrightarrow \mathbf{I}$*, defined by* $\mathbb{O}(Y) = 0$ *for all* $Y \in \mathcal{P}(E)$*, is mapped to the fuzzy quantifier defined by* $\mathcal{F}(\mathbb{O})(X) = 0$ *for all* $X \in \widetilde{\mathcal{P}}(E)$*;*
b. *If* $X \in \widetilde{\mathcal{P}}(E)$ *and there exists some* $e \in E$ *such that* $\mu_X(e) > 0$*, then* $\mathcal{F}(\exists)(X) > 0$*;*
c. *If* $X \in \widetilde{\mathcal{P}}(E)$ *and there exists* $e \in E$ *such that* $\mu_X(e) < 1$*, then* $\mathcal{F}(\sim\forall)(X) > 0$*, where* $\sim\forall : \mathcal{P}(E) \longrightarrow \mathbf{2}$ *is the quantifier defined by*

$$(\sim\forall)(Y) = \begin{cases} 1 & : \quad X \neq E \\ 0 & : \quad X = E \end{cases}$$

Then $\mathcal{F}$ *does not preserve convexity of unary quantifiers* $Q : \mathcal{P}(E) \longrightarrow \mathbf{I}$ *on finite base sets* $E \neq \varnothing$*. In particular, no DFS preserves convexity.*

This means that even if we restrict to the simple case of unary quantifiers on finite base sets and restrict our semantic postulates to the very elementary assumptions of the theorem, there is still no possible choice of a QFM which will preserve convexity in this restricted sense. Because contextuality is one of the most fundamental conditions, it seems better to weaken the requirements on the preservation of convexity rather than compromising contextuality or the other elementary conditions.

Noticing that the common examples of convex NL quantifiers are typically of the quantitative kind, it is straightforward to weaken the requirement of preserving convexity to quantitative convex quantifiers (see Def. 4.41 and Def. 4.42 for the assumed definitions of quantitativity). Unfortunately, this weakening is insufficient yet and the targeted class of convex quantifiers still too broad, as witnessed by the following theorem.

Theorem 6.17.

Suppose that $\mathcal{F}$ is a contextual QFM compatible with cylindrical extensions which satisfies the following properties: for all base sets $E \neq \varnothing$,

a. *the quantifier $\mathbb{O} : \mathcal{P}(E) \longrightarrow \mathbf{I}$, defined by $\mathbb{O}(Y) = 0$ for all $Y \in \mathcal{P}(E)$, is mapped to the fuzzy quantifier defined by $\mathcal{F}(\mathbb{O})(X) = 0$ for all $X \in \widetilde{\mathcal{P}}(E)$;*
b. *If $X \in \widetilde{\mathcal{P}}(E)$ and there exists some $e \in E$ such that $\mu_X(e) > 0$, then $\mathcal{F}(\exists)(X) > 0$;*
c. *If $X \in \widetilde{\mathcal{P}}(E)$ and there exists and there exists some $e \in E$ such that $\mu_X(e') = 0$ for all $e' \in E \setminus \{e\}$ and $\mu_X(e) < 1$, then $\mathcal{F}(\sim\exists)(X) > 0$, where $\sim\exists : \mathcal{P}(E) \longrightarrow \mathbf{2}$ is the quantifier defined by*

$$(\sim\exists)(Y) = \begin{cases} 1 & : \quad X = \varnothing \\ 0 & : \quad X \neq \varnothing \end{cases}$$

Then $\mathcal{F}$ does not preserve the convexity of quantitative semi-fuzzy quantifiers of arity $n > 1$ even on finite base sets. In particular, no DFS preserves the convexity of quantitative semi-fuzzy quantifiers of arity $n > 1$.

There remains a chance that certain models will preserve the convexity of quantitative semi-fuzzy quantifiers of arity $n = 1$. For simplicity, we shall investigate this preservation property for the case of finite domains only.

Definition 6.18.
A QFM $\mathcal{F}$ is said to weakly preserve convexity *if $\mathcal{F}$ preserves the convexity of quantitative one-place quantifiers on finite domains.*

In this case, we get positive results on the existence of models that satisfy the specified criterion. An example of a DFS which weakly preserves convexity, the model $\mathcal{M}_{CX}$, will be presented in Def. 7.56 on p. 201.

Let us now return to a special case of multi-place quantification. In general, we have negative results concerning the preservation of convexity for quantifiers of arity $n > 1$, even if these are quantitative (see Th. 6.17). However it is possible for a DFS to preserve convexity properties in the case of absolute two-place quantifiers, which are of obvious interest to natural language interpretation.

Theorem 6.19.
Suppose $Q : \mathcal{P}(E)^2 \longrightarrow \mathbf{I}$ is an absolute quantifier on a finite base set, i.e. there

exists a quantitative unary quantifier $Q' : \mathcal{P}(E) \longrightarrow \mathbf{I}$ *such that* $Q = Q'\cap$. *If a DFS* $\mathcal{F}$ *has the property of weakly preserving convexity and* Q *is convex in its arguments, then* $\mathcal{F}(Q)$ *is also convex in its arguments.*

Although the proportional type is not covered, the theorem demonstrates that weak preservation of convexity is strong enough to ensure a proper interpretation of many NL quantifiers of interest, e.g. **between 10 and 20**, **about 50** and others.

6.7 Conservativity

One of the pervasive properties of NL quantifiers is *conservativity*, see e.g. Keenan and Stavi [92, p. 275, eq. (40)], van Benthem [10, p. 445] and [9, p. 452] or Gamut [50, pp. 245-249].

Definition 6.20 (Conservativity).
We shall call $Q : \mathcal{P}(E)^2 \longrightarrow \mathbf{I}$ conservative *if*

$$Q(Y_1, Y_2) = Q(Y_1, Y_1 \cap Y_2)$$

for all $Y_1, Y_2 \in \mathcal{P}(E)$.

All two-valued or semi-fuzzy quantifiers introduced so far are conservative.[1] In particular, all proportional quantifiers are conservative by definition (see Def. 11.51 below). To give an example, if E is a set of persons, **married** $\in \mathcal{P}(E)$ is the subset of married persons, and **have_children** $\in \mathcal{P}(E)$ is the set of persons who have children, then the conservative semi-fuzzy quantifier **almost all** : $\mathcal{P}(E)^2 \longrightarrow \mathbf{I}$ satisfies

$$\begin{aligned}&\textbf{almost all}(\textbf{married}, \textbf{have\_children})\\&= \textbf{almost all}(\textbf{married}, \textbf{married} \cap \textbf{have\_children})\end{aligned}$$

i.e. the meanings of "Almost all married persons have children" and "Almost all married persons are married persons who have children" coincide. Like having extension, conservativity expresses an aspect of context insensitivity: If an element of the domain is irrelevant to the restriction (first argument) of a two-place quantifier, then it does not affect the quantification result at all. For example, every conservative $Q : \mathcal{P}(E)^2 \longrightarrow \mathbf{I}$ apparently satisfies $Q(X_1, X_2 \cap X_1) = Q(X_1, X_2 \cup \neg X_1)$. Hence in the case that $e \notin X_1$ for a given element e of the base set, it does not matter whether $e \in X_2$ or $e \notin X_2$. A corresponding fuzzy quantifier $\mathcal{F}(Q) : \widetilde{\mathcal{P}}(E)^2 \longrightarrow \mathbf{I}$ should at least possess the following property of *weak conservativity.*

[1] with the only exception of "only".

Definition 6.21 (Weak conservativity).
A fuzzy quantifier $\widetilde{Q} : \widetilde{\mathcal{P}}(E)^2 \longrightarrow \mathbf{I}$ *is said to be* weakly conservative *if*

$$\widetilde{Q}(X_1, X_2) = \widetilde{Q}(X_1, \mathrm{spp}(X_1) \cap X_2)\,,$$

for all $X_1, X_2 \in \widetilde{\mathcal{P}}(E)$*, where* $\mathrm{spp}(X_1)$ *is the support of* X_1*, see* (4.3).

This definition is sufficiently strong to capture the context insensitivity aspect of conservativity: an element $e \in E$ which is irrelevant to the restriction of the quantifier, i.e. $\mu_{X_1}(e) = 0$, has no effect on the quantification result, which is independent of $\mu_{X_2}(e)$.

Theorem 6.22.
Every DFS $\mathcal{F}$ *weakly preserves conservativity, i.e. if* $Q : \mathcal{P}(E)^2 \longrightarrow \mathbf{I}$ *is conservative, then* $\mathcal{F}(Q)$ *is weakly conservative.*

Definition 6.23 (Strong conservativity).
Let us say that a fuzzy quantifier $\widetilde{Q} : \widetilde{\mathcal{P}}(E)^2 \longrightarrow \mathbf{I}$ *is* strongly conservative *if*

$$\widetilde{Q}(X_1, X_2) = \widetilde{Q}(X_1, X_1 \mathbin{\widetilde{\cap}} X_2) \qquad \textit{for all } X_1, X_2 \in \widetilde{\mathcal{P}}(E).$$

In addition to the context insensitivity aspect, strong conservativity also reflects the definition of crisp conservativity in terms of intersection with the first argument.

Theorem 6.24.
Assume the QFM $\mathcal{F}$ *satisfies the following conditions: (a)* $\widetilde{\mathcal{F}}(\mathrm{id}_2) = \mathrm{id}_{\mathbf{I}}$*; (b)* $\widetilde{\neg}$ *is a strong negation operator; (c)* $\widetilde{\wedge}$ *is a t-norm; (d)* $\mathcal{F}$ *is compatible with internal meets, see Def. 4.23; (e)* $\mathcal{F}$ *is compatible with dualisation. Then* $\mathcal{F}$ *does not strongly preserve conservativity, i.e. there are conservative* $Q : \mathcal{P}(E)^2 \longrightarrow \mathbf{I}$ *such that* $\mathcal{F}(Q)$ *is not strongly conservative. In particular, no DFS strongly preserves conservativity.*

Hence strong preservation of conservativity cannot be ensured in a fuzzy framework, even under assumptions which are much weaker than the DFS axioms. However, the weak form of preserving conservativity already covers the most important aspect of context insensitivity, and it should be sufficient for most practical concerns that the considered models comply with conservativity in this sense.

It is instructive to see how conservativity interacts with the property of having extension, in case a quantifier satisfies both. Then for all crisp $Y_1, Y_2 \in \mathcal{P}(E)$, $Y_1 \neq \varnothing$,

$$\begin{aligned} Q_E(Y_1, Y_2) &= Q_E(Y_1, Y_1 \cap Y_2) \\ &= Q_{E'}(E', Y_1 \cap Y_2) \\ &= Q'_{E'}(Y_1 \cap Y_2)\,, \end{aligned}$$

where $E' = Y_1$, and $Q'_{E'} : \mathcal{P}(E') \longrightarrow \mathbf{I}$ is the unrestricted form of $Q_{E'}$ defined by $Q'_{E'}(Z) = Q_{E'}(E', Z)$ for all $Z \in \mathcal{P}(E')$. This example demonstrates that in the crisp case, restricted quantification based on a quantifier which is conservative and has extension can be reduced to unrestricted (unary) quantification on another domain (supplied by the first argument). The example thus explains why one-place, unrestricted quantification is important although natural language quantifiers are typically at least two-place. Of course, such a reduction is not possible in the fuzzy case because a fuzzy subset $Y_1 \in \widetilde{\mathcal{P}}(E)$ cannot serve as a domain (we have only admitted crisp base sets E).

6.8 Fuzzy Argument Insertion

When discussing crisp argument insertion (see p. 126), it has been remarked that adjectival restriction with fuzzy adjectives cannot be modelled directly: If $A \in \widetilde{\mathcal{P}}(E)$ is a *fuzzy* subset of E, then only $\mathcal{F}(Q)\triangleleft A$ is defined, but not $Q\triangleleft A$. However, one can ask if $\mathcal{F}(Q)\triangleleft A$ can be represented by a semi-fuzzy quantifier Q', i.e. if there is a Q' such that

$$\mathcal{F}(Q)\triangleleft A = \mathcal{F}(Q') . \tag{6.5}$$

The obvious choice of Q' is the following.

Definition 6.25.
Let $\mathcal{F}$ be a QFM, $Q : \mathcal{P}(E)^{n+1} \longrightarrow \mathbf{I}$ a semi-fuzzy quantifier and $A \in \widetilde{\mathcal{P}}(E)$. Then $Q \,\widetilde{\triangleleft}\, A : \mathcal{P}(E)^n \longrightarrow \mathbf{I}$ is defined by

$$Q \,\widetilde{\triangleleft}\, A = \mathcal{U}(\mathcal{F}(Q)\triangleleft A) ,$$

i.e. $Q \,\widetilde{\triangleleft}\, A(Y_1, \ldots, Y_n) = \mathcal{F}(Q)(Y_1, \ldots, Y_n, A)$ for all crisp $Y_1, \ldots, Y_n \in \mathcal{P}(E)$.

Notes 6.26.

- $Q \,\widetilde{\triangleleft}\, A$ is written with the 'tilde' notation $\widetilde{\triangleleft}$ in order to distinguish it from $Q\triangleleft A$ and emphasize that it depend on the chosen QFM $\mathcal{F}$.
- As already noted in [53, p. 54], $Q' = Q \,\widetilde{\triangleleft}\, A$ is the only choice of Q' which possibly satisfies (6.5), because any Q' which satisfies $\mathcal{F}(Q') = \mathcal{F}(Q)\triangleleft A$ also satisfies

$$Q' = \mathcal{U}(\mathcal{F}(Q')) = \mathcal{U}(\mathcal{F}(Q)\triangleleft A) = Q \,\widetilde{\triangleleft}\, A ,$$

  which is apparent from Th. 4.1.

Unfortunately, $Q \,\widetilde{\triangleleft}\, A$ is not guaranteed to fulfill (6.5) in a QFM (not even in a DFS). Let us therefore reformulate this equation as a postulate which ensures that $Q \,\widetilde{\triangleleft}\, A$ convey the intended meaning in a given QFM $\mathcal{F}$.

Definition 6.27.
Let $\mathcal{F}$ be a QFM. We say that $\mathcal{F}$ is compatible with fuzzy argument insertion *if for every semi-fuzzy quantifier $Q : \mathcal{P}(E)^n \longrightarrow \mathbf{I}$ of arity $n > 0$ and every $A \in \widetilde{\mathcal{P}}(E)$, $\mathcal{F}(Q \,\widetilde{\triangleleft}\, A) = \mathcal{F}(Q) \triangleleft A$.*

The main application of this property in natural language is that of adjectival restriction of a quantifier by means of a fuzzy adjective. For example, suppose E is a set of people, and **lucky** $\in \widetilde{\mathcal{P}}(E)$ is the fuzzy subset of those people in E who are lucky. Further suppose that **almost all** : $\mathcal{P}(E)^2 \longrightarrow \mathbf{I}$ is a semi-fuzzy quantifier which models "almost all". Finally, suppose the DFS $\mathcal{F}$ is chosen as the model of fuzzy quantification. We can then construct the semi-fuzzy quantifier $Q' = \mathbf{almost\ all} \cap \widetilde{\triangleleft}\, \mathbf{lucky}$. If $\mathcal{F}$ is compatible with fuzzy argument insertion, then the semi-fuzzy quantifier Q' is guaranteed to adequately model the composite expression "Almost all X_1's are lucky X_2's", because

$$\mathcal{F}(Q')(X_1, X_2) = \mathcal{F}(\mathbf{almost\ all})(X_1, X_2 \,\widetilde{\cap}\, \mathbf{lucky})$$

for all fuzzy arguments $X_1, X_2 \in \widetilde{\mathcal{P}}(E)$, which (relative to $\mathcal{F}$) is the proper expression for interpreting "Almost all X_1's are lucky X_2's" in the fuzzy case. Compatibility with fuzzy argument insertion is a very restrictive adequacy condition. We shall present the unique standard DFS which fulfills this condition on p. 201.

6.9 Chapter Summary

This chapter completes the investigations into the semantical properties of the proposed models of fuzzy quantification, which ultimately decide on the suitability of the proposal for linguistic modelling. While Chap. 4 was devoted to properties shared by every DFS and Chap. 5 to specific constructions and concepts defined on natural subgroups of the models, the present chapter finally discussed those properties which cannot be assumed of all prototypical models.

- Some of the properties are justified from the perspective of applications, but they exclude important models of theoretical interest. Requiring that all models be continuous in arguments, for example, will exclude the boundary case of those models which induce the smallest t-norm, because the drastic product is known to be discontinuous. Hence there is good reason for not including this condition into the list of absolute requirements which must be obeyed by all admissible models. Nevertheless, it is necessary to formalize the continuity requirement in order to permit the selection of robust models suited for practical applications.
- The second type of requirements comprises those problematic cases which either result in a partial inconsistency with some of the intended models

(e.g. with the standard DFSes), or even in a total inconsistency. The strategy then suggests itself of weakening the conflicting condition and reducing it to a core requirement still compatible with the basic axiom system.

In the chapter, we first considered the practical concern of robustness or stability against small changes in the inputs. Every practical model should absorb slight variations due to noise, quantization errors etc., which are typical of real-world applications. There are two kinds of input to the fuzzification mechanism, viz. the fuzzy quantifier and its arguments. Consequently, two facets of robustness must be distinguished, viz (a) robustness against variation in the quantifier, and (b), robustness against variation in the arguments. In both cases, it is necessary for robust system behaviour that the model at least satisfy the minimal requirement of continuity (i.e. small changes in the inputs will result in small changes of the quantification results). We therefore need models which are

- Q-continuous, i.e. continuous in the base quantifiers, in order to ensure the robustness of the quantification results under slight changes in the quantifier; and
- arg-continuous, i.e. the quantification results must be stable under slight changes in the membership functions of the quantifier's arguments.

The variation observed can either be non-systematic (random or resulting from imprecision) or systematic, in which case it reflects the varying interpretations of quantifiers and NL concepts across language users. In order to absorb these sources of variation in the inputs, it is essential for every practical model to obey these continuity conditions. Nonetheless, the robustness criteria should only be considered optional. As explained above, this will keep a number of interesting but discontinuous examples in the full class of models, like upper and lower specificity bounds.

Following the discussion of these robustness issues, we then investigated the compatibility of the models with the specificity order $\preceq_c$. In general, it seems reasonable to assume that whenever either the quantifier or the argument sets become fuzzier, the computed quantification results should become fuzzier as well. In order to express this condition in terms of the specificity order, the basic definition of $\preceq_c$ as a relation on scalars was extended to fuzzy arguments as well as semi-fuzzy and fuzzy quantifiers in the apparent way. This made it possible to state precisely what it means for a QFM to propagate fuzziness in its inputs, i.e. in the quantifiers and their arguments. The standard choice of fuzzy connectives min, max and $1 - x$ are known to be compatible with $\preceq_c$. Thus the standard DFSes possibly include models which propagate fuzziness (although other choices beyond standard models are also conceivable). As we shall see in the next chapter, there are indeed many standard DFSes which share this property. Not all standard models propagate fuzziness, though, as shown by the counterexamples described in Chap. 8. This chapter will also demonstrate that other demands on the models can outweigh the desideratum of propagating fuzziness. Thus, the propagation of fuzziness in arguments and quantifiers should not be considered for arbitrary models.

Next the compatibility requirement with respect to conjunctions and disjunctions of quantifiers was reconsidered. The earlier Th. 5.17 only established lower and upper bounds on the resulting quantifications. In the present chapter, it was shown that the compliance of the models with conjunctions and disjunctions cannot be assumed in the general case: This is only possible if the induced t-norm satisfy the law of contradiction. The bounded product $x_1 \mathbin{\widetilde{\wedge}_b} x_2 = \max(0, x_1 + x_2 - 1)$ is a possible example, but the standard conjunction $x_1 \wedge x_2 = \min(x_1, x_2)$ is not. Thus, in order to retain the standard models, the compositionality with conjunction and disjunction must be considered an optional requirement on the models.

Similar considerations apply in the case of another construction, which builds a new quantifier from a given one by unifying variables in different argument positions. This construction was motivated by the equivalence (biimplication) and antivalence (xor) connectives, whose definition involves multiple occurrences of variables. In order to describe such duplications of variables, we proposed a formal definition which generalizes the known constructions of argument permutations and cylindrical extensions. Apart from these unproblematic cases, the new construction also covers the desired multiple occurrences of variables. Based on this formalization, we were able to show that every model compatible with multiple occurrences of variables induces a t-norm which fulfills the law of contradiction. Again in order to retain the standard models, the compatibility with multiple occurrences of variables will be considered an optional requirement on the models.

Having considered these rather abstract constructions, the models were evaluated for their compliance with three linguistically motivated criteria. First of all, there is a more general form of monotonicity condition which no longer requires that the dependency of the quantifier on the argument of interest be uniformly increasing or decreasing (as in the unproblematic case of monotonicity properties), but also allows more general dependencies which are bell-shaped, trapezoidal etc. These quantifiers are called *continuous* in TGQ, but we are using a different terminology here and call these quantifiers *convex* in the considered argument, in order to avoid a possible confusion with continuity in its smoothness sense. A special case are unimodal quantifiers [185, p. 131]; these have a unique peak of full membership. For example, "about ten" can be modelled as a unimodal quantifier. In the chapter, we first discussed the convexity property for semi-fuzzy and fuzzy quantifiers. Abstracting from well-known properties of two-valued convex quantifiers, we established the closure property of convex quantifiers under the standard conjunction. It was further shown that every convex quantifier can be decomposed into a conjunction of a nonincreasing and a nondecreasing quantifier. Both results apply to semi-fuzzy as well as fuzzy quantifiers. Next we considered the intuitive expectation that the convex shape of quantifiers be preserved when applying a QFM. This investigation revealed that the preservation of convexity is difficult to achieve for models of fuzzy quantification. It cannot be guaranteed in the general case, even if a weaker system of basic axioms is assumed. In practice, it will be acceptable if a model of fuzzy quantification deform the convex shape of exotic quantifiers not expressed in NL, if only it will behave as expected for the cases of linguistic relevance. The requirement was therefore weakened to convex quantifiers of the quantitative variety. In addition, a restriction to unary quantifiers or convex absolute quantifiers like "about 50" was necessary to achieve compatibility with the basic framework. The resulting criterion of weakly preserving convexity is still very demanding. The model $\mathcal{M}_{CX}$ presented in Chap. 7 demonstrates that the requirement can indeed be fulfilled, but this property is confined to a small fragment of the models.

After this discussion of convexity properties, we turned to the definition and analysis of *conservative* quantifiers. Conservativity expresses an important aspect of context insensitivity, which is very common for simple quantifiers. Basically, a quantifier is considered conservative if only the restriction decides on its interpretation, and not the chosen domain. For example, the truth of "All men are mortal" depends on the set of considered men only. The remaining elements of the base set, which fall outside the considered subdomain of 'men', have no effect at all on the quantification results. The original notion of conservativity familiar from TGQ is readily extended to a definition for semi-fuzzy quantifiers, see Def. 6.20. The proper way of generalizing conservativity to fuzzy quantifiers is less obvious, though. In the chapter, we considered two alternative definitions which extend the original concept to the case of fuzzy arguments.

a. The proposed definition of *weak conservativity* merely targets at the domain insensitivity aspect of conservativity. The quantification results of weakly conservative quantifiers should thus be independent of those elements which are totally irrelevant to the restriction of the quantifier. This conception of conservativity combines easily with the proposed framework, i.e. all models will map conservative quantifiers to fuzzy quantifiers which are at least conservative in this weak sense.
b. The alternative proposal of *strong conservativity* is obtained when the crisp intersection in the original definition of conservativity is replaced with the induced fuzzy intersection $\widetilde{\cap}$ which will be used to directly combine the fuzzy arguments. Apart from the context insensitivity aspect, the resulting strong form of conservativity also parallels the definition of crisp conservativity in terms of an intersection with the first argument. There is negative evidence concerning the compliance of this alternative notion with the models of fuzzy quantification. As witnessed by Th. 6.24, the requirement of strong conservativity conflicts with fundamental assumptions on plausible choices of models, so we cannot assume it in the fuzzy framework. The weak form of conservativity will therefore be considered more appropriate when working with fuzzy arguments.

Finally we discussed the construction of fuzzy argument insertion. In order to motivate this criterion, we briefly reviewed adjectival restriction based on crisp adjectives like "married", which can be modelled by crisp argument insertions. From a linguistic point of view, it is desirable to achieve a compositional interpretation of adjectival restriction in accordance with Frege's principle of compositionality. Thus the interpretation of a composite expression should be a function of the meanings of its subexpressions. When applied to the current situation of adjectival restriction this means that the target quantifier $\widetilde{Q}'$ which results from adjectival restriction of Q by A should be representable in terms of a semi-fuzzy quantifier Q' which has $\widetilde{Q}' = \mathcal{F}(Q')$, and Q' should show a functional dependency on the original representation Q and the interpretation of the adjective A, i.e. $Q = c(Q, A)$, where c is the construction for adjectival restriction. In the crisp case, these goals were accomplished by letting $c(Q, A) = Q\tau_i \cap\!\triangleleft A\tau_i$; this restricts the i-th argument of Q to the crisp denotation A of the given adjective. In the important case of fuzzy adjectives, the strategy suggests itself to decompose fuzzy adjectival restriction in an analogous way, thus resolving it into an intersection, argument transpositions, and the crucial step of fuzzy argument insertion. However, fuzzy arguments are not admissible for semi-fuzzy quantifiers. Consequently a different approach is necessary to model the target construction of fuzzy argument insertion. To this end, we observed from Th. 4.1 that there is at most one conceivable choice of semi-fuzzy quantifier $Q' = Q \mathbin{\widetilde{\triangleleft}} A$ which possibly satisfies $\mathcal{F}(Q') = \mathcal{F}(Q)\triangleleft A$, viz $Q \mathbin{\widetilde{\triangleleft}} A = \mathcal{U}(\mathcal{F}(Q)\triangleleft A)$. However, it is not a matter of course that this choice of the semi-fuzzy quantifier $Q' = Q \mathbin{\widetilde{\triangleleft}} A$ will properly represent $\mathcal{F}(Q)\triangleleft A$ in the models. By explicitly requiring that the equality $\mathcal{F}(Q \mathbin{\widetilde{\triangleleft}} A) = \mathcal{F}(Q)\triangleleft A$ be valid, we obtain the criterion of *compatibility with fuzzy argument insertion*. A QFM compatible with this construction permits an explicit representation of the intermediate quantifiers which result from the insertion of a fuzzy argument into one of the argument slots, because it offers corresponding base descriptions $Q' = Q \mathbin{\widetilde{\triangleleft}} A$ on the level of semi-fuzzy quantifiers. The requirement of compatibility with fuzzy argument insertion is very restrictive. As shown in the next chapter, the DFS $\mathcal{M}_{CX}$ is in fact the only standard model with this property.

7

Models Defined in Terms of Three-Valued Cuts and Fuzzy-Median Aggregation

7.1 Motivation and Chapter Overview

By stating the DFS axioms, we have made explicit the linguistic expectations on models of fuzzy quantification. In the following chapters, we will turn attention from the discussion of semantic properties to the presentation of concrete models which instantiate the axiomatic framework. The existence of such concrete examples will prove the consistency of the proposed axioms. In addition, the constructions on which these examples are based and their internal structure might provide new insights into fuzzy quantification. Most importantly, the development of practical models and corresponding algorithms is absolutely mandatory to make the theory useful for applications. The following chapters are devoted to the investigation of increasingly broader classes of models, each of which will have its special characteristics. These classes will be defined in terms of parametrized mechanisms which build all models in the class from a generic constructive principle. For each considered construction, we will start by introducing the parametrized base mechanism, which spans a 'raw' class of totally unrestricted fuzzification mechanisms. By imposing conditions on the allowable choices of parameters, the unrestricted class of QFMs will then be pruned to the subclass of well-behaved models definable from the base mechanism.

In order to implement this basic strategy, we need a suitable starting point from which to abstract the required constructive principle. As we shall see below, some standard methods of fuzzy logic, and in particular α-cuts and the resolution principle, are not suited to define models of the theory. The problems with using α-cuts are caused by their lack of symmetry with respect to complementation. This observation will point attention to richer descriptions offered by three-valued sets. The definition of a three-valued cutting mechanism will overcome the asymmetry problem. In order to ensure a consistent processing of the resulting three-valued sets, we present a constructive principle for Kleene's three-valued logic, which is known to underly all standard models. The basic mechanism will then be fitted to the case of

I. Glöckner: *Fuzzy Quantifiers*, StudFuzz **193**, 179–220 (2006)
www.springerlink.com © Springer-Verlag Berlin Heidelberg 2006

semi-fuzzy quantifiers; this will be accomplished by fuzzy median aggregation. In this way, we obtain a fuzzy quantification result $Q_\gamma(X_1, \ldots, X_n) \in \mathbf{I}$ for each choice of the cutting parameter γ. The results spread over the cut levels must then be aggregated by applying an aggregation operator $\mathcal{B}$. We thus obtain a construction of fuzzification mechanisms in terms of three-valued cuts and fuzzy-median aggregation, By varying the aggregation operator $\mathcal{B}$, we obtain the class of $\mathcal{M}_{\mathcal{B}}$-QFMs. Next we will characterize the (well-behaved) $\mathcal{M}_{\mathcal{B}}$-DFSes in terms of necessary and sufficient conditions on the aggregation mapping $\mathcal{B}$. Some prototypical examples (including the most important cases) will also be presented. We will then assess the characteristics of the individual models. It will be shown that some of the considered properties are be shared by all $\mathcal{M}_{\mathcal{B}}$-DFSes. In addition, the extreme cases of models in terms of specificity will be identified. Finally, there will be a thorough discussion of a special $\mathcal{M}_{\mathcal{B}}$-model with unique semantic properties.[1]

7.2 Alpha-Cuts are not Suited to Define Models

In the proposed framework, an instance of fuzzy quantification can be described in terms of the following data, (a) the given semi-fuzzy quantifier $Q : \mathcal{P}(E)^n \longrightarrow \mathbf{I}$, and (b), a choice of fuzzy arguments $X_1, \ldots, X_n \in \widetilde{\mathcal{P}}(E)$. The problem to be solved is that of making the quantifier applicable to the fuzzy arguments, although it is (directly) defined for crisp arguments only. One idea that comes to mind is that of resorting to the familiar concept of α-cuts (or strict α-cuts) which resolve the fuzzy arguments into layers of crisp arguments which are organized by the cutting parameter α. For each considered layer, the resulting crisp arguments can then be supplied to the semi-fuzzy quantifier, which determines a corresponding quantification result for each cut level. In order to formalize this basic approach, let me first recall the usual definition of α-cuts and strict α-cuts:

Definition 7.1.
Let E be a given set, $X \in \widetilde{\mathcal{P}}(E)$ a fuzzy subset of E and $\alpha \in \mathbf{I}$. By $X_{\geq\alpha} \in \mathcal{P}(E)$ we denote the α-cut

$$X_{\geq\alpha} = \{e \in E : \mu_X(e) \geq \alpha\} .$$

Definition 7.2.
Let $X \in \widetilde{\mathcal{P}}(E)$ be given and $\alpha \in \mathbf{I}$. By $X_{>\alpha} \in \mathcal{P}(E)$ we denote the strict α-cut

$$X_{>\alpha} = \{e \in E : \mu_X(e) > \alpha\} .$$

The effect of α-cutting relative to $\mu_X(e)$ is displayed in Fig. 7.1.

[1] The proofs of the theorems cited in this chapter are presented in [55]. The main results have been published in [66].

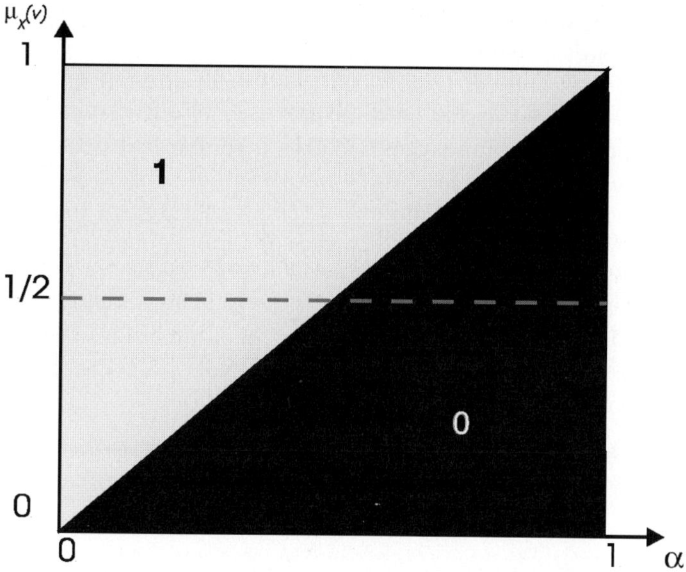

Fig. 7.1. α-cut as a function of α and $\mu_X(v)$

α-cuts and strict α-cuts are linked by negation, as witnessed by the following equalities. For all fuzzy subsets $X \in \widetilde{\mathcal{P}}(E)$ and $\alpha \in \mathbf{I}$,

$$(\neg X)_{\geq\alpha} = \neg(X_{>\neg\alpha}) \tag{7.1}$$

$$(\neg X)_{>\alpha} = \neg(X_{\geq\neg\alpha}) \,. \tag{7.2}$$

It is apparent from these equalities that neither α-cuts nor strict α-cuts are compatible with complementation.

We can now seize the above suggestion of evaluating the quantifier Q for the crisp arguments obtained at each cut level $\alpha \in \mathbf{I}$. The quantification results $f(\alpha) = Q((X_1)_{\geq\alpha}, \ldots, (X_n)_{\geq\alpha})$ obtained at the cut levels $\alpha \in \mathbf{I}$ must then be combined into a single scalar result. This process is delegated to a subsequent aggregation step. Probably the first choice of aggregation operation which comes to mind is that of integration. Let us hence consider the following attempt to define a QFM based on α-cuts and integration:

$$\mathcal{A}(Q)(X_1, \ldots, X_n) = \int_0^1 Q((X_1)_{\geq\alpha}, \ldots, (X_n)_{\geq\alpha}) \,\mathrm{d}\alpha \,,$$

for all semi-fuzzy quantifiers $Q : \mathcal{P}(E)^n \longrightarrow \mathbf{I}$ and $X_1, \ldots, X_n \in \widetilde{\mathcal{P}}(E)$. We should first notice that this is really only an 'attempt' to define a QFM, because the above integral may not exist (I have not made any assumptions on the 'well-behavedness' of Q, like measurability). Hence $\mathcal{A}$ is only partially defined.[2] However, even if we allow an arbitrary completion of $\mathcal{A}$ into a fully defined fuzzification mechanism, the resulting QFM $\mathcal{A}$ still fails to be a DFS. This is because $\mathcal{A}$ is subject to an additional flaw, apart from its 'definition gaps'. It is easily observed that $\mathcal{A}$ does not comply with the construction of internal complementation and thus violates a semantical postulate which is known from Th. 4.19 to be mandatory for all intended models.

From a broader perspective, the failure of the above attempt to define a model by integration over α-cuts is just an instance of a more general fact, and cannot be attributed to the choice of the integral which served as the aggregation operator. By contrast, it can be shown that it is impossible in general to define models of the DFS axioms from layers of α-cuts. This is because there exists no choice of aggregation operator which makes a DFS of the proposed base construction:

Theorem 7.3 (Nonexistence of alpha-cut based DFSes).
Define $\mathcal{C}$, $Q : \mathcal{P}(E)^n \longrightarrow \mathbf{I} \quad \mapsto \quad \mathcal{C}(Q) : \widetilde{\mathcal{P}}(E)^n \longrightarrow \widetilde{\mathcal{P}}(\mathbf{I})$ *by*

$$\mu_{\mathcal{C}(Q)(X_1,\ldots,X_n)}(\alpha) = Q((X_1)_{\geq\alpha}, \ldots, (X_n)_{\geq\alpha})$$

for $\alpha \in \mathbf{I}$. *There is no* $\mathcal{S} : \widetilde{\mathcal{P}}(\mathbf{I}) \longrightarrow \mathbf{I}$ *s.th.* $\mathcal{F} = \mathcal{S} \circ \mathcal{C}$ *satisfies the following conditions:*

- $\widetilde{\mathcal{F}}(\mathrm{id}_{\mathbf{2}}) = \mathrm{id}_{\mathbf{I}}$, *i.e. the identity truth function is mapped to its proper fuzzy analogue;*
- $\widetilde{\neg} = \widetilde{\mathcal{F}}(\neg)$ *is a strong negation operator;*
- $\mathcal{F}$ *is compatible with internal complementation.*

In particular, there exists no $\mathcal{S} : \widetilde{\mathcal{P}}(\mathbf{I}) \longrightarrow \mathbf{I}$ *such that* $\mathcal{F} = \mathcal{S} \circ \mathcal{C}$ *is a DFS.*

Note 7.4. The second part of the theorem is again apparent from Th. 4.19.

This failure of the α-cut based approach to define plausible models of fuzzy quantification is caused by their lack of symmetry with respect to complementation, see (7.1) and (7.2). Due to this negative evidence concerning the potential utility of α-cuts to decompose fuzzy quantification into several layers of quantification involving crisp arguments, I will now consider a different but related mechanism. The well-known *resolution principle* has successfully been applied in other contexts for the transfer of crisp concepts into corresponding concepts for fuzzy sets. Adapted to the present situation of semi-fuzzy quantifiers supplied with fuzzy arguments, the resolution principle becomes:

Definition 7.5 (Resolution principle).
Let $f : \mathcal{P}(E) \longrightarrow V$ *be a mapping. By* $\mathcal{R}(f)$ *we denote the mapping* $\mathcal{R}(f)$:

---

[2] To see this, consider a non-measurable function $f : \mathbf{I} \longrightarrow \mathbf{I}$ and let $E = \mathbf{I}$. Define $Q : \mathcal{P}(\mathbf{I}) \longrightarrow \mathbf{I}$ by $Q(X) = f(\inf X)$ for all $X \in \mathcal{P}(\mathbf{I})$. $\mathcal{A}$ is undefined on the fuzzy subset $X^* \in \widetilde{\mathcal{P}}(\mathbf{I})$ defined by $\mu_{X^*}(e) = e$ for all $e \in \mathbf{I}$, because $Q({X^*}_{\geq\alpha}) = f(\alpha)$.

$\widetilde{\mathcal{P}}(E) \longrightarrow \widetilde{\mathcal{P}}(V)$ *which to each* $X \in \widetilde{\mathcal{P}}(E)$ *associates the fuzzy set* $\mathcal{R}(f)(X)$ *defined by*

$$\mu_{\mathcal{R}(f)(X)}(v) = \sup\{\alpha \in \mathbf{I} : f(X_{\geq\alpha}) = v\}$$

for $v \in V$. *Then* $\mathcal{R}(f)$ *is said to result from* f *by applying the resolution principle.*

Note 7.6. I have stated the resolution principle for one-place functions only. Because $\mathcal{P}(E)^n \cong \mathcal{P}(E \times n)$ and $\widetilde{\mathcal{P}}(E)^n \cong \widetilde{\mathcal{P}}(E \times n)$, this generalises to n-place functions $f : \mathcal{P}(E)^n \longrightarrow V$ in the obvious way, i.e. by component-wise α-cutting.

By applying the resolution principle, then, a semi-fuzzy quantifier $Q : \mathcal{P}(E)^n \longrightarrow \mathbf{I}$ is mapped to that $\mathcal{R}(Q) : \widetilde{\mathcal{P}}(E)^n \longrightarrow \widetilde{\mathcal{P}}(\mathbf{I})$ which to each choice of $X_1, \ldots, X_n \in \widetilde{\mathcal{P}}(E)$ assigns the fuzzy subset $\mathcal{R}(Q)(X_1, \ldots, X_n) \in \widetilde{\mathcal{P}}(\mathbf{I})$ defined by

$$\mu_{\mathcal{R}(Q)(X_1,\ldots,X_n)}(v) = \sup\{\alpha \in \mathbf{I} : Q((X_1)_{\geq\alpha}, \ \ldots, \ (X_n)_{\geq\alpha}) = v\} \quad \text{for all } v \in \mathbf{I}.$$

Upon closer inspection, the resolution principle comes out as a special but important case of the general α-cut based approach discussed first, which has already been shown to be doomed to failure. As a corollary to the above theorem on α-cut based attempts to define DFSes, we hence obtain:

Theorem 7.7 (Nonexistence of models based on the resolution principle).
It is not possible to define models of the theory based on the resolution principle, i.e. no $\mathcal{S} : \mathcal{P}(\mathbf{I}) \longrightarrow \mathbf{I}$ *exists which makes* $\mathcal{F} = \mathcal{S} \circ \mathcal{R}$ *a DFS.*

By the *α-cut representation* of a choice of fuzzy arguments $(X_1, \ldots, X_n) \in \widetilde{\mathcal{P}}(E)^n$, I denote the fuzzy set $\mathrm{RES}(X_1, \ldots, X_n) \in \widetilde{\mathcal{P}}(\mathcal{P}(E))^n$ defined by

$$\mu_{\mathrm{RES}(X_1,\ldots,X_n)}((Y_1, \ldots, Y_n)) = \sup\{\alpha \in \mathbf{I} : (X_1)_{\geq\alpha} = Y_1, \ \ldots, \ (X_n)_{\geq\alpha} = Y_n\},$$

for all $Y_1, \ldots, Y_n \in \mathcal{P}(E)$. Of course, DFSes can be defined based on this α-cut representation of $X_1, \ldots, X_n$ (at least if DFSes exist at all, which will be proven below). But this is a trivial statement, since $(X_1, \ldots, X_n)$ can be recovered from its α-cut representation.

7.3 From Two-Valued Logic to Kleene's Logic, and Beyond

As witnessed by Th. 5.21, the $\{0, \frac{1}{2}, 1\}$-valued portion of the logic induced by a standard model coincides with Kleene's three-valued logic (see [95, p. 344];

also described e.g. in [97, p. 29]). This observation provides a suitable starting point for defining models of the theory. Kleene's logic can be constructed from two-valued logic by the following mechanism of generalizing a propositional function $f : \mathbf{2}^n \longrightarrow \mathbf{2}$ to the three-valued $\breve{f} : \{0, \frac{1}{2}, 1\}^n \longrightarrow \{0, \frac{1}{2}, 1\}$.

Suppose that $x_1, \ldots, x_n \in \{0, \frac{1}{2}, 1\}$ are given. Associate to each x_i a set $\mathcal{T}(x_i) \in \mathcal{P}(\mathbf{2})$ as follows:

$$\mathcal{T}(x_i) = \begin{cases} \{0\} & : \ x_i = 0 \\ \{0,1\} & : \ x_i = \frac{1}{2} \\ \{1\} & : \ x_i = 1 \end{cases}$$

The set $\mathcal{T}(x_i)$ collects the alternative interpretations of y_i as two-valued truth values: 0 and 1 are non-ambiguous and correspond to unique interpretations $\{0\}$ and $\{1\}$, respectively, while the $\frac{1}{2}$ represents absence of knowledge, and is hence ambiguous between all possible choices in $\{0, 1\}$. In terms of these sets of alternatives, we then define $\breve{f} : \{0, \frac{1}{2}, 1\}^n \longrightarrow \{0, \frac{1}{2}, 1\}$ by

$$\breve{f}(x_1, \ldots, x_n) = \begin{cases} 1 & : \ f(y_1, \ldots, y_n) = 1 \text{ for all } y_i \in \mathcal{T}(x_i), \ i = 1, \ldots, n \\ 0 & : \ f(y_1, \ldots, y_n) = 0 \text{ for all } y_i \in \mathcal{T}(x_i), \ i = 1, \ldots, n \\ \frac{1}{2} & : \ \text{else} \end{cases}$$

for all $x_1, \ldots, x_n \in \{0, \frac{1}{2}, 1\}$.

In this definition, the 'indeterminate' value $\frac{1}{2}$ is treated by considering both alternatives $0, 1$. The truth-values 0 and 1 do not induce any indeterminacy. Let us now recall the definition of the generalized fuzzy median, see Def. 5.9. By applying $\mathrm{m}_{\frac{1}{2}} : \mathcal{P}(\mathbf{I}) \longrightarrow \mathbf{I}$ to the supervaluation results, we can express the above definition more compactly, viz

$$\breve{f}(x_1, \ldots, x_n) = \mathrm{m}_{\frac{1}{2}} \{ f(y_1, \ldots, y_n) : y_i \in \mathcal{T}(x_i), i = 1, \ldots, n \} . \tag{7.3}$$

This median-based definition has the advantage that it can also be applied to functions $f : \mathbf{2}^n \longrightarrow \mathbf{I}$, which it maps to $\breve{f} : \{0, \frac{1}{2}, 1\}^n \longrightarrow \mathbf{I}$, and thus takes us beyond Kleene's logic.

7.4 Adaptation of the Mechanism to Quantifiers

In this book, we are more interested in quantifiers rather than propositional functions. Hence let us adapt the above mechanism to the case of three-valued quantifiers, which are then translated into quantifiers that accept three-valued arguments. To this end, I must first introduce some concepts related to three-valued subsets, which are necessary to express the definition of the target mechanism.

Three-valued subsets of a given base set E will be modelled in a way analogous to fuzzy subsets, i.e. we shall assume that each three-valued subset X of E

is uniquely characterised by its membership function $\nu_X : E \longrightarrow \{0, \frac{1}{2}, 1\}$ (I use the symbol ν_X rather than μ_X in order make unambiguous that the membership function is three-valued). The collection of three-valued subsets of a given E will be denoted $\breve{\mathcal{P}}(E)$; we shall assume that $\breve{\mathcal{P}}(E)$ is a set, and clearly we have $\breve{\mathcal{P}}(E) \cong \{0, \frac{1}{2}, 1\}^E$. As in the case of fuzzy subsets, it might be convenient to identify three-valued subsets and their membership functions, i.e. to stipulate $\breve{\mathcal{P}}(E) = \{0, \frac{1}{2}, 1\}^E$. However, I will again not enforce this identification.

I will assume that each crisp subset $X \in \mathcal{P}(E)$ can be viewed as a three-valued subset of E, and that each three-valued subset X of E can be viewed as a fuzzy subset of E. Hence at times, the same symbol X will be used to denote a particular crisp subset of E, as well as the corresponding three-valued and fuzzy subsets. If one chooses to identify membership functions and three-valued/fuzzy subsets, then the crisp subset X is distinct from its representation as a three-valued or fuzzy subset, which corresponds to its characteristic function χ_X. In this case, it is understood that the appropriate transformations (i.e., using characteristic function) are performed whenever X is substituted for a three-valued or fuzzy subset.

Let us now recall the construction of Kleene's logic presented in the previous section, which was based on the representation of three-valued $x \in \{0, \frac{1}{2}, 1\}$ by corresponding ambiguity ranges $\mathcal{T}(x) \in \mathcal{P}(\mathbf{2})$. This basic construction is easily adapted to three-valued subsets. In order to express the 'ambiguity' or 'indeterminacy' which originates from occurrences of the membership grade $\frac{1}{2}$, a closed range of crisp subsets of E will then be associated with each three-valued subset $X \in \breve{\mathcal{P}}(E)$. This range, denoted $\mathcal{T}(X)$, is intended to capture the alternative interpretations of the three-valued set X in terms of compatible two-valued sets.

Definition 7.8.
Suppose E is some set and $X \in \breve{\mathcal{P}}(E)$ is a three-valued subset of E. We associate with X crisp subsets $X^{\min}, X^{\max} \in \mathcal{P}(E)$, defined by

$$X^{\min} = \{e \in E : \nu_X(e) = 1\}$$
$$X^{\max} = \{e \in E : \nu_X(e) \geq \tfrac{1}{2}\}\,.$$

Based on $X^{\min}$ and $X^{\max}$, we associate with X the range of crisp sets $\mathcal{T}(X) \subseteq \mathcal{P}(E)$ defined by

$$\mathcal{T}(X) = \{Y \in \mathcal{P}(E) : X^{\min} \subseteq Y \subseteq X^{\max}\}\,.$$

Notes 7.9.

- If $e \in E$ is such that $\nu_X(e) = 1$, then all $Y \in \mathcal{T}(X)$ contain e; if $e \in E$ is such that $\nu_X(e) = 0$, then no $Y \in \mathcal{T}(X)$ contains e, and for those $e \in E$ with $\nu_X = \frac{1}{2}$, $\mathcal{T}(X)$ contains all combinations of the alternatives $e \in Y$, $e \notin Y$.

- The pairs $(X^{\min}, X^{\max})$, $X^{\min} \subseteq X^{\max}$ form a representation of the three-valued sets X because every X can be recovered from $(X^{\min}, X^{\max})$ in the apparent way, viz

$$\nu_X(e) = \begin{cases} 1 & : \quad e \in X^{\min} \\ 0 & : \quad e \notin X^{\max} \\ \frac{1}{2} & : \quad \text{else} \end{cases}$$

  for all $e \in E$.

Based on this concept, which resolves each three-valued subset into its ambiguity range of compatible crisp subsets, I can now modify the above mechanism for extending two-valued propositional functions to three-valued arguments, and fit its core idea to the case of three-valued quantifiers. It is then convenient to abbreviate

$$\mathcal{T}(X_1, \ldots, X_n) = \mathcal{T}(X_1) \times \cdots \times \mathcal{T}(X_n)\,.$$

I can then define

$$\begin{aligned} &\breve{Q}(X_1, \ldots, X_n) \\ &= \begin{cases} 1 & : \quad Q(Y_1, \ldots, Y_n) = 1 \text{ for all } (Y_1, \ldots, Y_n) \in \mathcal{T}(X_1, \ldots, X_n) \\ 0 & : \quad Q(Y_1, \ldots, Y_n) = 0 \text{ for all } (Y_1, \ldots, Y_n) \in \mathcal{T}(X_1, \ldots, X_n) \\ \frac{1}{2} & : \quad \text{else} \end{cases} \end{aligned}$$

for all $X_1, \ldots, X_n \in \breve{\mathcal{P}}(E)$. This construction extends the three-valued quantifier $Q : \mathcal{P}(E)^n \longrightarrow \{0, \frac{1}{2}, 1\}$ to a quantifier $\breve{Q} : \breve{\mathcal{P}}(E)^n \longrightarrow \{0, \frac{1}{2}, 1\}$, which now accepts three-valued arguments. Again, we can profit from the generalized fuzzy median $\mathrm{m}_{\frac{1}{2}}$ and rephrase this according to the same idea which underlies equation (7.3). Stated this way, the mechanism can be applied to arbitrary semi-fuzzy quantifiers as well, which produce continuous rather than three-valued outputs. A given semi-fuzzy quantifier $Q : \mathcal{P}(E)^n \longrightarrow \mathbf{I}$ is then mapped to a quantifier $\breve{Q} : \breve{\mathcal{P}}(E)^n \longrightarrow \mathbf{I}$, which generalizes its base quantifier Q to the case of three-valued arguments:

Definition 7.10.
To all semi-fuzzy quantifiers $Q : \mathcal{P}(E)^n \longrightarrow \mathbf{I}$, we associate a corresponding $\breve{Q} : \breve{\mathcal{P}}(E)^n \longrightarrow \mathbf{I}$ defined by

$$\breve{Q}(X_1, \ldots, X_n) = \mathrm{m}_{\frac{1}{2}}\{Q(Y_1, \ldots, Y_n) : Y_1 \in \mathcal{T}(X_1), \ldots, Y_n \in \mathcal{T}(X_n)\}$$

for all $X_1, \ldots, X_n \in \breve{\mathcal{P}}(E)$.

The resulting $\breve{Q}$ now accepts three-valued arguments. In the general case, though, we must deal with unrestricted fuzzy arguments which are not tied to membership grades in $\{0, \frac{1}{2}, 1\}$. In the following, I therefore suggest a reduction of fuzzy quantification to the case of three-valued arguments.

7.5 Three-Valued Cuts of Fuzzy Subsets

We are now able to evaluate quantifying statements which involve three-valued arguments. In order to further generalize this construction and support arbitrary fuzzy arguments, I propose the use of a cutting mechanism which reduces the fuzzy arguments of the quantifier to corresponding three-valued arguments, and control this reduction by a cutting parameter $\gamma \in \mathbf{I}$. We know from Th. 7.3 that α-cuts (i.e. simple two-valued cuts) are not suited to effect the reduction, due to their lack of symmetry with respect to complementation. However, it is apparent how a reduction to three-valued sets (rather than the crisp sets of α-cuts) can be performed in a symmetrical fashion, and hence achieves the desired compatibility with negation. Let me first state the definition for individual membership grades in the unit interval, which are cut to membership grades in $\{0, \frac{1}{2}, 1\}$ which correspond to 'false', 'unknown' and 'true', respectively.

Definition 7.11.
For every $x \in \mathbf{I}$ and $\gamma \in \mathbf{I}$, the three-valued cut of x at γ is defined by

$$t_\gamma(x) = \begin{cases} 1 & : \quad x \geq \frac{1}{2} + \frac{1}{2}\gamma \\ \frac{1}{2} & : \quad \frac{1}{2} - \frac{1}{2}\gamma < x < \frac{1}{2} + \frac{1}{2}\gamma \\ 0 & : \quad x \leq \frac{1}{2} - \frac{1}{2}\gamma \end{cases}$$

if $\gamma > 0$, and

$$t_0(x) = \begin{cases} 1 & : \quad x > \frac{1}{2} \\ \frac{1}{2} & : \quad x = \frac{1}{2} \\ 0 & : \quad x < \frac{1}{2} \end{cases}$$

in the case that $\gamma = 0$.

Note 7.12. The cutting parameter γ can be conceived of a degree of 'cautiousness' because the larger γ becomes, the closer $t_\gamma(x)$ will approach the 'undecided' result of $\frac{1}{2}$. Hence $t_{\gamma'}(x) \preceq_c t_\gamma(x)$ whenever $\gamma \leq \gamma'$.

The three-valued cut mechanism for scalars can be extended to three-valued cuts of fuzzy subsets, by applying it element-wise to the membership functions.

Definition 7.13.
Suppose E is some set and $X \in \widetilde{\mathcal{P}}(E)$ a subset of E. The three-valued cut of X at $\gamma \in \mathbf{I}$ is the three-valued subset $\mathrm{T}_\gamma(X) \in \breve{\mathcal{P}}(E)$ defined by

$$\nu_{\mathrm{T}_\gamma(X)}(e) = t_\gamma(\mu_X(e)),$$

for all $e \in E$.

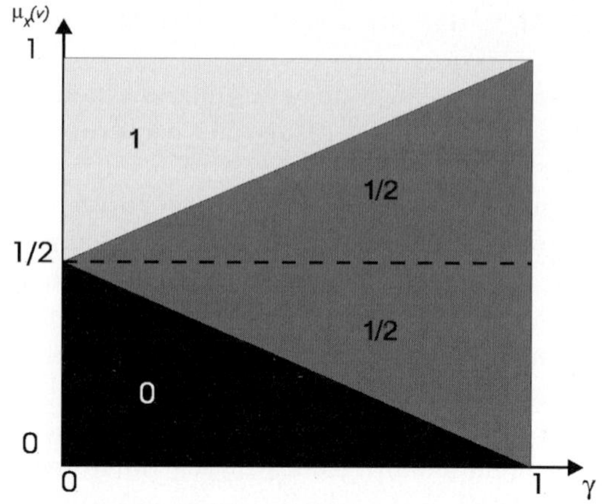

Fig. 7.2. Three-valued cut as a function of γ and $\mu_X(v)$

Note 7.14. The effect of three-valued cutting is displayed in Fig. 7.2.

Usually the three-valued cut will not be applied in isolation. By contrast, it will be composed with the construction introduced in Def. 7.8, and hence resolved into a pair of crisp sets $((\mathrm{T}_\gamma(X))^{\min}, (\mathrm{T}_\gamma(X))^{\max})$, or alternatively into the crisp range $\mathcal{T}(\mathrm{T}_\gamma(X))$ determined by the three-valued set which results from the cut operation. Due to this practice, which generally combines the cut operation with the subsequent resolution, it is convenient to introduce the succinct notation $X_\gamma^{\min}$ and $X_\gamma^{\max}$ for the above pair of crisp sets, and also introduce an abbreviation $\mathcal{T}_\gamma(X)$ for the cut range $\mathcal{T}(\mathrm{T}_\gamma(X))$ which represents the three-valued cut at the 'cautiousness level' $\gamma \in \mathbf{I}$ by a set of crisp alternatives.

Definition 7.15.

Let E be some set, $X \in \widetilde{\mathcal{P}}(E)$ a fuzzy subset of E and $\gamma \in \mathbf{I}$. $X_\gamma^{\min}, X_\gamma^{\max} \in \mathcal{P}(E)$ and $\mathcal{T}_\gamma(X) \subseteq \mathcal{P}(E)$ are defined by

$$\begin{aligned}
X_\gamma^{\min} &= (\mathrm{T}_\gamma(X))^{\min}\\
X_\gamma^{\max} &= (\mathrm{T}_\gamma(X))^{\max}\\
\mathcal{T}_\gamma(X) &= \mathcal{T}(\mathrm{T}_\gamma(X)) = \{Y : X_\gamma^{\min} \subseteq Y \subseteq X_\gamma^{\max}\}\,.
\end{aligned}$$

Hence if $\gamma > 0$, then

$$\begin{aligned}
X_\gamma^{\min} &= X_{\geq \frac{1}{2}+\frac{1}{2}\gamma}\\
X_\gamma^{\max} &= X_{> \frac{1}{2}-\frac{1}{2}\gamma}\,,
\end{aligned}$$

and in the case that $\gamma = 0$,

$$X_0^{\min} = X_{>\frac{1}{2}}$$
$$X_0^{\max} = X_{\geq\frac{1}{2}} .$$

Notes 7.16.

- Let us recall that the three-valued set which results from the three-valued cut of X at $\gamma \in \mathbf{I}$ can be represented by the pair $(X_\gamma^{\min}, X_\gamma^{\max})$, and further notice that both $X_\gamma^{\min}$ and $X_\gamma^{\max}$ are defined by α-cuts. Hence every three-valued cut can be represented by a pair of α-cuts, which in turn also determines the corresponding range of crisp sets, $\mathcal{T}_\gamma(X)$.
- The γ-indexed family $(X_\gamma^{\min}, X_\gamma^{\max})_{\gamma\in\mathbf{I}}$ faithfully represents X because X can recovered from this representation by means of

$$\mu_X(e) = \begin{cases} \frac{1}{2} + \frac{1}{2}\sup\{\gamma \in \mathbf{I} : e \in X_\gamma^{\min}\} & : \quad e \in X_0^{\min} \\ \frac{1}{2} - \frac{1}{2}\sup\{\gamma \in \mathbf{I} : e \notin X_\gamma^{\max}\} & : \quad e \notin X_0^{\min} \end{cases}$$

  for all $e \in E$. Hence every fuzzy subset X can be resolved into a family of pairs $(X_\gamma^{\min}, X_\gamma^{\max})_{\gamma\in\mathbf{I}}$ which represent the three-valued cuts of X at all choices of $\gamma \in \mathbf{I}$.
- In turn, X is also faithfully represented by the family of cut ranges $(\mathcal{T}_\gamma(X))_{\gamma\in\mathbf{I}}$, because $X_\gamma^{\min}$ and $X_\gamma^{\max}$ can be recovered from these ranges by forming the intersection $X_\gamma^{\min} = \cap\,\mathcal{T}_\gamma(X)$ and the union $X_\gamma^{\max} = \cup\,\mathcal{T}_\gamma(X)$. Hence when resorting to the cut ranges, we also achieve the resolution property for three-valued cuts.
- Abusing language, I will usually identify the pair of crisp sets $(X_\gamma^{\min}, X_\gamma^{\max})$, and the cut range $\mathcal{T}_\gamma(X)$ with the three-valued cut $\mathrm{T}_\gamma(X)$, and refer to both as 'the three-valued cut of X at γ' although in a precise sense, the three-valued cut is only represented by these modelling devices.

It is also instructive to observe how the choice of the 'cautiousness parameter' γ effects the results of the cutting operation on a given fuzzy subset X. If $\gamma = 0$, the set of indeterminates (i.e. of those $e \in E$ such that $e \in X_\gamma^{\max} \setminus X_\gamma^{\min}$) contains only those $e \in E$ with $\mu_X(e) = \frac{1}{2}$; all other elements of E are mapped to the closest truth value in $\{0, 1\}$. As γ increases, the set of indeterminates is increasing. For $\gamma = 1$, then, the level of maximal cautiousness is reached where all elements of E except those with $\mu_X(e) \in \{0, 1\}$ are interpreted as indeterminates (see also Fig. 7.2).

Some obvious properties of $X_\gamma^{\min}$ and $X_\gamma^{\max}$ are summarized in the following theorem.

Theorem 7.17.
Suppose that E *is a given set and* $X, X' \in \widetilde{\mathcal{P}}(E)$ *are fuzzy subsets of* E. *Then for all* $\gamma \in \mathbf{I}$,

a. $(\neg X)_\gamma^{\min} = \neg(X_\gamma^{\max})$ *and* $(\neg X)_\gamma^{\max} = \neg(X_\gamma^{\min})$;
b. $(X \cap X')_\gamma^{\min} = X_\gamma^{\min} \cap X'^{\min}_\gamma$ *and* $(X \cap X')_\gamma^{\max} = X_\gamma^{\max} \cap X'^{\max}_\gamma$;
c. $(X \cup X')_\gamma^{\min} = X_\gamma^{\min} \cup X'^{\min}_\gamma$ *and* $(X \cup X')_\gamma^{\max} = X_\gamma^{\max} \cup X'^{\max}_\gamma$.

This demonstrates that the three-valued cutting mechanism is compatible with all Boolean operations on the arguments of a quantifier, i.e.

$$\begin{aligned}
\mathcal{T}_\gamma((\neg X)) &= \{\neg Y : Y \in \mathcal{T}_\gamma(X)\}\\
\mathcal{T}_\gamma((X \cap X')) &= \{Y \cap Y' : Y \in \mathcal{T}_\gamma(X), Y' \in \mathcal{T}_\gamma(X')\}\\
\mathcal{T}_\gamma((X \cup X')) &= \{Y \cup Y' : Y \in \mathcal{T}_\gamma(X), Y' \in \mathcal{T}_\gamma(X')\},
\end{aligned}$$

for all $X, X' \in \widetilde{\mathcal{P}}(E)$ and $\gamma \in \mathbf{I}$.

7.6 The Fuzzy-Median Based Aggregation Mechanism

In this section we can put the pieces together and combine the interpretation mechanism for three-valued arguments, which takes Q to $\breve{Q}$, and the three-valued cutting operation. The three-valued cuts at a given level γ will then be applied to reduce given choices of fuzzy arguments to the simpler situation of three-valued arguments, which can already be handled by $\breve{Q}$. Hence in order to evaluate a quantifying statement based on a semi-fuzzy quantifier $Q : \mathcal{P}(E)^n \longrightarrow \mathbf{I}$ at a given level of cautiousness γ, we apply the above mechanism to Q, yielding $\breve{Q} : \breve{\mathcal{P}}(E)^n \longrightarrow \mathbf{I}$, and then insert the three-valued cuts of the fuzzy argument sets at γ. In fact, both mechanisms, i.e. quantifier generalisation and three-valued cut, are exclusively used in this specific combination. It is therefore beneficial to introduce a shorthand notation for the composition of the two mechanisms, and subsequently treat the resulting combination as an atomic construction which can be used to define models. This approach is summarized in the following definition.

Definition 7.18.
For every $\gamma \in \mathbf{I}$, *we denote by* $(\bullet)_\gamma$ *the QFM defined by*

$$Q_\gamma(X_1, \ldots, X_n) = \mathrm{m}_{\frac{1}{2}}\{Q(Y_1, \ldots, Y_n) : Y_i \in \mathcal{T}_\gamma(X_i)\},$$

for all semi-fuzzy quantifiers $Q : \mathcal{P}(E)^n \longrightarrow \mathbf{I}$.

Notes 7.19.

- It should be apparent how the definition of $Q_\gamma(X_1, \ldots, X_n)$ evolves from the composition of $\breve{Q}$ introduced in Def. 7.10, and the three valued cuts defined by Def. 7.13.
- Considering the resulting construction in total, the basic idea that can be identified is that of treating the crisp ranges which correspond to the three-valued cuts as providing a number of alternatives to be checked.

Hence in order to evaluate a semi-fuzzy quantifier Q at a certain cut level γ, we have to consider all choices of $Q(Y_1, \ldots, Y_n)$, where $Y_i \in \mathcal{T}_\gamma(X_i)$. The collection of supervaluation results obtained in this way must then be aggregated to a single result in the unit interval, which is achieved by applying the generalized fuzzy median.

- If the base set E is finite, then the number of *distinct* three-valued cuts of $X_1, \ldots, X_n \in \widetilde{\mathcal{P}}(E)$ is finite as well. In this case, $f(\gamma) = \mathcal{Q}_\gamma(X_1, \ldots, X_n)$ is a step function, i.e. piece-wise constant in γ, except for a finite number of discontinuities. It is this property which makes the concept of three-valued cuts and the definition of $(\bullet)_\gamma$ still suited for efficient implementation: results need to be computed only for the (usually small) number of distinct three-valued cuts. More details on the computational aspect can be found in Chap. 11.

The next theorem presents a result concerning the monotonicity properties of $f(\gamma) = \mathcal{Q}_\gamma(X_1, \ldots, X_n)$, when viewed as a function of $\gamma \in \mathbf{I}$.

Theorem 7.20.
Let $Q : \mathcal{P}(E)^n \longrightarrow \mathbf{I}$ be a semi-fuzzy quantifier and $X_1, \ldots, X_n \in \widetilde{\mathcal{P}}(E)$.

a. If $\mathcal{Q}_0(X_1, \ldots, X_n) > \frac{1}{2}$, then $\mathcal{Q}_\gamma(X_1, \ldots, X_n)$ is monotonically nonincreasing in γ and $\mathcal{Q}_\gamma(X_1, \ldots, X_n) \geq \frac{1}{2}$ for all $\gamma \in \mathbf{I}$.
b. If $\mathcal{Q}_0(X_1, \ldots, X_n) = \frac{1}{2}$, then $\mathcal{Q}_\gamma(X_1, \ldots, X_n) = \frac{1}{2}$ for all $\gamma \in \mathbf{I}$.
c. If $\mathcal{Q}_0(X_1, \ldots, X_n) < \frac{1}{2}$, then $\mathcal{Q}_\gamma(X_1, \ldots, X_n)$ is monotonically nondecreasing in γ and $\mathcal{Q}_\gamma(X_1, \ldots, X_n) \leq \frac{1}{2}$ for all $\gamma \in \mathbf{I}$.

Note 7.21. These monotonicity properties reflect the intuition that as soon as one becomes more cautious (i.e. the 'level of cautiousness' γ increases), the results obtained will become less specific, in the sense of being closer to $\frac{1}{2}$.

Owing to this fuzzy-median based aggregation mechanism, we are now able to interpret fuzzy quantifiers for any fixed choice of the cutting parameter $\gamma \in \mathbf{I}$. However, none of the QFMs $(\bullet)_\gamma$ is a DFS yet, because the required information is spread over various cut levels, and the fuzzy median suppresses too much structure at each individual cut level. Nonetheless, these QFMs prove useful in defining models of fuzzy quantification. We simply need to abstract from individual cut levels, and simultaneously consider the results obtained at all levels of cautiousness γ.

7.7 The Unrestricted Class of $\mathcal{M}_{\mathcal{B}}$-QFMs

In order to implement this basic idea of supplying the assumed QFM with the total amount of information, which is spread over cautiousness levels, we consider the γ-indexed family $(\mathcal{Q}_\gamma(X_1, \ldots, X_n))_{\gamma \in \mathbf{I}}$. Unlike the individual QFMs $(\bullet)_\gamma$, which fail to be a DFS because they only see one choice of γ at a

for all $z \in \mathbf{I}$. *Then*

$$Q_\gamma(X) = f(\gamma)$$

for all $\gamma \in \mathbf{I}$.

Note 7.30. Combining this with the above observations, we now have evidence that $\mathbb{B}$ is the minimal set of mappings $f : \mathbf{I} \longrightarrow \mathbf{I}$ which embeds all possible operands of the assumed aggregation mapping. Hence $\mathbb{B}$ precisely characterises the domain of suitable aggregation operators.

Let us now return to the original idea of abstracting from the above definition of $\mathcal{M}$, and formulating a generic construction of QFMs by aggregating the results of $Q_\gamma(X_1, \ldots, X_n)$ obtained for all choices of the cautiousness parameter. We know from the last theorem that a corresponding aggregation operator must be defined on $\mathbb{B}$, because $(Q_\gamma(X_1, \ldots, X_n))_{\gamma \in \mathbf{I}} \in \mathbb{B}$, and no smaller set $A \subseteq \mathbf{I}^\mathbf{I}$ will suffice.
Hence consider an aggregation operator $\mathcal{B} : \mathbb{B} \longrightarrow \mathbf{I}$. By composing $\mathcal{B}$ with the γ-indexed cut mechanism, we can now construct a QFM, denoted $\mathcal{M}_\mathcal{B}$, which corresponds to the operator $\mathcal{B}$. This basic suggestion can be formalized as follows.

Definition 7.31.
Let $\mathcal{B} : \mathbb{B} \longrightarrow \mathbf{I}$ *be given. The QFM* $\mathcal{M}_\mathcal{B}$ *is defined by*

$$\mathcal{M}_\mathcal{B}(Q)(X_1, \ldots, X_n) = \mathcal{B}((Q_\gamma(X_1, \ldots, X_n))_{\gamma \in \mathbf{I}}),$$

for all semi-fuzzy quantifiers $Q : \mathcal{P}(E)^n \longrightarrow \mathbf{I}$ *and* $X_1, \ldots, X_n \in \widetilde{\mathcal{P}}(E)$.

By the class of $\mathcal{M}_\mathcal{B}$-QFMs I mean the class of all QFMs $\mathcal{M}_\mathcal{B}$ defined in this way. Because no conditions whatsoever were imposed on the aggregation mapping $\mathcal{B}$, it is apparent that the resulting fuzzification mechanisms constitute a 'raw', totally unrestricted class of $\mathcal{M}_\mathcal{B}$-QFMs, which apart from the well-behaved models, also contains many QFMs which violate the axiomatic requirements. In order to shrink down this 'raw' class to the reasonable cases of $\mathcal{M}_\mathcal{B}$-DFSes, thus identifying its plausible models, I will now address the problem of formalizing the required conditions on admissible choices of the aggregation mapping.

7.8 Characterisation of $\mathcal{M}_\mathcal{B}$-DFSes

In this section, I will identify the class of well-behaved models among the 'raw' $\mathcal{M}_\mathcal{B}$-QFMs, based on properties that are visible in the aggregation mapping. To this end, I first introduce some constructions on $\mathbb{B}$. These will permit us to capture the relevant behaviour of the aggregation mapping in a system of conditions imposed on $\mathcal{B}$. In turn, the proposed conditions will then be shown to be necessary and sufficient for $\mathcal{M}_\mathcal{B}$ to be a DFS, i.e. they precisely characterise the admissible choices of $\mathcal{B}$. To begin with, let me define the required constructions on the domain of the aggregation operator.

Definition 7.32.

Suppose $f : \mathbf{I} \longrightarrow \mathbf{I}$ is a monotonic mapping (i.e., either nondecreasing or nonincreasing). The mappings $f^\flat$, $f^\sharp : \mathbf{I} \longrightarrow \mathbf{I}$ are defined by:

$$f^\sharp = \begin{cases} \lim\limits_{y \searrow x} f(y) & : \quad x < 1 \\ f(1) & : \quad x = 1 \end{cases}$$

$$f^\flat = \begin{cases} \lim\limits_{y \nearrow x} f(y) & : \quad x > 0 \\ f(0) & : \quad x = 0 \end{cases}$$

for all $x \in \mathbf{I}$.

Notes 7.33.

- Let me remark that $f^\sharp$ and $f^\flat$ are well-defined, i.e. the limits in the above expressions exist, regardless of f. This is because f is known to be bounded and monotonic.
- It is further apparent that if $f \in \mathbb{B}$, then $f^\sharp \in \mathbb{B}$ and $f^\flat \in \mathbb{B}$.

I will now introduce several coefficients which describe certain aspects of a mapping $f : \mathbf{I} \longrightarrow \mathbf{I}$.

Definition 7.34. *Suppose that $f : \mathbf{I} \longrightarrow \mathbf{I}$ is a monotonic mapping (i.e., either nondecreasing or nonincreasing). We define*

$$f_0^* = \lim_{\gamma \searrow 0} f(\gamma) \tag{7.5}$$

$$f_*^0 = \inf\{\gamma \in \mathbf{I} : f(\gamma) = 0\} \tag{7.6}$$

$$f_*^{\frac{1}{2}} = \inf\{\gamma \in \mathbf{I} : f(\gamma) = \tfrac{1}{2}\} \tag{7.7}$$

$$f_1^* = \lim_{\gamma \nearrow 1} f(\gamma) \tag{7.8}$$

$$f_*^{1\uparrow} = \sup\{\gamma \in \mathbf{I} : f(\gamma) = 1\}\,. \tag{7.9}$$

As usual, let us stipulate that $\sup \varnothing = 0$ *and* $\inf \varnothing = 1$.

Notes 7.35.

- All limits in the definition of these coefficients exist due to the assumed monotonicity of $f : \mathbf{I} \longrightarrow \mathbf{I}$.
- The conditions imposed on legal choices of $\mathcal{B}$ will only utilize one of these coefficients, $f_*^{\frac{1}{2}}$. The remaining coefficients will permit the subsequent definition of example models.

Based on these concepts, I can now state several axioms governing the behaviour of reasonable choices of $\mathcal{B}$.

Definition 7.36. *Let an aggregation mapping $\mathcal{B} : \mathbb{B} \longrightarrow \mathbf{I}$ be given. For all $f, g \in \mathcal{B}$, we define the following conditions on $\mathcal{B}$:*

$$\mathcal{B}(f) = f(0) \quad \text{if } f \text{ is constant, i.e. } f(x) = f(0) \text{ for all } x \in \mathbf{I} \tag{B-1}$$

$$\mathcal{B}(1 - f) = 1 - \mathcal{B}(f) \tag{B-2}$$

$$\text{If } \widehat{f}(\mathbf{I}) \subseteq \{0, \tfrac{1}{2}, 1\}, \text{ then} \tag{B-3}$$

$$\mathcal{B}(f) = \begin{cases} \frac{1}{2} + \frac{1}{2} f_*^{\frac{1}{2}} & : \ f \in \mathbb{B}^+ \\ \frac{1}{2} & : \ f \in \mathbb{B}^{\frac{1}{2}} \\ \frac{1}{2} - \frac{1}{2} f_*^{\frac{1}{2}} & : \ f \in \mathbb{B}^- \end{cases}$$

$$\mathcal{B}(f^\sharp) = \mathcal{B}(f^\flat) \tag{B-4}$$

$$\text{If } f \leq g, \text{ then } \mathcal{B}(f) \leq \mathcal{B}(g) \tag{B-5}$$

Note 7.37. Let me briefly comment on the meaning of the individual conditions. (B-1) states that $\mathcal{B}$ preserves constants. In particular, all $\mathcal{M}_\mathcal{B}$-QFMs such that $\mathcal{B}$ satisfies (B-1) coincide on three-valued argument sets. (B-2) expresses that $\mathcal{B}$ is compatible with the standard negation $1 - x$. (B-3) ensures that all conforming $\mathcal{M}_\mathcal{B}$-QFMs coincide on three-valued quantifiers. (B-4) expresses some kind of insensitivity property of $\mathcal{B}$, which turns out to be crucial with respect to $\mathcal{M}_\mathcal{B}$ satisfying *functional application* (Z-6). Finally, (B-5) expresses that $\mathcal{B}$ is monotonic, i.e. application of $\mathcal{B}$ preserves inequalities.

Taken as a whole, the proposed system of conditions on $\mathcal{B}$ precisely captures the requirements on reasonable choices of $\mathcal{B}$. To see this, we first notice that every $\mathcal{B} : \mathbb{B} \longrightarrow \mathbf{I}$ which satisfies (B-1) to (B-5) makes $\mathcal{M}_\mathcal{B}$ a standard DFS:

Theorem 7.38.
The conditions (B-1)–(B-5) *on $\mathcal{B} : \mathbb{B} \longrightarrow \mathbf{I}$ are sufficient for $\mathcal{M}_\mathcal{B}$ to be a standard DFS.*

Next I consider the converse issue that every choice of $\mathcal{B} : \mathbb{B} \longrightarrow \mathbf{I}$ which makes $\mathcal{M}_\mathcal{B}$ a DFS, in fact satisfies the proposed set of conditions. Again, I have a positive result here.

Theorem 7.39.
The conditions (B-1)–(B-5) *on $\mathcal{B} : \mathbb{B} \longrightarrow \mathbf{I}$ are necessary for $\mathcal{M}_\mathcal{B}$ to be a DFS.*

Combining this with Th. 7.38, it is easily seen that

Theorem 7.40.
Every $\mathcal{M}_\mathcal{B}$-DFS is a standard DFS.

Let us further notice that

Theorem 7.41.
The conditions (B-1)–(B-5) *are independent.*

Hence none of these conditions can be expressed in terms of the remaining conditions. In other words, I have succeeded in formalizing a minimal set of conditions, and the conditions indeed *separate* the distinct factors that guide reasonable choices of $\mathcal{B}$. The mutual independence of the conditions also avoids redundant effort in proofs.

7.9 A Simplified Construction Based on Mappings $\mathcal{B}' : \mathbb{H} \longrightarrow \mathbf{I}$

In the following, a simplified representation of $\mathcal{M}_{\mathcal{B}}$-DFSes will be developed, which facilitates the presentation and investigation of specific models. To this end, let us first observe that $\mathcal{B}$ is completely determined by its behaviour on $\mathbb{B}^+ \cup \mathbb{B}^{\frac{1}{2}}$, provided that it satisfies (B-2). This is apparent because for every $f \in \mathbb{B}^-$, we obtain that $1 - f \in \mathbb{B}^+$ and hence $\mathcal{B}(f) = 1 - \mathcal{B}(1 - f)$. Noticing that $\mathrm{c}_{\frac{1}{2}} = 1 - \mathrm{c}_{\frac{1}{2}}$, we can further conclude that in every $\mathcal{B}$ which satisfies (B-2), it holds that

$$\mathcal{B}(\mathrm{c}_{\frac{1}{2}}) = 1 - \mathcal{B}(1 - \mathrm{c}_{\frac{1}{2}}) = 1 - \mathcal{B}(\mathrm{c}_{\frac{1}{2}}),$$

i.e. $\mathcal{B}(\mathrm{c}_{\frac{1}{2}}) = \frac{1}{2}$. We can hence strengthen the above remark, and state that $\mathcal{B}$ is even fully determined by its behaviour on $\mathbb{B}^+$ only. If we further assume that $\mathcal{B}$ satisfies (B-5), then we also know that $\mathcal{B}(f) \geq \mathcal{B}(\mathrm{c}_{\frac{1}{2}}) = \frac{1}{2}$ for all $f \in \mathbb{B}^+$. This means that we can restrict the range of possible $\mathcal{B}(f)$ to the upper half of the unit interval, $[\frac{1}{2}, 1]$.

In order to develop the simplified constructions, let us now recall from Th. 7.39 that the considered conditions (B-2) and (B-5) are necessary for $\mathcal{M}_{\mathcal{B}}$ to be a DFS. Hence no models of interest are lost if we focus on those $\mathcal{B} : \mathbb{B} \longrightarrow \mathbf{I}$ which satisfy (B-2) and result in $\mathcal{B}(f) \geq \frac{1}{2}$ for all $f \in \mathbb{B}^+$. In this case, I can give a more concise description of the models:

Definition 7.42.
By $\mathbb{H} \subseteq \mathbf{I}^{\mathbf{I}}$ we denote the set of nonincreasing mappings $f : \mathbf{I} \longrightarrow \mathbf{I}$, $f \neq \mathrm{c}_0$, i.e.

$$\mathbb{H} = \{f \in \mathbf{I}^{\mathbf{I}} : \ f \ \textit{nonincreasing and} \ f(0) > 0\ \}.$$

We can then associate with each $\mathcal{B}' : \mathbb{H} \longrightarrow \mathbf{I}$ a $\mathcal{B} : \mathbb{B} \longrightarrow \mathbf{I}$ as follows:

$$\mathcal{B}(f) = \begin{cases} \frac{1}{2} + \frac{1}{2}\mathcal{B}'(2f - 1) & : \quad f \in \mathbb{B}^+ \\ \frac{1}{2} & : \quad f \in \mathbb{B}^{\frac{1}{2}} \\ \frac{1}{2} - \frac{1}{2}\mathcal{B}'(1 - 2f) & : \quad f \in \mathbb{B}^- \end{cases} \tag{7.10}$$

It is apparent that each $\mathcal{B}$ constructed in this way again satisfies (B-2) and results in $\mathcal{B}(f) \geq \frac{1}{2}$ for all $f \in \mathbb{B}^+$. Conversely, all suitable choices of $\mathcal{B}$ can be represented in terms of this new construction, as I will now state.

Theorem 7.43.
Consider $\mathcal{B} : \mathbb{B} \longrightarrow \mathbf{I}$ and suppose that $\mathcal{B}$ satisfies (B-2) *and results in $\mathcal{B}(f) \geq \frac{1}{2}$ for all $f \in \mathbb{B}^+$. Then $\mathcal{B}$ can be defined in terms of a mapping $\mathcal{B}' : \mathbb{H} \longrightarrow \mathbf{I}$ according to equation* (7.10)*. $\mathcal{B}'$ is defined by*

$$\mathcal{B}'(f) = 2\mathcal{B}(\tfrac{1}{2} + \tfrac{1}{2}f) - 1\,. \tag{7.11}$$

By the above reasoning, these conditions are satisfied by every $\mathcal{B} : \mathbb{B} \longrightarrow \mathbf{I}$ which fulfills (B-2) and (B-5); in particular, every choice of $\mathcal{B}$ which makes $\mathcal{M}_{\mathcal{B}}$ a DFS can be represented in this simplified format. We can therefore restrict attention on properties of mappings $\mathcal{B}' : \mathbb{H} \longrightarrow \mathbf{I}$, and still cover all of the desired models.

7.10 Characterisation of the Models in Terms of Conditions on $\mathcal{B}'$

Let us now characterise the $\mathcal{M}_{\mathcal{B}}$-models in terms of conditions on the core aggregation mapping $\mathcal{B}'$, which are fitted to the simplified construction.

Definition 7.44. *Suppose $\mathcal{B}' : \mathbb{H} \longrightarrow \mathbf{I}$ is given. For all $f, g \in \mathbb{H}$, we define the following conditions on $\mathcal{B}'$:*

$$\mathcal{B}'(f) = f(0) \quad \textit{if } f \textit{ is constant, i.e. } f(x) = f(0) \textit{ for all } x \in \mathbf{I} \tag{C-1}$$

$$\textit{If } f(\mathbf{I}) \subseteq \{0, 1\}, \textit{ then } \mathcal{B}'(f) = f_*^0, \tag{C-2}$$

$$\mathcal{B}'(f^\sharp) = \mathcal{B}'(f^\flat) \quad \textit{if } \widehat{f}((0,1]) \neq \{0\} \tag{C-3}$$

$$\textit{If } f \leq g, \textit{ then } \mathcal{B}'(f) \leq \mathcal{B}'(g) \tag{C-4}$$

These conditions on $\mathcal{B}'$ are straightforward from the known conditions (B-1)–(B-5) that must be imposed on the corresponding $\mathcal{B}$. The following results on the (C-1)–(C-4) are therefore apparent from the previous results on the 'B-conditions':

Theorem 7.45.
The conditions (C-1)–(C-4) *on $\mathcal{B}' : \mathbb{H} \longrightarrow \mathbf{I}$ are sufficient for $\mathcal{M}_{\mathcal{B}}$ to be a standard DFS.*

Theorem 7.46.
The conditions (C-1)–(C-4) *on* $\mathcal{B}' : \mathbb{H} \longrightarrow \mathbf{I}$ *are necessary for* $\mathcal{M}_\mathcal{B}$ *to be a DFS.*

Theorem 7.47.
The conditions (C-1)–(C-4) *are independent.*

Note 7.48. In the above theorems on the 'C-conditions', it is understood that $\mathcal{M}_\mathcal{B}$ be constructed from $\mathcal{B}'$ according to (7.10) and Def. 7.31.

To sum up, the simplified construction of the relevant $\mathcal{M}_\mathcal{B}$-QFMs in terms of a mapping $\mathcal{B}' : \mathbb{H} \longrightarrow \mathbf{I}$ results in a succinct characterisation of the target models. Based on this characterisation, it has now become easy to verify that a 'DFS candidate' $\mathcal{M}_\mathcal{B}$ indeed qualifies as one of the intended models, by performing a few simple checks on the aggregation operator $\mathcal{B} : \mathbb{B} \longrightarrow \mathbf{I}$ or $\mathcal{B}' : \mathbb{H} \longrightarrow \mathbf{I}$. In fact, knowing the precise contraints that govern admissible choices of $\mathcal{B}'$, also guided my search into examples of models.

7.11 Examples of $\mathcal{M}_\mathcal{B}$-Models

In the following, I will present some examples of $\mathcal{M}_\mathcal{B}$-QFMs. For convenience, the new construction based on $\mathcal{B}'$ will be utilized, in order to simplify the presentation of these models and achieve succinct definitions. Let us first reconsider the model $\mathcal{M}$, which motivated the generalization to the class of $\mathcal{M}_\mathcal{B}$-QFMs, and fit its definition into the new format. In terms of the improved notation which is now available, the original definition of $\mathcal{M}$ can be rephrased as follows.

Theorem 7.49.
$\mathcal{M}$ is the $\mathcal{M}_\mathcal{B}$-QFM defined by

$$\mathcal{B}'_\int(f) = \int_0^1 f(x)\,\mathrm{d}x\,,$$

for all $f \in \mathbb{H}$.

Now turning to new examples, I first introduce the following model $\mathcal{M}_U$.

Definition 7.50.
By $\mathcal{M}_U$ we denote the $\mathcal{M}_\mathcal{B}$-QFM defined by

$$\mathcal{B}'_U(f) = \max(f_*^{1\uparrow}, f_1^*)$$

for all $f \in \mathbb{H}$, where the coefficients f_1^* and $f_*^{1\uparrow}$ are defined by equalities (7.8) *and* (7.9)*, resp.*

As I now state, $\mathcal{M}_U$ is indeed a plausible model. In fact, this can be asserted for a broader type of $\mathcal{M}_\mathcal{B}$-QFMs, of which $\mathcal{M}_U$ is only a special example:

Theorem 7.51.
Suppose $\oplus : \mathbf{I}^2 \longrightarrow \mathbf{I}$ is an s-norm and $\mathcal{B}' : \mathbb{H} \longrightarrow \mathbf{I}$ is defined by

$$\mathcal{B}'(f) = f_*^{1\uparrow} \oplus f_1^* \,, \qquad \text{(Th. 7.51.a)}$$

for all $f \in \mathbb{H}$, where the coefficients $f_*^{1\uparrow}$ and f_1^* are defined by (7.9) and (7.8), respectively. Further suppose that $\mathcal{B} : \mathbb{B} \longrightarrow \mathbf{I}$ is defined in terms of $\mathcal{B}'$ according to equation (7.10), and that $\mathcal{M}_\mathcal{B}$ is the QFM defined in terms of $\mathcal{B}$ according to Def. 7.31. The QFM $\mathcal{M}_\mathcal{B}$ is a standard DFS.

In particular, $\mathcal{M}_U$ is a standard DFS. We will see later on that $\mathcal{M}_U$ plays a special role among the models of the $\mathcal{M}_\mathcal{B}$-type, by establishing a lower bound on these models with respect to the specificity order.
In [53, Def. 44, p. 63], another example $\mathcal{M}^*$ was presented, which is defined as follows.

Definition 7.52.
By $\mathcal{M}^*$ we denote the $\mathcal{M}_\mathcal{B}$-QFM defined by

$$\mathcal{B}^{*\prime}(f) = f_*^0 \cdot f_0^* \,,$$

for all $f \in \mathbb{H}$, where the coefficients f_*^0 and f_0^* are defined by (7.6) and (7.5), resp.

The following model, based on a similar construction, will be of some interest to the later analysis of the class of $\mathcal{M}_\mathcal{B}$-DFSes.

Definition 7.53.
By $\mathcal{M}_S$ we denote the $\mathcal{M}_\mathcal{B}$-QFM defined by

$$\mathcal{B}'_S(f) = \min(f_*^0, f_0^*)$$

for all $f \in \mathbb{H}$, using the same abbreviations as for $\mathcal{M}^*$.

Abstracting from the underlying construction, we may assert the following.

Theorem 7.54.
Suppose $\mathcal{B}' : \mathbb{H} \longrightarrow \mathbf{I}$ is defined by

$$\mathcal{B}'(f) = f_*^0 \odot f_0^*$$

for all $f \in \mathbb{H}$, where $\odot : \mathbf{I}^2 \longrightarrow \mathbf{I}$ is a t-norm. Further suppose that the QFM $\mathcal{M}_\mathcal{B}$ is defined in terms of $\mathcal{B}'$ according to (7.10) and Def. 7.31. Then $\mathcal{M}_\mathcal{B}$ is a standard DFS.

In particular, $\mathcal{M}^*$ and $\mathcal{M}_S$ are standard DFSes. Like $\mathcal{M}_U$, the model $\mathcal{M}_S$ is also of special relevance to the $\mathcal{M}_\mathcal{B}$-type. We shall see later that it plays the opposite role, and constitutes an upper bound on the models under the specificity order.
In [53, Def. 45, p. 64], a further type of model $\mathcal{M}_*$ was presented.

Definition 7.55.
By $\mathcal{M}_*$ we denote the $\mathcal{M}_{\mathcal{B}}$-QFM defined by

$$\mathcal{B}'_*(f) = \sup\{x \cdot f(x) : x \in \mathbf{I}\},$$

for all $f \in \mathbb{H}$.

By slightly altering this construction, I discovered a model which is now known to be of key relevance to DFS theory.

Definition 7.56.
By $\mathcal{M}_{CX}$ we denote the $\mathcal{M}_{\mathcal{B}}$-QFM defined by

$$\mathcal{B}'{}_{CX}(f) = \sup\{\min(x, f(x)) : x \in \mathbf{I}\}$$

for all $f \in \mathbb{H}$.

By abstracting from the constructive principle underlying these examples, we obtain the following more general result:

Theorem 7.57.
Suppose $\odot : \mathbf{I}^2 \longrightarrow \mathbf{I}$ is a continuous t-norm and $\mathcal{B}' : \mathbb{H} \longrightarrow \mathbf{I}$ is defined by

$$\mathcal{B}'(f) = \sup\{\gamma \odot f(\gamma) : \gamma \in \mathbf{I}\} \qquad \text{(Th. 7.57.a)}$$

for all $f \in \mathbb{H}$. Further suppose that $\mathcal{B} : \mathbb{B} \longrightarrow \mathbf{I}$ is defined in terms of $\mathcal{B}'$ according to equation (7.10)*. Then the QFM $\mathcal{M}_{\mathcal{B}}$, defined by Def. 7.31, is a standard DFS.*

Note 7.58. In particular, $\mathcal{M}_{CX}$ and $\mathcal{M}_*$ are standard models.

As remarked above, $\mathcal{M}_{CX}$ is a DFS with unique properties, and a separate section has therefore been devoted to its discussion. Before focusing on this particular model in Sect. 7.13, however, it is first necessary to develop the criteria on $\mathcal{B}$ or $\mathcal{B}'$ which permit us to check whether a given DFS $\mathcal{M}_{\mathcal{B}}$ satisfies an adequacy property of interest. For example, it would be helpful if we could perform specificity comparisons solely based on the known aggregation mappings $\mathcal{B}'$; if we could test $\mathcal{M}_{\mathcal{B}}$ for Q-continuity and arg-continuity by looking at $\mathcal{B}'$ only, etc. In turn, these conditions will then permit a deeper investigation of $\mathcal{M}_{CX}$ and other prominent models of the theory.

7.12 Properties of $\mathcal{M}_{\mathcal{B}}$-Models

Let us now take a closer look at $\mathcal{M}_{\mathcal{B}}$-DFSes. The present goal is to expose the internal structure of these models and relate it to the properties of interest, like the continuity requirements or propagation of fuzziness. I am also interested in identifying properties shared by all $\mathcal{M}_{\mathcal{B}}$-DFSes which unveil some characteristic aspects of this class of models. In addition, it is useful to locate models with special properties, and in particular the boundary cases with respect to the specificity order. Knowing these extreme cases delimits the space of possible models, which is opened by the proposed base construction. These special examples hence provide some insight into the behaviour that must be expected by other models of interest, which are known to be in-between the extreme poles.

To begin with, I state a theorem which simplifies the comparison of $\mathcal{M}_{\mathcal{B}}$-DFSes with respect to specificity.

Theorem 7.59.
Suppose $\mathcal{B}'_1, \mathcal{B}'_2 : \mathbb{H} \longrightarrow \mathbf{I}$ are given. Further suppose that $\mathcal{B}_1, \mathcal{B}_2 \in \mathbb{H}$ are the mappings associated with $\mathcal{B}'_1$ and $\mathcal{B}'_2$, respectively, according to equation (7.10), and $\mathcal{M}_{\mathcal{B}_1}$, $\mathcal{M}_{\mathcal{B}_2}$ are the corresponding QFMs defined by Def. 7.31. Then $\mathcal{M}_{\mathcal{B}_1} \preceq_c \mathcal{M}_{\mathcal{B}_2}$ if and only if $\mathcal{B}'_1 \leq \mathcal{B}'_2$.

Based on the above theorem and the examples of models introduced so far, it is now easily shown that $\preceq_c$ is a genuine partial order on $\mathcal{M}_{\mathcal{B}}$-DFSes.

Theorem 7.60.
$\preceq_c$ is not a total order on $\mathcal{M}_{\mathcal{B}}$-DFSes.

Note 7.61. In particular, the standard DFSes are only partially ordered by $\preceq_c$.

In practice, this means that we cannot simply compile a list of the example models (or other models) ordered by specificity, because some of these might fail to be comparable under $\preceq_c$. However, it is still possible to investigate extreme cases of $\mathcal{M}_{\mathcal{B}}$-DFSes with respect to the specificity order. For example, it is not hard to prove that the DFS $\mathcal{M}_U$ defined by Def. 7.50 represents a boundary case of $\mathcal{M}_{\mathcal{B}}$-DFSes in terms of specificity:

Theorem 7.62.
$\mathcal{M}_U$ is the least specific $\mathcal{M}_{\mathcal{B}}$-DFS.

Let us now consider the question of the existence of most specific $\mathcal{M}_{\mathcal{B}}$-DFSes. We first observe that

Theorem 7.63.
All $\mathcal{M}_{\mathcal{B}}$-DFSes are specificity consistent.

It is then immediate from Th. 5.13 that there exists a least upper specificity bound $\mathcal{F}_{\text{lub}}$ on every nonempty collection of $\mathcal{M}_{\mathcal{B}}$-models. As I will now show, $\mathcal{F}_{\text{lub}}$ is in fact an $\mathcal{M}_{\mathcal{B}}$-DFS.

Theorem 7.64.
Let $\mathbb{F}$ be a given nonempty collection of $\mathcal{M}_{\mathcal{B}}$-DFSes. The least upper specificity bound $\mathcal{F}_{\text{lub}}$ of $\mathbb{F}$ is an $\mathcal{M}_{\mathcal{B}}$-DFS.

Hence arbitrary collections of $\mathcal{M}_{\mathcal{B}}$-QFMs have a least upper specificity bound, which is again an $\mathcal{M}_{\mathcal{B}}$-DFS. If we start from the collection of *all* $\mathcal{M}_{\mathcal{B}}$-DFSes, then we obtain the model $\mathcal{M}_S$ defined in Def. 7.53, which represents the other extreme case of $\mathcal{M}_{\mathcal{B}}$-DFS in terms of specificity:

Theorem 7.65.
$\mathcal{M}_S$ is the most specific $\mathcal{M}_{\mathcal{B}}$-DFS.

Next I will discuss the two continuity requirements that were identified in Sect. 6.2, and develop the required criteria for deciding upon the Q-continuity and arg-continuity of a given $\mathcal{M}_{\mathcal{B}}$-DFS. Let us first make some general observations how the continuity conditions are related to $(\bullet)_\gamma$.

Theorem 7.66.
Let $Q, Q' : \mathcal{P}(E)^n \longrightarrow \mathbf{I}$ be given. Then $d(Q_\gamma, Q'_\gamma) \leq d(Q, Q')$ for all $\gamma \in \mathbf{I}$.

We can use this inequality to formulate a condition on $\mathcal{B}' : \mathbb{H} \longrightarrow \mathbf{I}$ which is necessary and sufficient for $\mathcal{M}_{\mathcal{B}}$ to be Q-continuous. To this end, we first define a metric $d : \mathbb{H} \times \mathbb{H} \longrightarrow \mathbf{I}$ by

$$d(f, g) = \sup\{|f(\gamma) - g(\gamma)| : \gamma \in \mathbf{I}\} \tag{7.12}$$

for all $f, g \in \mathbb{H}$.

Theorem 7.67.
Suppose $\mathcal{M}_{\mathcal{B}}$ is an $\mathcal{M}_{\mathcal{B}}$-DFS and $\mathcal{B}'$ is the corresponding mapping $\mathcal{B}' : \mathbb{H} \longrightarrow \mathbf{I}$.
$\mathcal{M}_{\mathcal{B}}$ is Q-continuous if and only if for all $\varepsilon > 0$ there exists $\delta > 0$ such that $|\mathcal{B}'(f) - \mathcal{B}'(g)| < \varepsilon$ whenever $f, g \in \mathbb{H}$ satisfy $d(f, g) < \delta$.

In the case of continuity in arguments, we need a different distance measure on $\mathbb{H}$. Hence let us define $d' : \mathbb{H} \times \mathbb{H} \longrightarrow \mathbf{I}$ by

$$d'(f, g) = \sup\{\inf\{\gamma' : \gamma' \in \mathbf{I}, \max(f(\gamma'), g(\gamma')) \leq \min(f(\gamma), g(\gamma))\} - \gamma : \gamma \in \mathbf{I}\}, \tag{7.13}$$

for all $f, g \in \mathbb{H}$. Let us notice that d' is only a 'pseudo-metric'. It is symmetric and satisfies the triangle inequality, but fails to be a metric in the strict sense because $d'(f, g) = 0$ does not entail that $f = g$. However, those f, g with $d'(f, g) = 0$ are treated alike by reasonable choices of $\mathcal{B}'$. It is this fact which justifies the use of d' in lieu of a metric, which permits a successful reduction of arg-continuity to a criterion decidable from the aggregation mapping. The next theorem established this desired criterion, and hence shows how the arg-continuous $\mathcal{M}_{\mathcal{B}}$ models can be characterised in terms of the underlying mapping $\mathcal{B}' : \mathbb{H} \longrightarrow \mathbf{I}$.

Theorem 7.68.
Suppose $\mathcal{B}' : \mathbb{H} \longrightarrow \mathbf{I}$ satisfies (C-2), (C-3) *and* (C-4). *Further suppose that $\mathcal{M}_{\mathcal{B}}$ is defined in terms of $\mathcal{B}'$ according to* (7.10) *and Def. 7.31. Then the following conditions are equivalent:*

Note 7.78. This is different from general standard DFSes which by Th. 4.1 are guaranteed to coincide for two-valued arguments only.

A similar result can be established for three-valued quantifiers.

Theorem 7.79.
All $\mathcal{M}_{\mathcal{B}}$-DFSes coincide on three-valued semi-fuzzy quantifiers, i.e. in the case that Q assumes results in the restricted range $\{0, \frac{1}{2}, 1\}$ only, and can thus be expressed as $Q : \mathcal{P}(E)^n \longrightarrow \{0, \frac{1}{2}, 1\}$.

Again, this is different from general standard DFSes, which are guaranteed to coincide only for two-valued quantifiers, see Th. 5.22.

7.13 A Standard Model with Unique Properties: $\mathcal{M}_{CX}$

We already know from the previous section that the model $\mathcal{M}_{CX}$ is both Q-continuous and arg-continuous. It is thus a practical model and suited for applications. In addition, $\mathcal{M}_{CX}$ propagates fuzziness in quantifiers and in arguments, just like every $\mathcal{M}_{\mathcal{B}}$-DFS. However, what exactly is it that makes $\mathcal{M}_{CX}$ so special? The present section is devoted exclusively to this issue and will substantiate that $\mathcal{M}_{CX}$ is indeed unique among $\mathcal{M}_{\mathcal{B}}$-DFSes. In fact, the particular qualities held by this model are distinguished even among the full class of standard models of fuzzy quantification.

In Sect. 4.10, I have introduced the construction of fuzzy argument insertion, and I explained that only those models which conform to this construction can adequately represent adjectival restriction by a fuzzy adjective, as in "Most X's are lucky Y's".

Compatibility with fuzzy argument insertion is therefore highly desirable from a linguistic standpoint, because it ensures that this general type of adjectival restriction can be interpreted in a compositional way. As concerns fuzzy argument insertion, the following positive result has been proven for $\mathcal{M}_{CX}$:

Theorem 7.80.
The DFS $\mathcal{M}_{CX}$ is compatible with fuzzy argument insertion.

It is this property held by the model which explains the distinguished role of $\mathcal{M}_{CX}$. In fact, it can be shown that $\mathcal{M}_{CX}$ is the *unique* standard DFS which satisfies this condition.

Theorem 7.81.
$\mathcal{M}_{CX}$ is the only standard DFS which is compatible with fuzzy argument insertion.

Hence $\mathcal{M}_{CX}$ is foremost among the standard models. Due to the special importance of $\mathcal{M}_{CX}$, it is probably worthwhile to attempt an axiomatization of the model in terms of a system of conditions which uniquely identify $\mathcal{M}_{CX}$.

In fact, the desired axiomatization is rather obvious, because we only need to augment the previous axiomatization of the standard models, which has been presented in Def. 5.23, by the requirement of being compatible with fuzzy argument insertion:

Definition 7.82.
Consider a QFM $\mathcal{F}$. For all semi-fuzzy quantifiers $Q : \mathcal{P}(E)^n \longrightarrow \mathbf{I}$, I introduce the following conditions.

Correct generalisation	$\mathcal{U}(\mathcal{F}(Q)) = Q \quad \textit{if } n \leq 1$	(M-1)
Projection quantifiers	$\mathcal{F}(Q) = \widetilde{\pi}_e \quad \textit{if there exists } e \in E \textit{ s.th. } Q = \pi_e$	(M-2)
Dualisation	$\mathcal{F}(Q\square) = \mathcal{F}(Q)\square \quad n > 0$	(M-3)
Internal joins	$\mathcal{F}(Q\cup) = \mathcal{F}(Q)\cup \quad n > 0$	(M-4)
Preservation of monotonicity	*If Q is nonincreasing in n-th arg, then* $\mathcal{F}(Q)$ *is nonincreasing in n-th arg,* $n > 0$	(M-5)
Functional application	$\mathcal{F}(Q \circ \overset{n}{\underset{i=1}{\times}} \widehat{f_i}) = \mathcal{F}(Q) \circ \overset{n}{\underset{i=1}{\times}} \hat{\hat{f}}_i$ *where* $f_1, \ldots, f_n : E' \longrightarrow E,\ E' \neq \varnothing$.	(M-6)
Fuzzy argument insertion	$\mathcal{F}(Q \mathbin{\widetilde{\triangleleft}} A) = \mathcal{F}(Q) \triangleleft A$ *for all* $A \in \widetilde{\mathcal{P}}(E),\ n > 0$	(M-7)

We can then assert that the above conditions (M-1)–(M-7) indeed achieve an axiomatisation of the model $\mathcal{M}_{CX}$.

Theorem 7.83.
$\mathcal{M}_{CX}$ is the unique QFM which satisfies (M-1)–(M-7).

Note 7.84. By the above reasoning, this theorem is really straightforward and an immediate consequence of Th. 5.25 and Th. 7.81.

As witnessed by the theorem, we have succeeded to uniquely identify $\mathcal{M}_{CX}$ in the full space of possible QFMs, solely based on its observable behaviour. This was possible because the structure of $\mathcal{M}_{CX}$ shows up in the distinguishing property (M-7), which is sufficient to characterise $\mathcal{M}_{CX}$ within the class of standard models. Because the system (M-1)–(M-7) also comprises a minimal set of conditions, this makes a rather satisfying result from a theoretical point of view.

In addition to being compatible with fuzzy argument insertion, $\mathcal{M}_{CX}$ also exhibits a number of other remarkable characteristics. At present, no other models are known which possess any of these distinguished properties. I hence suspect that the other properties, too, are unique to $\mathcal{M}_{CX}$, but there is no proof of this yet, and more research should be directed at these issues.

Let us first consider the preservation of convexity properties.

Theorem 7.85.
The DFS $\mathcal{M}_{CX}$ weakly preserves convexity.

Theorem 7.86.
Suppose $\mathcal{M}_{\mathcal{B}}$ is an $\mathcal{M}_{\mathcal{B}}$-DFS. If $\mathcal{M}_{\mathcal{B}}$ weakly preserves convexity, then $\mathcal{M}_{CX} \preceq_c \mathcal{M}_{\mathcal{B}}$.

The above theorems substantiate that $\mathcal{M}_{CX}$ is the least specific $\mathcal{M}_{\mathcal{B}}$-DFS which weakly preserves convexity. As remarked above, there is no evidence yet that other models exist which fulfill this criterion, and it might hence express another distinctive property of $\mathcal{M}_{CX}$.

Let us further observe that $\mathcal{M}_{CX}$ is a concrete implementation of a so-called 'substitution approach' to fuzzy quantification [173], i.e. the fuzzy quantifier is modelled by constructing an equivalent logical formula. This is apparent if we rewrite $\mathcal{M}_{CX}$ as follows.

Theorem 7.87.
For every $Q : \mathcal{P}(E)^n \longrightarrow \mathbf{I}$ and $X_1, \ldots, X_n \in \widetilde{\mathcal{P}}(E)$,

$$\begin{aligned}
&\mathcal{M}_{CX}(Q)(X_1, \ldots, X_n)\\
&= \sup\{\widetilde{Q}^L_{V,W}(X_1, \ldots, X_n) : V, W \in \mathcal{P}(E)^n, V_1 \subseteq W_1, \ldots, V_n \subseteq W_n\}\\
&= \inf\{\widetilde{Q}^U_{V,W}(X_1, \ldots, X_n) : V, W \in \mathcal{P}(E)^n, V_1 \subseteq W_1, \ldots, V_n \subseteq W_n\}
\end{aligned}$$

where

$$\widetilde{Q}^L_{V,W}(X_1, \ldots, X_n) \tag{7.15}$$
$$= \min(\widetilde{\Xi}_{V,W}(X_1, \ldots, X_n), L(Q, V, W)) \tag{7.16}$$
$$\widetilde{Q}^U_{V,W}(X_1, \ldots, X_n) \tag{7.17}$$
$$= \max(1 - \widetilde{\Xi}_{V,W}(X_1, \ldots, X_n), U(Q, V, W)) \tag{7.18}$$
$$\widetilde{\Xi}_{V,W}(X_1, \ldots, X_n) \tag{7.19}$$
$$= \min_{i=1}^{n} \min(\inf\{\mu_{X_i}(e) : e \in V_i\}, \inf\{1 - \mu_{X_i}(e) : e \notin W_i\}) \tag{7.20}$$
$$L(Q, V, W) \tag{7.21}$$
$$= \inf\{Q(Y_1, \ldots, Y_n) : V_i \subseteq Y_i \subseteq W_i, \ \textit{all } i\} \tag{7.22}$$
$$U(Q, V, W) \tag{7.23}$$
$$= \sup\{Q(Y_1, \ldots, Y_n) : V_i \subseteq Y_i \subseteq W_i, \ \textit{all } i\}\,. \tag{7.24}$$

Notes 7.88.

- To see how this is related to the substitution approach, let us notice that in the common case of a finite base set, inf and sup reduce to the logical connectives $\wedge = \max$ and $\vee = \min$ as usual. The infimum and supremum are only needed to express the infinitary conjunctions and disjunctions that are required for quantification on infinite domains. In addition, it was necessary to allow occurrences of continuous-valued constants

Table 7.1. Computation of $L(Q, V, W)$ and $U(Q, V, W)$ (example)

V	W	$\widetilde{\Xi}_{V,W}(X)$	$L(Q, V, W)$	$U(Q, V, W)$
$\varnothing$	$\varnothing$	$\neg p \wedge \neg q$	0	0
$\varnothing$	$\{a\}$	$\neg q$	0	1
$\varnothing$	$\{b\}$	$\neg p$	0	$\frac{2}{3}$
$\varnothing$	$\{a, b\}$	1	0	1
$\{a\}$	$\{a\}$	$p \wedge \neg q$	1	1
$\{a\}$	$\{a, b\}$	p	1	1
$\{b\}$	$\{b\}$	$\neg p \wedge q$	$\frac{2}{3}$	$\frac{2}{3}$
$\{b\}$	$\{a, b\}$	q	$\frac{2}{3}$	1
$\{a, b\}$	$\{a, b\}$	$p \wedge q$	1	1

$L(Q, V, W), U(Q, V, W) \in \mathbf{I}$ in the resulting formulas because the fuzzification mechanism is applied to general semi-fuzzy quantifiers, not only to two-valued quantifiers.

- From a different point of view, the above representation of $\mathcal{M}_{CX}$ demonstrates that the model can be defined independently of the cut ranges and the median-based aggregation mechanism. By contrast, it can be rephrased into a compact form, which reduces fuzzy quantification in $\mathcal{M}_{CX}$ to the evaluation of a Boolean formula involving fuzzy connectives and continuous-valued propositional variables.

Before discussing any implications of the above reformulation of $\mathcal{M}_{CX}$ in terms of the substitution approach, I will first give an example of the proposed construction, in order to facilitate understanding of the above reformulation of $\mathcal{M}_{CX}$, and to elucidate the structure of the constructed formula and the function of the involved coefficients. Hence let us consider a simple two-element base set $E = \{a, b\}$ and a semi-fuzzy quantifier defined on E, say

$$Q(Y) = \begin{cases} 0 & : \quad Y = \varnothing \\ \frac{2}{3} & : \quad Y = \{b\} \\ 1 & : \quad Y = \{a\} \vee Y = \{a, b\} \end{cases}$$

for all $Y \in \mathcal{P}(\{a, b\})$. Let us now construct the quantification formula step by step, in order to gain some understanding how it is built from the base quantifier. Hence let a fuzzy argument set $X \in \widetilde{\mathcal{P}}(\{a, b\})$ be given. It is convenient to abbreviate $p = \mu_X(a)$, $q = \mu_X(b)$. Table 7.1 lists all possible choices of $V, W \in \mathcal{P}(\{a, b\})$ with $V \subseteq W$, along with the corresponding quantities $\widetilde{\Xi}_{V,W}(X)$ which express the compatibility of X with the closed range of subsets $\{Y \in \mathcal{P}(E) : V \subseteq Y \subseteq W\}$, as well as the lower and upper bounds $L(Q, V, W)$ and $U(Q, V, W)$ on the quantification results achieved by Q in this given range of subsets. Based on this table, we can then fill in the formulas which define the upper and lower bound quantifiers $\widetilde{Q}^U_{V,W}(X)$ and $\widetilde{Q}^L_{V,W}(X)$, as shown in Table 7.2. Recalling the first construction of the quantification formula described in Th. 7.87, which resorts to the lower bound quantifiers

Table 7.2. Computation of $\widetilde{Q}^L_{V,W}(X)$ and $\widetilde{Q}^U_{V,W}(X)$ (example)

V	W	$\widetilde{Q}^L_{V,W}(X)$	$\widetilde{Q}^U_{V,W}(X)$
$\varnothing$	$\varnothing$	$\neg p \wedge \neg q \wedge 0$	$p \vee q \vee 0$
$\varnothing$	$\{a\}$	$\neg q \wedge 0$	$q \vee 1$
$\varnothing$	$\{b\}$	$\neg p \wedge 0$	$p \vee \frac{2}{3}$
$\varnothing$	$\{a,b\}$	$1 \wedge 0$	$0 \vee 1$
$\{a\}$	$\{a\}$	$p \wedge \neg q \wedge 1$	$\neg p \vee q \vee 1$
$\{a\}$	$\{a,b\}$	$p \wedge 1$	$\neg p \vee 1$
$\{b\}$	$\{b\}$	$\neg p \wedge q \wedge \frac{2}{3}$	$p \vee \neg q \vee \frac{2}{3}$
$\{b\}$	$\{a,b\}$	$q \wedge \frac{2}{3}$	$\neg q \vee 1$
$\{a,b\}$	$\{a,b\}$	$p \wedge q \wedge 1$	$\neg p \vee \neg q \vee 1$

$\widetilde{Q}^L_{V,W}$, we can now build the desired formula which reduces $\mathcal{M}_{CX}(Q)(X)$ to a propositional expression involving continuous-valued variables:

$$\mathcal{M}_{CX}(Q)(X) = \max\{\neg p \wedge \neg q \wedge 0, \neg q \wedge 0, \neg p \wedge 0, 1 \wedge 0, p \wedge \neg q \wedge 1, p \wedge 1, \neg p \wedge q \wedge \tfrac{2}{3}, q \wedge \tfrac{2}{3}, p \wedge q \wedge 1\}\,.$$

After some element-wise simplification, this becomes

$$\mathcal{M}_{CX}(Q)(X) = \max\{0, 0, 0, 0, p \wedge \neg q, p, \neg p \wedge q \wedge \tfrac{2}{3}, q \wedge \tfrac{2}{3}, p \wedge q\}\,.$$

By eliminating all occurrences of the identity 0 of max, and by utilizing the law of absorption, the above expression can then be reduced to

$$\mathcal{M}_{CX}(Q)(X) = p \vee (q \wedge \tfrac{2}{3})\,.$$

Of course, we could also have applied the second construction of the quantification formula proposed in Th. 7.87, which is based on the upper bound quantifiers $\widetilde{Q}^U_{V,W}$. In this case, too, we start by building the 'raw' formula which expresses $\mathcal{M}_{CX}(Q)(X)$, now based on the second construction. We then obtain the conjunction,

$$\mathcal{M}_{CX}(Q)(X) = \min\{p \vee q \vee 0, q \vee 1, p \vee \tfrac{2}{3}, 0 \vee 1, \neg p \vee q \vee 1, \neg p \vee 1, p \vee \neg q \vee \tfrac{2}{3}, \neg q \vee 1, \neg p \vee \neg q \vee 1\}\,.$$

Now effecting some element-wise simplification, the raw formula becomes

$$\mathcal{M}_{CX}(Q)(X) = \min\{p \vee q, 1, p \vee \tfrac{2}{3}, 1, 1, 1, p \vee \neg q \vee \tfrac{2}{3}, 1, 1\}\,.$$

Again by removing occurrences of the identity 1 of min, and by taking benefit of the law of absorption in order to eliminate some of the subexpressions, we can further simplify the latter result into

$$\mathcal{M}_{CX}(Q)(X) = (p \vee q) \wedge (p \vee \tfrac{2}{3})\,.$$

Now recalling the law of distributivity, we can extract the variable p from the two disjunctions, thus rewriting the above equation as

$$\mathcal{M}_{CX}(Q)(X) = p \vee (q \wedge \tfrac{2}{3}) .$$

Hence the result of the second construction indeed coincides with that obtained from the first construction, as claimed by the theorem. This completes the example, which demonstrated the utility of the substitution formulas to the modelling of fuzzy quantification, and I will now discuss some further observations which are apparent from the reformulation of $\mathcal{M}_{CX}$.

The above representation of $\mathcal{M}_{CX}$, which is made explicit by Th. 7.87, also reveals the robustness of this model against variation in the quantifier and its arguments. This is important because in practice, there often remains some uncertainty concerning the precise choice of numeric membership grades which best model the target NL quantifier and the arguments (i.e. the target NL concepts which are to be modelled by suitable fuzzy sets).

- Let us first observe from the above representation that $\mathcal{M}_{CX}(Q)(X_1, \ldots, X_n)$ is not only arg-continuous (i.e. a smooth function of the membership grades $\mu_{X_i}(e)$ attained by the arguments), but indeed extremely stable against slight changes of the membership grades. This is because for all $i \in \{1, \ldots, n\}$ and $e \in E$,

$$\left| \frac{\partial \mathcal{M}_{CX}(Q)(X_1, \ldots, X_n)}{\partial \mu_{X_i}(e)} \right| \leq 1$$

  whenever the partial derivative is defined, which is obvious from the above representation. A change in the arguments which does not exceed a given Δ thus cannot change the quantification results by more than Δ. This indicates that with $\mathcal{M}_{CX}$, the precise choice of membership grades for the arguments (which might be to some degree arbitrary), is rather uncritical.
- A similar remark can also be made concerning the robustness of the quantification results $\mathcal{M}_{CX}(Q)(X_1, \ldots, X_n)$ against variation in the chosen *quantifier*. In this case, we observe that for all $(Y_1, \ldots, Y_n) \in \mathcal{P}(E)^n$,

$$\left| \frac{\partial \mathcal{M}_{CX}(Q)(X_1, \ldots, X_n)}{\partial Q(Y_1, \ldots, Y_n)} \right| \leq 1$$

  whenever the partial derivative is defined, which is again obvious from the above representation. In fact, a change of $Q(Y_1, \ldots, Y_n)$ by some $\Delta > 0$ can maximally change the quantification results by Δ. Hence the precise choice of the semi-fuzzy quantifier is uncritical with $\mathcal{M}_{CX}$ as well.

This means that the imprecision in the choice of numeric membership grades for quantifier and arguments is not amplified in any way when applying $\mathcal{M}_{CX}$, but either kept at its original level, or even suppressed.

Summarizing, I have managed to relate the model $\mathcal{M}_{CX}$ to Yager's suggestion of modelling fuzzy quantification by constructing a suitable logical formula [173], known as the 'substitution approach'. In addition, I have discussed some important implications that are apparent from the representation of $\mathcal{M}_{CX}$ in terms of a fuzzy propositional formula. In the following, I will continue along these lines, thus connecting the model to existing work on fuzzy quantification. To this end, I first recall the definition of the Sugeno integral, which is very closely tied to the model.

Definition 7.89.
Suppose $Q : \mathcal{P}(E) \longrightarrow \mathbf{I}$ is a nondecreasing semi-fuzzy quantifier and $X \in \widetilde{\mathcal{P}}(E)$. The Sugeno integral $(S) \int X \, dQ$ *is defined by*

$$(S) \int X \, dQ = \sup\{\min(\alpha, Q(X_{\geq\alpha})) : \alpha \in \mathbf{I}\} .$$

In fact, $\mathcal{M}_{CX}$ properly generalizes the Sugeno integral:

Theorem 7.90.
Suppose $Q : \mathcal{P}(E) \longrightarrow \mathbf{I}$ is nondecreasing. Then for all $X \in \widetilde{\mathcal{P}}(E)$,

$$(S) \int X \, dQ = \mathcal{M}_{CX}(Q)(X) .$$

Hence $\mathcal{M}_{CX}(Q)$ coincides with the well-known Sugeno integral whenever the latter is defined. Unlike the integral, though, $\mathcal{M}_{CX}$ is defined for arbitrary semi-fuzzy quantifiers, and hence extends the Sugeno integral from monotonic measures to the 'hard' cases of unrestricted multi-place and non-monotonic quantifiers.
In order to relate this result with previous approaches to fuzzy quantification, let us now formally define the quantity $\mu_{[j]}(X)$ already mentioned in the introduction.

Definition 7.91.
Let a finite base set $E \neq \varnothing$ of cardinality $|E| = m$ be given. For a fuzzy subset $X \in \widetilde{\mathcal{P}}(E)$, we denote by $\mu_{[j]}(X) \in \mathbf{I}$, $j = 1, \ldots, m$, the j-th largest membership value of X (including duplicates).
More formally, consider an ordering of the elements of E such that $E = \{e_1, \ldots, e_m\}$ and $\mu_X(e_1) \geq \cdots \geq \mu_X(e_m)$. Then define $\mu_{[j]}(X) = \mu_X(e_j)$. It is apparent that the results do not depend on the chosen ordering if ambiguities exist.
Let us further stipulate that $\mu_{[0]}(X) = 1$ and that $\mu_{[j]}(X) = 0$ whenever $j > m$.

As a corollary to the above theorem, we then obtain (cf. [21]):

Theorem 7.92.
Suppose $E \neq \varnothing$ is a finite base set, $q : \{0, \ldots, |E|\} \longrightarrow \mathbf{I}$ is a nondecreasing mapping and $Q : \mathcal{P}(E) \longrightarrow \mathbf{I}$ is defined by $Q(Y) = q(|Y|)$ for all $Y \in \mathcal{P}(E)$. Then for all $X \in \widetilde{\mathcal{P}}(E)$,

$$\mathcal{M}_{CX}(Q)(X) = \max\{\min(q(j), \mu_{[j]}(X)) : 0 \leq j \leq |E|\} .$$

Notes 7.93.

- Hence $\mathcal{M}_{CX}$ consistently generalises the basic FG-count approach of [174, 197], which is restricted to quantitative and nondecreasing one-place quantifiers.
- From a computational point of view, the theorem shows that in the special case of nondecreasing quantitative unary quantifiers on finite domains, $\mathcal{M}_{CX}(Q)(X)$ can be determined from a fuzzy cardinality measure on X, namely from the well-known FG-count(X). As will be shown later in Th. 11.18, it is possible to generalize this result to arbitrary quantitative unary quantifiers on finite domains. In this generic case, though, the FG-count must be replaced with a different measure of fuzzy cardinality, the fuzzy interval cardinality $\|X\|_{\mathrm{iv}}$, which will be introduced in Def. 11.16 below.

In closing this section on $\mathcal{M}_{CX}$, I finally discuss two additional properties of the model which make it particular suited for real-world applications. The first property is concerned with an aspect of parsimony. Typical applications of fuzzy logic do not fully exploit the available range of continuous truth values. In many situations, it is sufficient to restrict attention to a modest number of truth values/membership grades. For example, a repertoire of five possible grades

$$\Upsilon = \{0, \tfrac{1}{4}, \tfrac{1}{2}, \tfrac{3}{4}, 1\} \tag{7.25}$$

might be adequate, which correspond to the ordinal scale

'false', 'quite false', 'undecided', 'quite true', 'true'. (7.26)

The model $\mathcal{M}_{CX}$ combines well with such ordinal scales of truth values, and hence supports a parsimoneous use of membership grades. This is because $\mathcal{M}_{CX}$ does not 'invent' any new truth values, provided that the restricted scale is closed under negation:

Theorem 7.94.
Let $\Upsilon \subset \mathbf{I}$ be a given set with the following properties:

- *Υ is finite;*
- *if $v \in \Upsilon$, then $1 - v \in \Upsilon$;*
- *$\{0, 1\} \subseteq \Upsilon$.*

Further suppose that $Q : \mathcal{P}(E)^n \longrightarrow \Upsilon$ is a semi-fuzzy quantifier with quantification results in Υ and that $X_1, \ldots, X_n \in \widetilde{\mathcal{P}}(E)$ are Υ-valued fuzzy subsets of E, i.e. based on membership functions $\mu_{X_i} : E \longrightarrow \Upsilon$, $i = 1, \ldots, n$. Then it also holds that $\mathcal{M}_{CX}(Q)(X_1, \ldots, X_n) \in \Upsilon$.

Hence in the above example of the five element set Υ defined by (7.25), fuzzy quantification based on $\mathcal{M}_{CX}$ will always yield results in $\{0, \frac{1}{4}, \frac{1}{2}, \frac{3}{4}, 1\}$, provided that both the semi-fuzzy quantifier and its arguments assume values in this restricted range only.

Due to its special conformance with linguistic expectations, $\mathcal{M}_{CX}$ is the preferred choice for all applications that need to implement natural language semantics. However, the model also has practical virtues, especially its robustness against random or systematic variation in the model's inputs.

8

Models Defined in Terms of Upper and Lower Bounds on Three-Valued Cuts

8.1 Motivation and Chapter Overview

The previous chapter introduced a rich class of standard models for fuzzy quantification. In addition, we identified a distinguished model which is the preferred choice for all applications which emphasize the natural language aspect of fuzzy quantifiers and need to capture their intuitive semantics. This positive result on the model $\mathcal{M}_{CX}$ might suggest that we can stop the exploration of further models, now that the universal, perfect case of a standard DFS has been discovered. Things are not that easy, though, because there are situations in which different requirements on the models prevail over the desideratum of linguistic adequacy, and these requirements are not necessarily satisfied by the model $\mathcal{M}_{CX}$, nor by any other choice of an $\mathcal{M}_{\mathcal{B}}$-DFS. Such demands may arise, for example, in applications of fuzzy quantifiers like information aggregation, data fusion and multi-criteria decisionmaking. These applications usually involve the computation of a result ranking, or a weighting of alternatives, which in turn permits the selection of the best choice. In this kind of situation, it can be the prime concern to ensure a fine-grained differentiation of the available options, and the model of fuzzy quantification must thus be as discriminating as possible.

As will be shown below, the $\mathcal{M}_{\mathcal{B}}$-models are disadvantageous in this respect. Surprisingly, the reason for their suboptimal discriminating power is buried in the fact that every $\mathcal{M}_{\mathcal{B}}$-DFS propagates fuzziness. There is an apparent trade-off here because propagation of fuzziness captures very basic expectations on plausible models – it is simply hard to understand that the results should become more specific when the inputs (quantifier or argument) get fuzzier. However, there is a price to be paid for the propagation of fuzziness: As the input becomes less specific, the result of an $\mathcal{M}_{\mathcal{B}}$-DFS is likely to attain the least specific value of $\frac{1}{2}$, see Th. 8.26 and Th. 8.33 below. In certain applications, it might therefore be preferable to sacrifice the propagation of fuzziness, in favour of a model with enhanced discriminatory force – a model which remains useful when the input is overly fuzzy. Secondly, the investiga-

I. Glöckner: *Fuzzy Quantifiers*, StudFuzz **193**, 221–235 (2006)
www.springerlink.com © Springer-Verlag Berlin Heidelberg 2006

tion of models beyond the $\mathcal{M}_\mathcal{B}$ type will also be useful to relate existing work on fuzzy quantification to the axiomatic solution proposed in this work. The previous chapter has shown that the Sugeno integral and hence the 'basic' FG-count approach can be fitted into the proposed framework. By investigating a richer class of models, it might be possible to prove a similar result for the Choquet integral and hence the 'basic' OWA approach. Thirdly, the study of a richer class of models is of great theoretical interest, because it will improve our understanding of the full class of standard models. Specifically, the study of prototypical standard models which do not propagate fuzziness in quantifiers and/or arguments will reveal the structure and properties of models which are rather distinct from $\mathcal{M}_\mathcal{B}$-DFSes. The bottom line is that broadening the considered class of models will also broaden the knowledge about fuzzy quantification in general.

In order to achieve the desired generalization of $\mathcal{M}_\mathcal{B}$-type models, the use of the fuzzy median in the definition of $Q_\gamma(X_1, \ldots, X_n)$ will be replaced with a more general construction. We get an idea of how to proceed if we simply expand the definition of the generalized fuzzy median and rewrite $Q_\gamma(X_1, \ldots, X_n)$ as

$$\begin{aligned} Q_\gamma(X_1, \ldots, X_n) = \ & \mathrm{med}_{\frac{1}{2}}(\sup\{Q(Y_1, \ldots, Y_n) : Y_i \in \mathcal{T}_\gamma(X_i)\}, \\ & \inf\{Q(Y_1, \ldots, Y_n) : Y_i \in \mathcal{T}_\gamma(X_i)\}) . \end{aligned} \tag{8.1}$$

(This reformulation is justified by Def. 5.9 and Def. 7.18). The fuzzy median can then be replaced with other connectives, e.g. the arithmetic mean $(x + y)/2$. If we view $\sup\{Q(Y_1, \ldots, Y_n) : Y_i \in \mathcal{T}_\gamma(X_i)\}$ and $\inf\{Q(Y_1, \ldots, Y_n) : Y_i \in \mathcal{T}_\gamma(X_i)\}$ as mappings that depend on γ, then we can even eliminate the pointwise application of the connective and define more 'holistic' mechanisms.

In the subsequent chapter, we will discuss the corresponding extension of $\mathcal{M}_\mathcal{B}$-DFSes. Following the usual scheme of developing classes of models, we shall first motivate the new base construction which spans the unrestricted class of $\mathcal{F}_\xi$-QFMs. Again, the full class of raw mechanisms will then be pruned to the reasonable cases, by characterizing the $\mathcal{F}_\xi$-DFSes in terms of conditions imposed on the aggregation mapping ξ. After presenting examples of models, we will develop the necessary formal machinery to assess the relevant properties of $\mathcal{F}_\xi$-DFSes. Finally, the resulting techniques will be tested on the example models. As will emerge from this analysis, the novel class of models is rich enough to accomplish the three objectives which necessitated the quest for additional models. First of all, the two criteria of propagating fuzziness will no longer be tied to all models, and now become independent attributes which may or may not be possessed by the considered examples. The second goal will also be achieved, by identifying a prototypical model in the new class, which generalizes the Choquet integral. Finally, the introduction of the new class will also have a number of theoretical ramifications, and it sheds some light into the unknown of fuzzy quantification.

8.2 The Unrestricted Class of $\mathcal{F}_\xi$-QFMs

As mentioned above, it was decided to build the new class from a generalisation of the construction for $\mathcal{M}_\mathcal{B}$-type models. Hence let us reconsider the definition of $Q_\gamma(X_1, \ldots, X_n)$ which underlies the construction of the known models. As demonstrated by equation (8.1) above, the application of the fuzzy median connective $\mathrm{med}_{\frac{1}{2}}$ can be separated from a prior computation of the supremum and infimum $\sup\{Q(Y_1, \ldots, Y_n) : Y_i \in \mathcal{T}_\gamma(X_i)\}$ and $\inf\{Q(Y_1, \ldots, Y_n) : Y_i \in \mathcal{T}_\gamma(X_i)\}$, respectively. By abstracting from the cutting parameter γ in these expressions, we arrive at the following definition of upper and lower bound mappings, which delimit the possible quantification results in the cut ranges:

Definition 8.1.
Let a semi-fuzzy quantifier $Q : \mathcal{P}(E)^n \longrightarrow \mathbf{I}$ and fuzzy arguments $X_1, \ldots, X_n$ be given. The upper bound mapping $\top_{Q,X_1,\ldots,X_n} : \mathbf{I} \longrightarrow \mathbf{I}$ and the lower bound mapping $\bot_{Q,X_1,\ldots,X_n} : \mathbf{I} \longrightarrow \mathbf{I}$ are defined by

$$\begin{aligned}\top_{Q,X_1,\ldots,X_n}(\gamma) &= \sup\{Q(Y_1, \ldots, Y_n) : Y_1 \in \mathcal{T}_\gamma(X_1), \ldots, Y_n \in \mathcal{T}_\gamma(X_n)\}\\ \bot_{Q,X_1,\ldots,X_n}(\gamma) &= \inf\{Q(Y_1, \ldots, Y_n) : Y_1 \in \mathcal{T}_\gamma(X_1), \ldots, Y_n \in \mathcal{T}_\gamma(X_n)\}\,.\end{aligned}$$

The following properties of $\top_{Q,X_1,\ldots,X_n}$ and $\bot_{Q,X_1,\ldots,X_n}$ are apparent:

Theorem 8.2.
Suppose $Q : \mathcal{P}(E)^n \longrightarrow \mathbf{I}$ is a semi-fuzzy quantifier and $X_1, \ldots, X_n \in \widetilde{\mathcal{P}}(E)$ a choice of fuzzy arguments. Then

1. *$\top_{Q,X_1,\ldots,X_n}$ is monotonically nondecreasing;*
2. *$\bot_{Q,X_1,\ldots,X_n}$ is monotonically nonincreasing;*
3. *$\bot_{Q,X_1,\ldots,X_n} \le \top_{Q,X_1,\ldots,X_n}$.*

We can therefore define the domain $\mathbb{T}$ of aggregation operators $\xi : \mathbb{T} \longrightarrow \mathbf{I}$ which combine the results of $\top_{Q,X_1,\ldots,X_n}$ and $\bot_{Q,X_1,\ldots,X_n}$ as follows.

Definition 8.3.
$\mathbb{T} \subseteq \mathbf{I}^\mathbf{I} \times \mathbf{I}^\mathbf{I}$ is defined by

$$\mathbb{T} = \{(\top, \bot) :\ \top : \mathbf{I} \longrightarrow \mathbf{I} \textit{ nondecreasing}, \ \bot : \mathbf{I} \longrightarrow \mathbf{I} \textit{ nonincreasing}, \ \bot \le \top\}\,.$$

It is apparent from Th. 8.2 that $(\top_{Q,X_1,\ldots,X_n}, \bot_{Q,X_1,\ldots,X_n}) \in \mathbb{T}$, regardless of the semi-fuzzy quantifier $Q : \mathcal{P}(E)^n \longrightarrow \mathbf{I}$ and the chosen fuzzy arguments $X_1, \ldots, X_n \in \widetilde{\mathcal{P}}(E)$. In addition, it can be shown that $\mathbb{T}$ is the minimal set which embeds all such pairs of mappings.

Theorem 8.4.
Let $(\top, \bot) \in \mathbb{T}$ be given. We define semi-fuzzy quantifiers $Q', Q'', Q : \mathcal{P}(\mathbf{2} \times \mathbf{I}) \longrightarrow \mathbf{I}$ by

$$Q'(Y) = \top(\sup Y') \tag{8.2}$$
$$Q''(Y) = \bot(\inf Y'') \tag{8.3}$$
$$Q(Y) = \begin{cases} Q''(Y) & : & Y' = \varnothing \\ Q'(Y) & : & \textit{else} \end{cases} \tag{8.4}$$

where

$$Y' = \{z \in \mathbf{I} : (0, z) \in Y\} \tag{8.5}$$
$$Y'' = \{z \in \mathbf{I} : (1, z) \in Y\} \tag{8.6}$$

for all $Y \in \mathcal{P}(\mathbf{2} \times \mathbf{I})$.
Further suppose that the fuzzy subset $X \in \widetilde{\mathcal{P}}(\mathbf{2} \times \mathbf{I})$ *is defined by*

$$\mu_X(c, z) = \begin{cases} \frac{1}{2} - \frac{1}{2}z & : & c = 0 \\ \frac{1}{2} + \frac{1}{2}z & : & c = 1 \end{cases} \tag{8.7}$$

for all $(c, z) \in \mathbf{2} \times \mathbf{I}$.
Then $\top = \top_{Q,X}$ *and* $\bot = \bot_{Q,X}$.

Based on the aggregation operator $\xi : \mathbb{T} \longrightarrow \mathbf{I}$, we define a corresponding QFM $\mathcal{F}_\xi$ in the obvious way.

Definition 8.5.
For every mapping $\xi : \mathbb{T} \longrightarrow \mathbf{I}$*, the QFM* $\mathcal{F}_\xi$ *is defined by*

$$\mathcal{F}_\xi(Q)(X_1, \dots, X_n) = \xi(\top_{Q,X_1,\dots,X_n}, \bot_{Q,X_1,\dots,X_n})\,, \tag{8.8}$$

for all semi-fuzzy quantifiers $Q : \mathcal{P}(E)^n \longrightarrow \mathbf{I}$ *and all fuzzy subsets* $X_1, \dots, X_n \in \widetilde{\mathcal{P}}(E)$.

The class of QFMs so defined will be called the class of $\mathcal{F}_\xi$-QFMs. Apparently, it contains a number of models that do not fulfill the DFS axioms. We therefore impose five elementary conditions on the aggregation mapping ξ which provide a characterisation of the well-behaved models, i.e. of the class of $\mathcal{F}_\xi$-DFSes.

8.3 Characterisation of $\mathcal{F}_\xi$-Models

Definition 8.6. *For all* $(\top, \bot) \in \mathbb{T}$*, we impose the following conditions on aggregation mappings* $\xi : \mathbb{T} \longrightarrow \mathbf{I}$.

$$\textit{If } \top = \bot, \textit{ then } \xi(\top, \bot) = \top(0) \tag{X-1}$$
$$\xi(1 - \bot, 1 - \top) = 1 - \xi(\top, \bot) \tag{X-2}$$
$$\textit{If } \top = \mathrm{c}_1 \textit{ and } \bot(\mathbf{I}) \subseteq \{0, 1\}, \textit{ then } \xi(\top, \bot) = \tfrac{1}{2} + \tfrac{1}{2}\bot_*^0 \tag{X-3}$$
$$\xi(\top^\flat, \bot) = \xi(\top^\sharp, \bot) \tag{X-4}$$
$$\textit{If } (\top', \bot') \in \mathbb{T} \textit{ such that } \top \le \top' \textit{ and } \bot \le \bot', \textit{ then } \xi(\top, \bot) \le \xi(\top', \bot') \tag{X-5}$$

As stated in the following theorems, the conditions imposed on ξ capture exactly the requirements that make $\mathcal{F}_\xi$ a DFS. Let us show first that (X-1) to (X-5) are sufficient for $\mathcal{F}_\xi$ to be a standard model.

Theorem 8.7.
If $\xi : \mathbb{T} \longrightarrow \mathbf{I}$ satisfies (X-1) to (X-5), then $\mathcal{F}_\xi$ is a standard DFS.

Theorem 8.8.
The conditions (X-1) to (X-5) on $\xi : \mathbb{T} \longrightarrow \mathbf{I}$ are necessary for $\mathcal{F}_\xi$ to be a DFS.

Hence the 'X-conditions' are necessary and sufficient for $\mathcal{F}_\xi$ to be a DFS, and all $\mathcal{F}_\xi$-DFSes are indeed standard DFSes. The criteria can also be shown to be independent. To carry out the independence proof, it was useful to relate $\mathcal{M}_\mathcal{B}$-QFMs to the broader class of $\mathcal{F}_\xi$-QFMs:

Theorem 8.9.
Suppose $\mathcal{B} : \mathbb{B} \longrightarrow \mathbf{I}$ is a given aggregation mapping. Then $\mathcal{M}_\mathcal{B} = \mathcal{F}_\xi$, where $\xi : \mathbb{T} \longrightarrow \mathbf{I}$ is defined by

$$\xi(\top, \bot) = \mathcal{B}(\mathrm{med}_{\frac{1}{2}}(\top, \bot)) \tag{8.9}$$

for all $(\top, \bot) \in \mathbb{T}$, and $\mathrm{med}_{\frac{1}{2}}(\top, \bot)$ abbreviates

$$\mathrm{med}_{\frac{1}{2}}(\top, \bot)(\gamma) = \mathrm{med}_{\frac{1}{2}}(\top(\gamma), \bot(\gamma)),$$

for all $\gamma \in \mathbf{I}$.

Hence all $\mathcal{M}_\mathcal{B}$-QFMs are $\mathcal{F}_\xi$-QFMs, and all $\mathcal{M}_\mathcal{B}$-DFSes are $\mathcal{F}_\xi$-DFSes. Sometimes we should be aware of the relationship between the 'B-conditions' and the 'X-conditions' in the case of $\mathcal{M}_\mathcal{B}$-QFMs. The next theorem helps us to show that the 'X-conditions' are independent, because the 'B-conditions' have already been shown to be independent in [55]:

Theorem 8.10.
Suppose $\mathcal{B} : \mathbb{B} \longrightarrow \mathbf{I}$ is given and $\xi : \mathbb{T} \longrightarrow \mathbf{I}$ is defined by equation (8.9). Then

1. (B-1) *is equivalent to* (X-1);
2. (B-2) *is equivalent to* (X-2);
3. *a)* (B-3) *entails* (X-3);
   b) the conjunction of (X-2) *and* (X-3) *entails* (B-3);
4. *a)* (B-4) *entails* (X-4);
   b) the conjunction of (X-2) *and* (X-4) *entails* (B-4);
5. (B-5) *is equivalent to* (X-5).

Note 8.11. The theorem also proved useful in other contexts, e.g. to show that the $\mathcal{M}_\mathcal{B}$-DFSes are exactly those $\mathcal{F}_\xi$-DFSes that propagate fuzziness in both quantifiers and arguments.

Theorem 8.12.
The conditions (X-1) to (X-5) are independent.

8.4 Examples of $\mathcal{F}_\xi$-Models

Let us now give examples of 'genuine' $\mathcal{F}_\xi$-DFSes (i.e. models that go beyond the special case of $\mathcal{M}_\mathcal{B}$-DFSes).

Definition 8.13.
The QFM $\mathcal{F}_{\mathrm{owa}} = \mathcal{F}_{\xi_{\mathrm{owa}}}$ *is defined in terms of* $\xi_{\mathrm{owa}} : \mathbb{T} \longrightarrow \mathbf{I}$ *by*

$$\xi_{\mathrm{owa}}(\top, \bot) = \tfrac{1}{2} \int_0^1 \top(\gamma)\,\mathrm{d}\gamma + \tfrac{1}{2} \int_0^1 \bot(\gamma)\,\mathrm{d}\gamma\,,$$

for all $(\top, \bot) \in \mathbb{T}$.

Note 8.14. Both integrals are guaranteed to exist because $\top$ and $\bot$ are monotonic and bounded.

Theorem 8.15.
$\mathcal{F}_{\mathrm{owa}}$ *is a standard DFS.*

The DFS $\mathcal{F}_{\mathrm{owa}}$ is of special interest because of its close relationship to the well-known Choquet integral.

Definition 8.16.
Suppose $Q : \mathcal{P}(E) \longrightarrow \mathbf{I}$ *is a nondecreasing semi-fuzzy quantifier and* $X \in \widetilde{\mathcal{P}}(E)$. *The* Choquet integral $(Ch) \int X\,dQ$ *is defined by*

$$(Ch) \int X\,dQ = \int_0^1 Q(X_{\geq\alpha})\,\mathrm{d}\alpha\,.$$

Theorem 8.17.
Suppose $Q : \mathcal{P}(E) \longrightarrow \mathbf{I}$ *is nondecreasing. Then for all* $X \in \widetilde{\mathcal{P}}(E)$,

$$(Ch) \int X\,dQ = \mathcal{F}_{\mathrm{owa}}(Q)(X)\,.$$

Hence $\mathcal{F}_{\mathrm{owa}}$ coincides with the Choquet integral on fuzzy quantifiers whenever the latter is defined. Recalling the notation $\mu_{[j]}(X)$ introduced in Def. 7.91, we obtain the following corollary to the above theorem (cf. [21]):

Theorem 8.18.
Suppose $E \neq \varnothing$ *is a finite base set,* $q : \{0, \ldots, |E|\} \longrightarrow \mathbf{I}$ *is a nondecreasing mapping such that* $q(0) = 0$, $q(|E|) = 1$, *and* $Q : \mathcal{P}(E) \longrightarrow \mathbf{I}$ *is defined by* $Q(Y) = q(|Y|)$ *for all* $Y \in \mathcal{P}(E)$. *Then for all* $X \in \widetilde{\mathcal{P}}(E)$,

$$\mathcal{F}_{\mathrm{owa}}(Q)(X) = \sum_{j=1}^{|E|} (q(j) - q(j-1)) \cdot \mu_{[j]}(X)\,,$$

i.e. $\mathcal{F}_{\mathrm{owa}}$ *consistently generalises Yager's OWA approach [179].*

Definition 8.19.
The QFM $\mathcal{F}_S$ is defined in terms of $\xi_S : \mathbb{T} \longrightarrow \mathbf{I}$ by

$$\xi_S(\top, \bot) = \begin{cases} \min(\top_1^*, \frac{1}{2} + \frac{1}{2}\bot_*^{\leq \frac{1}{2}}) & : \quad \bot(0) > \frac{1}{2} \\ \max(\bot_1^*, \frac{1}{2} - \frac{1}{2}\top_*^{\geq \frac{1}{2}}) & : \quad \top(0) < \frac{1}{2} \\ \frac{1}{2} & : \quad else \end{cases}$$

for all $(\top, \bot) \in \mathbb{T}$, where the coefficients $f_*^{\leq \frac{1}{2}}, f_*^{\geq \frac{1}{2}} \in \mathbf{I}$ are defined by

$$f_*^{\leq \frac{1}{2}} = \inf\{\gamma \in \mathbf{I} : f(\gamma) \leq \tfrac{1}{2}\} \tag{8.10}$$

$$f_*^{\geq \frac{1}{2}} = \inf\{\gamma \in \mathbf{I} : f(\gamma) \geq \tfrac{1}{2}\}, \tag{8.11}$$

for all $f : \mathbf{I} \longrightarrow \mathbf{I}$.

Theorem 8.20.
$\mathcal{F}_S$ is a standard DFS.

A third model of interest is the following QFM $\mathcal{F}_A$:

Definition 8.21.
The QFM $\mathcal{F}_A$ is defined in terms of $\xi_A : \mathbb{T} \longrightarrow \mathbf{I}$ by

$$\xi_A(\top, \bot) = \begin{cases} \min(\bot_0^*, \frac{1}{2} + \frac{1}{2}\bot_*^0) & : \quad \bot_0^* > \frac{1}{2} \\ \max(\top_0^*, \frac{1}{2} - \frac{1}{2}\top_*^{1\downarrow}) & : \quad \top_0^* < \frac{1}{2} \\ \frac{1}{2} & : \quad else \end{cases}$$

for all $(\top, \bot) \in \mathbb{T}$.

Theorem 8.22.
$\mathcal{F}_A$ is a standard DFS.

8.5 Properties of the $\mathcal{F}_\xi$-Models

Turning to properties of $\mathcal{F}_\xi$-DFSes, let us first investigate the precise conditions under which an $\mathcal{F}_\xi$-DFS propagates fuzziness in quantifiers and/or in arguments.

Definition 8.23.
We say that $\xi : \mathbb{T} \longrightarrow \mathbf{I}$ propagates fuzziness *if and only if*

$$\xi(\top, \bot) \preceq_c \xi(\top', \bot')$$

whenever $(\top, \bot), (\top', \bot') \in \mathbb{T}$ with $\top \preceq_c \top'$ and $\bot \preceq_c \bot'$.

In particular, the two conditions of propagating fuzziness are independent in the case of $\mathcal{F}_\xi$-DFSes, as stated in the following corollary.

Theorem 8.38.
The conditions of propagating fuzziness in quantifiers and in arguments are independent for $\mathcal{F}_\xi$-DFSes.

Finally, we can characterize the subclass of $\mathcal{M}_\mathcal{B}$-DFSes which are exactly those $\mathcal{F}_\xi$-DFSes that propagate fuzziness in both quantifiers and arguments.

Theorem 8.39.
Suppose an $\mathcal{F}_\xi$-DFS propagates fuzziness in both quantifiers and arguments. Then $\mathcal{F}_\xi$ is an $\mathcal{M}_\mathcal{B}$-DFS.

Note 8.40. The converse implication is already known from Th. 7.75 and Th. 7.76.

Next we shall investigate the exact conditions under which an $\mathcal{F}_\xi$-QFM is Q-continuous or arg-continuous. To be able to discuss Q-continuous $\mathcal{F}_\xi$-QFMs, we introduce a metric $d : \mathbb{T} \times \mathbb{T} \longrightarrow \mathbf{I}$. For all nondecreasing mappings $\top, \top' : \mathbf{I} \longrightarrow \mathbf{I}$, we define

$$d(\top, \top') = \sup\{|\top(\gamma) - \top'(\gamma)| : \gamma \in \mathbf{I}\} . \tag{8.12}$$

We proceed similarly for nondecreasing mappings $\bot, \bot' : \mathbf{I} \longrightarrow \mathbf{I}$. In this case,

$$d(\bot, \bot') = \sup\{|\bot(\gamma) - \bot'(\gamma)| : \gamma \in \mathbf{I}\} . \tag{8.13}$$

Finally, we define $d : \mathbb{T} \times \mathbb{T} \longrightarrow \mathbf{I}$ by

$$d((\top, \bot), (\top', \bot')) = \max(d(\top, \top'), d(\bot, \bot')) , \tag{8.14}$$

for all $(\top, \bot), (\top', \bot') \in \mathbb{T}$. It is apparent that d is indeed a metric. We will utilize d to express a condition on ξ which characterises the Q-continuous $\mathcal{F}_\xi$-QFMs.

Theorem 8.41.
Let $\xi : \mathbb{T} \longrightarrow \mathbf{I}$ be a given mapping which satisfies (X-5). *Then the following conditions are equivalent:*

a. *$\mathcal{F}_\xi$ is Q-continuous;*
b. *for all $\varepsilon > 0$, there exists $\delta > 0$ such that $|\xi(\top, \bot) - \xi(\top', \bot')| < \varepsilon$ whenever $(\top, \bot), (\top', \bot') \in \mathbb{T}$ satisfy $d((\top, \bot), (\top', \bot')) < \delta$.*

If ξ is sufficiently well-behaved, then the above condition can be simplified to the following criterion, which is easier to check.

Theorem 8.42.
Suppose $\xi : \mathbb{T} \longrightarrow \mathbf{I}$ satisfies (X-2) *and* (X-5). *Then the following conditions are equivalent:*

a. $\mathcal{F}_\xi$ *is Q-continuous;*
b. for all $\varepsilon > 0$*, there exists* $\delta > 0$ *such that* $\xi(\top', \bot) - \xi(\top, \bot) < \varepsilon$ *whenever* $(\top, \bot), (\top', \bot) \in \mathbb{T}$ *satisfy* $d(\top, \top') < \delta$ *and* $\top \leq \top'$.

I have the following results for the examples of $\mathcal{F}_\xi$-DFSes.

Theorem 8.43.
$\mathcal{F}_{\mathrm{owa}}$ *is Q-continuous.*

Theorem 8.44.
$\mathcal{F}_S$ *is not Q-continuous.*

Theorem 8.45.
$\mathcal{F}_A$ *is not Q-continuous.*

As concerns continuity in arguments, we first need to introduce another distance measure $d' : \mathbb{T} \times \mathbb{T} \longrightarrow \mathbf{I}$, which can be used to characterise the arg-continuous $\mathcal{F}_\xi$-QFMs in terms of conditions on ξ. For all nondecreasing mappings $\top, \top' : \mathbf{I} \longrightarrow \mathbf{I}$, we define

$$d'(\top, \top') = \sup\{\inf\{\gamma' \min(\top(\gamma'), \top'(\gamma')) \geq \max(\top(\gamma), \top'(\gamma))\} - \gamma : \gamma \in \mathbf{I}\} . \tag{8.15}$$

Similarly for nonincreasing mappings $\bot, \bot' : \mathbf{I} \longrightarrow \mathbf{I}$,

$$d'(\bot, \bot') = \sup\{\inf\{\gamma' : \max(\bot(\gamma'), \bot'(\gamma')) \leq \min(\bot(\gamma), \bot'(\gamma))\} - \gamma : \gamma \in \mathbf{I}\} . \tag{8.16}$$

Finally, we define $d' : \mathbb{T} \times \mathbb{T} \longrightarrow \mathbf{I}$ by

$$d'((\top, \bot), (\top', \bot')) = \max(d'(\top, \top'), d'(\bot, \bot')) , \tag{8.17}$$

for all $(\top, \bot), (\top', \bot') \in \mathbb{T}$. It is easily checked that d' is a 'pseudo-metric', i.e. it is symmetric and satisfies the triangle inequality, but $d'((\top, \bot), (\top', \bot')) = 0$ does not imply that $(\top, \bot) = (\top', \bot')$. However, d' is a metric modulo $\sharp\flat$, i.e. on the equivalence classes of $(\top, \bot) \sim (\top', \bot') \Leftrightarrow (\top^{\flat\sharp}, \bot^{\flat\sharp}) = (\top'^{\flat\sharp}, \bot'^{\flat\sharp})$. Hence $d'((\top, \bot), (\top', \bot')) = 0$ entails that $(\top, \bot) \sim (\top', \bot')$, i.e. $\xi(\top, \bot) = \xi(\top', \bot')$ whenever ξ satisfies (X-2), (X-4) and (X-5). Based on d', we can now assert the following.

Theorem 8.46.
Suppose $\xi : \mathbb{T} \longrightarrow \mathbf{I}$ *satisfies* (X-2), (X-4) *and* (X-5)*. Then the following conditions are equivalent:*

a. $\mathcal{F}_\xi$ *is arg-continuous;*
b. for all $(\top, \bot) \in \mathbb{T}$ *and all* $\varepsilon > 0$*, there exists* $\delta > 0$ *such that* $|\xi(\top, \bot) - \xi(\top', \bot')| < \varepsilon$ *whenever* $(\top', \bot') \in \mathbb{T}$ *satisfies* $d'((\top, \bot), (\top', \bot')) < \delta$.

In some cases, the following sufficient condition can shorten the proof that a given $\mathcal{F}_\xi$ is arg-continuous.

Theorem 8.47.
Suppose $\xi : \mathbb{T} \longrightarrow \mathbf{I}$ satisfies (X-2) *and* (X-5) *Then $\mathcal{F}_\xi$ is arg-continuous if the following condition holds: For all $\varepsilon > 0$ there exists $\delta > 0$ such that $\xi(\top', \bot) - \xi(\top, \bot) < \varepsilon$ whenever $(\top, \bot), (\top', \bot) \in \mathbb{T}$ satisfy $d'(\top, \top') < \delta$ and $\top \leq \top'$.*

Based on these theorems, it is easy to prove the following.

Theorem 8.48.
$\mathcal{F}_{\mathrm{owa}}$ is arg-continuous.

Theorem 8.49.
$\mathcal{F}_S$ is not arg-continuous.

Theorem 8.50.
$\mathcal{F}_A$ is not arg-continuous.

Hence $\mathcal{F}_{\mathrm{owa}}$ is continuous both in quantifiers and arguments; which is important for applications. The second example, $\mathcal{F}_S$, fails at both continuity conditions and is thus not practical. (We will see below that $\mathcal{F}_S$ is of theoretical interest because it represents a boundary case of $\mathcal{F}_\xi$-DFS).
Let us also investigate the specificity of $\mathcal{F}_\xi$-DFSes. The following theorem facilitates the proof that a given $\mathcal{F}_\xi$-QFM is less specific than another $\mathcal{F}_\xi$-QFM by relating the specificity order on $\mathcal{F}_\xi$ to the specificity order on ξ:

Theorem 8.51.
Let $\xi, \xi' : \mathbb{T} \longrightarrow \mathbf{I}$ be given mappings. Then the following conditions are equivalent:

a. $\mathcal{F}_\xi \preceq_c \mathcal{F}_{\xi'}$;
b. $\xi \preceq_c \xi'$.

In the case of $\mathcal{F}_\xi$-models that propagate fuzziness in quantifiers, it is sufficient to check a simpler condition.

Theorem 8.52.
Let $\xi, \xi' : \mathbb{T} \longrightarrow \mathbf{I}$ be given mappings which satisfy (X-1) *to* (X-5) *and suppose that ξ, ξ' have the additional property that $\xi(\top, \bot) = \xi'(\top, \bot) = \frac{1}{2}$ whenever $(\top, \bot) \in \mathbb{T}$ with $\top(0) \geq \frac{1}{2} \geq \bot(0)$. Then the following conditions are equivalent:*

a. $\mathcal{F}_\xi \preceq_c \mathcal{F}_{\xi'}$;
b. for all $(\top, \bot) \in \mathbb{T}$ *with* $\bot(0) > \frac{1}{2}$, $\xi(\top, \bot) \leq \xi'(\top, \bot)$.

As regards least specific $\mathcal{F}_\xi$-DFSes, we can prove the following:

Theorem 8.53.
$\mathcal{M}_U$ is the least specific $\mathcal{F}_\xi$-DFS.

Turning to the issue of most specific models, let us first state a theorem for establishing or rejecting specificity consistency. This is useful because specificity consistency is tightly coupled to the existence of least upper specificity bounds, see Th. 5.13.

Theorem 8.54.
Consider a pair of mappings $\xi, \xi' : \mathbb{T} \longrightarrow \mathbf{I}$. *The QFMs* $\mathcal{F}_\xi$ *and* $\mathcal{F}_{\xi'}$ *are specificity consistent if and only if* ξ, ξ' *are specificity consistent, i.e. for all* $(\top, \bot) \in \mathbb{T}$, *either* $\{\xi(\top, \bot), \xi'(\top, \bot)\} \subseteq [0, \frac{1}{2}]$ *or* $\{\xi(\top, \bot), \xi'(\top, \bot)\} \subseteq [\frac{1}{2}, 1]$.

An investigation of a possible most specific $\mathcal{F}_\xi$-DFS reveals the following.

Theorem 8.55.
The class of $\mathcal{F}_\xi$*-DFSes is not specificity consistent.*

Hence by Th. 5.13, a 'most specific $\mathcal{F}_\xi$-DFS' does not exist. However, we obtain a positive result if we restrict attention to the class of $\mathcal{F}_\xi$-DFSes which propagate fuzziness in quantifiers or arguments. This is apparent from the following observation.

Theorem 8.56.
Suppose $\mathbb{F}$ *is a collection of* $\mathcal{F}_\xi$*-DFSes* $\mathcal{F}_\xi \in \mathbb{F}$ *with the property that* $\xi(\top, \bot) = \frac{1}{2}$ *whenever* $(\top, \bot) \in \mathbb{T}$ *is such that* $\top(0) \geq \frac{1}{2} \geq \bot(0)$. *Then* $\mathbb{F}$ *is specificity consistent.*

We then have the following corollaries.

Theorem 8.57.
The class of $\mathcal{F}_\xi$*-DFSes that propagate fuzziness in quantifiers is specificity consistent.*

Theorem 8.58.
The class of $\mathcal{F}_\xi$*-DFSes that propagate fuzziness in arguments is specificity consistent.*

By Th. 5.13, the $\mathcal{F}_\xi$-models that propagate fuzziness in quantifiers have a least upper specificity bound which, as it turns out, also propagates fuzziness in quantifiers.

Theorem 8.59.
$\mathcal{F}_S$ *is the most specific* $\mathcal{F}_\xi$*-DFS that propagates fuzziness in quantifiers.*

Similarly, we can conclude from Th. 8.58 that there is a most specific $\mathcal{F}_\xi$-DFS that propagates fuzziness in arguments.

Theorem 8.60.
$\mathcal{F}_A$ *is the most specific* $\mathcal{F}_\xi$*-DFS that propagates fuzziness in arguments.*

9

The Full Class of Models Defined in Terms of Three-Valued Cuts

9.1 Motivation and Chapter Overview

In this chapter, a further step will be taken to extend the class of known models. While the construction of $\mathcal{F}_\xi$-QFM is based on upper and lower bounds on the results obtained for the three-valued cuts and a subsequent aggregation step, we will now define fuzzification mechanisms directly in terms of these supervaluation sets, to which an aggregation mapping Ω is then applied. In order to identify the useful models in this class, we then make explicit the necessary and sufficient conditions on the aggregation mapping Ω that make $\mathcal{F}_\Omega$ a DFS. In addition, the required theory will be developed which permits us to check interesting properties of $\mathcal{F}_\Omega$-models. It is shown that the new class of models is genuinely broader than $\mathcal{F}_\xi$-DFSes. However, the new class does not introduce any new 'practical' models because those $\mathcal{F}_\Omega$-DFSes which are Q-continuous (and thus of potential interest to applications) are in fact $\mathcal{F}_\xi$-DFSes. These findings justify the introduction of $\mathcal{F}_\xi$-QFMs. It will also be shown that the full class of standard models which propagate fuzziness both in quantifiers and arguments is genuinely broader than the class of $\mathcal{M}_\mathcal{B}$-DFSes. But again, all models outside the known range of models are not Q-continuous. We shall further investigate a subclass of $\mathcal{F}_\Omega$-QFMs, the class of $\mathcal{F}_\omega$-QFMs. These QFMs can be expressed in terms of a simpler construction which excludes some of the original $\mathcal{F}_\Omega$-QFMs. However, the considered subclass still contains all well-behaved models, i.e. the $\mathcal{F}_\Omega$-DFSes and $\mathcal{F}_\omega$-DFSes coincide. The $\mathcal{F}_\omega$ type will be needed later on to establish the relationship between the models defined in terms of three-valued cuts and those defined in terms of the extension principle.

9.2 The Unrestricted Class of $\mathcal{F}_\Omega$-QFMs

Let us now extend the class of $\mathcal{F}_\xi$-QFMs to the full class of QFMs definable in terms of three-valued cuts of the argument sets. Hence consider a semi-fuzzy

I. Glöckner: *Fuzzy Quantifiers*, StudFuzz **193**, 237–258 (2006)
www.springerlink.com
© Springer-Verlag Berlin Heidelberg 2006

quantifier $Q : \mathcal{P}(E)^n \longrightarrow \mathbf{I}$ and a choice of fuzzy arguments $X_1, \ldots, X_n \in \widetilde{\mathcal{P}}(E)$. In order to spot a starting point for the desired generalization, we recall the definition of $\top_{Q,X_1,\ldots,X_n}$ and $\bot_{Q,X_1,\ldots,X_n}$. Apparently, the upper and lower bound mappings can be decomposed into (a) the three-valued cut mechanism, and (b) a subsequent inf/sup-based aggregation:

$$\begin{aligned} &\top_{Q,X_1,\ldots,X_n}(\gamma) \\ &= \sup\{Q(Y_1, \ldots, Y_n) : (Y_1, \ldots, Y_n) \in \mathcal{T}_\gamma(X_1, \ldots, X_n)\} \\ &= \sup \mathcal{S}_{Q,X_1,\ldots,X_n}(\gamma) \end{aligned} \tag{9.1}$$

and

$$\begin{aligned} &\bot_{Q,X_1,\ldots,X_n}(\gamma) \\ &= \inf\{Q(Y_1, \ldots, Y_n) : (Y_1, \ldots, Y_n) \in \mathcal{T}_\gamma(X_1, \ldots, X_n)\} \\ &= \inf \mathcal{S}_{Q,X_1,\ldots,X_n}(\gamma) \end{aligned} \tag{9.2}$$

for all $\gamma \in \mathbf{I}$, provided we define the supervaluation set $\mathcal{S}_{Q,X_1,\ldots,X_n}(\gamma)$ as follows.

Definition 9.1.
For every semi-fuzzy quantifier $Q : \mathcal{P}(E)^n \longrightarrow \mathbf{I}$ and all $X_1, \ldots, X_n \in \widetilde{\mathcal{P}}(E)$, the mapping $\mathcal{S}_{Q,X_1,\ldots,X_n} : \mathbf{I} \longrightarrow \mathcal{P}(\mathbf{I})$ is defined by

$$\mathcal{S}_{Q,X_1,\ldots,X_n}(\gamma) = \{Q(Y_1, \ldots, Y_n) : (Y_1, \ldots, Y_n) \in \mathcal{T}_\gamma(X_1, \ldots, X_n)\},$$

for all $\gamma \in \mathbf{I}$.

Some basic properties of the set of all supervaluation results $\mathcal{S}_{Q,X_1,\ldots,X_n}$ are stated in this theorem.

Theorem 9.2.
Consider a semi-fuzzy quantifier $Q : \mathcal{P}(E)^n \longrightarrow \mathbf{I}$ and choice of fuzzy subsets $X_1, \ldots, X_n \in \widetilde{\mathcal{P}}(E)$. Then

a. $\mathcal{S}_{Q,X_1,\ldots,X_n}(0) \neq \varnothing$;
b. $\mathcal{S}_{Q,X_1,\ldots,X_n}(\gamma) \subseteq \mathcal{S}_{Q,X_1,\ldots,X_n}(\gamma')$ whenever $\gamma, \gamma' \in \mathbf{I}$ with $\gamma \leq \gamma'$.

It is thus apparent that all possible choices of $\mathcal{S}_{Q,X_1,\ldots,X_n}$ are contained in the following set $\mathbb{K}$.

Definition 9.3.
$\mathbb{K} \subseteq \mathcal{P}(\mathbf{I})^{\mathbf{I}}$ is defined by

$$\mathbb{K} = \{\mathcal{S} \in \mathcal{P}(\mathbf{I})^{\mathbf{I}} : \mathcal{S}(0) \neq \varnothing \text{ and } \mathcal{S}(\gamma) \subseteq \mathcal{S}(\gamma') \text{ whenever } \gamma \leq \gamma'\}.$$

As we shall now state, $\mathbb{K}$ is the minimal set which contains all possible choices for $\mathcal{S}_{Q,X_1,\ldots,X_n}$. To this end, we first have to introduce coefficients $s(z) \in \mathbf{I}$ associated with $\mathcal{S} \in \mathbb{K}$, which will play an essential role throughout this chapter.

Definition 9.4.
Consider $S \in \mathbb{K}$. We associate with S a mapping $s : \mathbf{I} \longrightarrow \mathbf{I}$ defined by

$$s(z) = \inf\{\gamma \in \mathbf{I} : z \in S(\gamma)\}\,,$$

for all $z \in \mathbf{I}$.

It is convenient to define a notation for the $s(z)$'s obtained from a given quantifier and arguments.

Definition 9.5.
For every semi-fuzzy quantifier $Q : \mathcal{P}(E)^n \longrightarrow \mathbf{I}$ and all $X_1, \ldots, X_n \in \widetilde{\mathcal{P}}(E)$, we denote the mapping s obtained from $S_{Q,X_1,\ldots,X_n}$ by applying Def. 9.4 by $s_{Q,X_1,\ldots,X_n} : \mathbf{I} \longrightarrow \mathbf{I}$. The resulting mapping is thus defined by

$$s_{Q,X_1,\ldots,X_n}(z) = \inf\{\gamma \in \mathbf{I} : z \in S_{Q,X_1,\ldots,X_n}(\gamma)\}\,,$$

for all $z \in \mathbf{I}$.

As we shall see later, all $\mathcal{F}_\Omega$-DFSes can be defined in terms of $s_{Q,X_1,\ldots,X_n}$.

Theorem 9.6.
Let $S \in \mathbb{K}$ be given and define $Q : \mathcal{P}(\mathbf{2} \times \mathbf{I}) \longrightarrow \mathbf{I}$ by

$$Q(Y) = Q_{\inf Y'}(Y'') \tag{9.3}$$

for all $Y \in \mathcal{P}(\mathbf{2} \times \mathbf{I})$, where

$$Y' = \{y \in \mathbf{I} : (0, y) \in Y\} \tag{9.4}$$
$$Y'' = \{y \in \mathbf{I} : (1, y) \in Y\} \tag{9.5}$$

and the $Q_z : \mathcal{P}(\mathbf{I}) \longrightarrow \mathbf{I}$, $z \in \mathbf{I}$ are defined by

$$Q_z(Y'') = \begin{cases} z & : \sup Y'' > s(z) \\ z_0 & : \text{else} \end{cases} \tag{9.6}$$

for all $Y'' \in \mathcal{P}(\mathbf{I})$ if $z \notin S(s(z))$, and

$$Q_z(Y'') = \begin{cases} z & : \sup Y'' \geq s(z) \\ z_0 & : \text{else} \end{cases} \tag{9.7}$$

in the case that $z \in S(s(z))$. z_0 is an arbitrary element

$$z_0 \in S(0)\,, \tag{9.8}$$

which exists by Th. 9.2. Further suppose that $X \in \widetilde{\mathcal{P}}(\mathbf{2} \times \mathbf{I})$ is defined by

$$\mu_X(a, y) = \begin{cases} \frac{1}{2} & : a = 0 \\ \frac{1}{2} - \frac{1}{2}y & : a = 1 \end{cases} \tag{9.9}$$

for all $a \in \mathbf{2}$, $y \in \mathbf{I}$. Then $S_{Q,X} = S$.

Hence $\mathbb{K}$ is exactly the set of all $S = S_{Q,X_1,\dots,X_n}$ obtained for arbitrary choices of quantifiers and arguments. In order to obtain a quantification result from $S_{Q,X_1,\dots,X_n}$, we will apply an aggregation operator $\Omega : \mathbb{K} \longrightarrow \mathbf{I}$ in the obvious way.

Definition 9.7.
Consider an aggregation operator $\Omega : \mathbb{K} \longrightarrow \mathbf{I}$. The corresponding QFM $\mathcal{F}_\Omega$ is defined by

$$\mathcal{F}_\Omega(Q)(X_1, \dots, X_n) = \Omega(S_{Q,X_1,\dots,X_n}),$$

for all semi-fuzzy quantifiers $Q : \mathcal{P}(E)^n \longrightarrow \mathbf{I}$ and fuzzy arguments $X_1, \dots, X_n \in \widetilde{\mathcal{P}}(E)$.

By the class of $\mathcal{F}_\Omega$-QFMs, we denote the collection of all QFMs defined in this way. As usual, we must impose conditions to shrink the full class of $\mathcal{F}_\Omega$ to its subclass of $\mathcal{F}_\Omega$-DFSes.

9.3 Characterisation of the $\mathcal{F}_\Omega$-Models

Definition 9.8.
For all $S \in \mathbb{K}$, we define $S^\sharp, S^\flat \in \mathbb{K}$ as follows.

$$S^\sharp = \begin{cases} \cap\{S(\gamma') : \gamma' > \gamma\} & : \gamma < 1 \\ \mathbf{I} & : \gamma = 1 \end{cases} \qquad S^\flat = \begin{cases} S(0) & : \gamma = 0 \\ \cup\{S(\gamma') : \gamma' < \gamma\} & : \gamma > 0 \end{cases}$$

for all $\gamma \in \mathbf{I}$.

Note 9.9. The definition is slightly asymmetric; we have departed from the usual scheme of defining $S^\sharp(1) = S(1)$ in this case because the present definition of $S^\sharp(1) = \mathbf{I}$ allows more compact conditions on Ω, and eventually for shorter proofs.

Let us further stipulate a definition of $S \sqsubseteq S'$ which will serve to express a monotonicity condition on Ω.

Definition 9.10.
For all $S, S' \in \mathbb{K}$, let us say that $S \sqsubseteq S'$ if and only if the following two conditions are valid for all $\gamma \in \mathbf{I}$:

1. *for all $z \in S(\gamma)$, there exists $z' \in S'(\gamma)$ with $z' \geq z$;*
2. *for all $z' \in S'(\gamma)$, there exists $z \in S(\gamma)$ with $z \leq z'$.*

It is apparent from this definition that $\sqsubseteq$ is reflexive and transitive, but not necessarily antisymmetric (i.e. $S \sqsubseteq S'$ and $S' \sqsubseteq S$ does not imply that $S = S'$). Hence $\sqsubseteq$ is a preorder.

We are now ready to state the conditions on reasonable choices of $\Omega : \mathbb{K} \longrightarrow \mathbf{I}$, in analogy to the conditions (B-1)–(B-5) for $\mathcal{M}_\mathcal{B}$-models and to the conditions (X-1)–(X-5) for the $\mathcal{F}_\xi$-type.

Definition 9.11.
Consider $\Omega : \mathbb{K} \longrightarrow \mathbf{I}$. We impose the following conditions on Ω. For all $S \in \mathbb{K}$,

If there exists $a \in \mathbf{I}$ with $S(\gamma) = \{a\}$ for all $\gamma \in \mathbf{I}$, then $\Omega(S) = a$. (Ω-1)

If $S'(\gamma) = \{1 - z : z \in S(\gamma)\}$ for all $\gamma \in \mathbf{I}$, then $\Omega(S') = 1 - \Omega(S)$. ($\Omega$-2)

If $1 \in S(0)$ and $S(\gamma) \subseteq \{0,1\}$ for all $\gamma \in \mathbf{I}$, then $\Omega(S) = \frac{1}{2} + \frac{1}{2}s(0)$. ($\Omega$-3)

$\Omega(S) = \Omega(S^\sharp)$ (Ω-4)

If $S' \in \mathbb{K}$ satisfies $S \sqsubseteq S'$, then $\Omega(S) \leq \Omega(S')$. ($\Omega$-5)

Note 9.12. The only condition which is slighly different from the usual scheme is (Ω-4). The departure from requiring $\Omega(S^\sharp) = \Omega(S^\flat)$ turned out to shorten the proofs. The latter equation is entailed by the above conditions, however.

Theorem 9.13.
The conditions (Ω-1)–(Ω-5) on $\Omega : \mathbb{K} \longrightarrow \mathbf{I}$ are sufficient for $\mathcal{F}_\Omega$ to be a standard DFS.

In the following, we consider another construction which elucidates the exact properties of $S \in \mathbb{K}$ that a conforming choice of Ω can rely on.

Definition 9.14.
For all $S \in \mathbb{K}$, we define $S^\ddagger \in \mathbb{K}$ by

$$S^\ddagger(\gamma) = \{z \in \mathbf{I} : \textit{there exist } z', z'' \in S(\gamma) \textit{ with } z' \leq z \leq z''\}$$

for all $\gamma \in \mathbf{I}$.

Note 9.15. It is apparent that indeed $S^\ddagger \in \mathbb{K}$. The effect of applying ‡ to S is that of 'filling the gaps' in the interior of S. The resulting $S^\ddagger$ will be a closed, half-open, or open interval.

The importance of this construction with respect to $\mathcal{F}_\Omega$-QFMs stems from the invariance of well-behaved $\mathcal{F}_\Omega$-QFMs with respect to the gap-filling operation:

Theorem 9.16.
Suppose $\Omega : \mathbb{K} \longrightarrow \mathbf{I}$ is a given mapping such that $\mathcal{F}_\Omega$ satisfies (Z-5). Then

$$\Omega(S) = \Omega(S^\ddagger)\,,$$

for all $S \in \mathbb{K}$.

This means that a well-behaved choice of Ω may only depend on $\sup S(\gamma)$, $\inf S(\gamma)$, and the knowledge whether $\sup S(\gamma) \in S(\gamma)$ and $\inf S(\gamma) \in S(\gamma)$. Apart from this, the 'interior structure' of $S(\gamma)$ is irrelevant to the determination of $\Omega(S)$.

The above gap-filling operation is useful for proving that (Ω-5) be necessary for $\mathcal{F}_\Omega$ to satisfy (Z-5). The other 'Ω-conditions' are easily shown to be necessary for $\mathcal{F}_\Omega$ to be a DFS, and require only minor adjustments of the corresponding proofs for $\mathcal{F}_\xi$-QFMs that were presented in [57].

Theorem 9.17.
The conditions (Ω-1)–(Ω-5) *on* $\Omega : \mathbb{K} \longrightarrow \mathbf{I}$ *are necessary for* $\mathcal{F}_\Omega$ *to be a DFS.*

Hence the 'Ω-conditions' are necessary and sufficient for $\mathcal{F}_\Omega$ to be a DFS, and all $\mathcal{F}_\Omega$-DFSes are indeed standard models. In order to prove that the criteria are independent, let us relate $\mathcal{F}_\xi$-QFMs to their apparent superclass of $\mathcal{F}_\Omega$-QFMs.

Theorem 9.18.
Consider an aggregation mapping $\xi : \mathbb{T} \longrightarrow \mathbf{I}$. *Then* $\mathcal{F}_\xi = \mathcal{F}_\Omega$, *where* $\Omega : \mathbb{K} \longrightarrow \mathbf{I}$ *is defined by*

$$\Omega(S) = \xi(\top_S, \bot_S)\,, \tag{9.10}$$

for all $S \in \mathbb{K}$, *and* $(\top_S, \bot_S) \in \mathbb{T}$ *is defined by*

$$\top_S(\gamma) = \sup S(\gamma) \tag{9.11}$$
$$\bot_S(\gamma) = \inf S(\gamma) \tag{9.12}$$

for all $\gamma \in \mathbf{I}$.

This is apparent. Hence all $\mathcal{F}_\xi$-QFMs are $\mathcal{F}_\Omega$-QFMs and all $\mathcal{F}_\xi$-DFSes are $\mathcal{F}_\Omega$-DFSes. The next theorem permits to reduce the independence proof of the conditions on Ω to the independence proof of the conditions imposed on ξ.

Theorem 9.19.
Suppose $\xi : \mathbb{T} \longrightarrow \mathbf{I}$ *is given and* $\Omega : \mathbb{K} \longrightarrow \mathbf{I}$ *is defined by* (9.10). *Then*

a. (X-1) *is equivalent to* (Ω-1)*;*
b. (X-2) *is equivalent to* (Ω-2)*;*
c. (X-3) *is equivalent to* (Ω-3)*;*
d. 1. *the conjunction of* (X-2), (X-4) *and* (X-5) *implies* (Ω-4)*;*
   2. (Ω-4) *implies* (X-4)*;*
e. (X-5) *is equivalent to* (Ω-5).

Based on this theorem and the known independence of the conditions (X-1)–(X-5), it is now easy to prove the desired result concerning independence.

Theorem 9.20.
The conditions (Ω-1)–(Ω-5) *imposed on* $\Omega : \mathbb{K} \longrightarrow \mathbf{I}$ *are independent.*

9.4 The Unrestricted Class of $\mathcal{F}_\omega$-QFMs

As has been remarked above, every $\mathcal{F}_\Omega$-DFS can be defined in terms of the mapping $s_{Q,(X_1,\ldots,X_n)} : \mathbf{I} \longrightarrow \mathbf{I}$ and this usually makes a simpler representation. It therefore makes sense to introduce the class of QFMs definable in terms of $s_{Q,(X_1,\ldots,X_n)} : \mathbf{I} \longrightarrow \mathbf{I}$.

Theorem 9.21.
Suppose $Q : \mathcal{P}(E)^n \longrightarrow \mathbf{I}$ is a semi-fuzzy quantifier and $X_1, \ldots, X_n \in \widetilde{\mathcal{P}}(E)$ a choice of fuzzy arguments. Then ${s_{Q,X_1,\ldots,X_n}}^{-1}(0) \neq \varnothing$, i.e. there exists $z_0 \in \mathbf{I}$ with $s_{Q,X_1,\ldots,X_n}(z_0) = 0$.

Hence all possible choices of $s_{Q,X_1,\ldots,X_n}$ are contained in the following set $\mathbb{L}$.

Definition 9.22.
$\mathbb{L} \subseteq \mathbf{I}^{\mathbf{I}}$ is defined by

$$\mathbb{L} = \{s \in \mathbf{I}^{\mathbf{I}} : s^{-1}(0) \neq \varnothing\}\,.$$

The following theorem states that $\mathbb{L}$ is the minimum subset of $\mathbf{I}^{\mathbf{I}}$ which contains all possible mappings $s_{Q,X_1,\ldots,X_n}$:

Theorem 9.23.
For all $s \in \mathbb{L}$, let us define $S : \mathbf{I} \longrightarrow \mathcal{P}(\mathbf{I})$ by

$$S(\gamma) = \{z \in \mathbf{I} : \gamma \geq s(z)\} \tag{9.13}$$

for all $\gamma \in \mathbf{I}$. It is apparent that $S \in \mathbb{K}$. Let us further suppose that $Q : \mathcal{P}(\mathbf{2} \times \mathbf{I}) \longrightarrow \mathbf{I}$ is defined by (9.3) and that $X \in \widetilde{\mathcal{P}}(\mathbf{2} \times \mathbf{I})$ is the fuzzy subset defined by (9.9). Then $s_{Q,X} = s$.

In order to define quantification results based on $s_{Q,X_1,\ldots,X_n}$, we need an aggregation mapping $\omega : \mathbb{L} \longrightarrow \mathbf{I}$. The corresponding QFM $\mathcal{F}_\omega$ is defined in the usual way.

Definition 9.24.
Let a mapping $\omega : \mathbb{L} \longrightarrow \mathbf{I}$ be given. By $\mathcal{F}_\omega$ we denote the QFM defined by

$$\mathcal{F}_\omega(Q)(X_1, \ldots, X_n) = \omega(s_{Q,X_1,\ldots,X_n})\,,$$

for all semi-fuzzy quantifiers $Q : \mathcal{P}(E)^n \longrightarrow \mathbf{I}$ and all $X_1, \ldots, X_n \in \widetilde{\mathcal{P}}(E)$.

9.5 The Classes of $\mathcal{F}_\omega$-Models and $\mathcal{F}_\Omega$-Models Coincide

It is obvious from the definition of $s_{Q,X_1,\ldots,X_n}$ in terms of $S_{Q,X_1,\ldots,X_n}$ that all $\mathcal{F}_\omega$-QFMs are $\mathcal{F}_\Omega$-QFMs, using the apparent choice of $\Omega : \mathbb{K} \longrightarrow \mathbf{I}$,

$$\Omega(S) = \omega(s) \tag{9.14}$$

where $s(z) = \inf\{\gamma \in \mathbf{I} : z \in S(\gamma)\}$, see Def. 9.5. It is then clear from Def. 9.7 and Def. 9.24 that

$$\mathcal{F}_\Omega = \mathcal{F}_\omega \,. \tag{9.15}$$

The converse is not true, i.e. it is not the case that all $\mathcal{F}_\Omega$-QFMs are $\mathcal{F}_\omega$-QFMs. However, if an $\mathcal{F}_\Omega$-QFMs is sufficiently 'well-behaved', then it is also an $\mathcal{F}_\omega$-QFM. In particular, this is the case for an $\mathcal{F}_\Omega$-DFS.

Theorem 9.25.

a. If $\Omega : \mathbb{K} \longrightarrow \mathbf{I}$ satisfies (Ω-4), then $\mathcal{F}_\Omega = \mathcal{F}_\omega$, provided we define $\omega : \mathbb{L} \longrightarrow \mathbf{I}$ by

$$\omega(s) = \Omega(S) \tag{9.16}$$

for all $s \in \mathbb{L}$, where

$$S(\gamma) = \{z \in \mathbf{I} : \gamma \geq s(z)\} \tag{9.17}$$

for all $\gamma \in \mathbf{I}$.
b. If $\Omega : \mathbb{K} \longrightarrow \mathbf{I}$ does not satisfy (Ω-4), then $\mathcal{F}_\Omega$ is not an $\mathcal{F}_\omega$-QFM.

Therefore an $\mathcal{F}_\Omega$-QFM is an $\mathcal{F}_\omega$-QFM if and only if it satisfies (Ω-4). Let us recall that by Th. 9.17, (Ω-4) is necessary for $\mathcal{F}_\Omega$ to be a DFS. This means that we do not lose any models of interest if we restrict attention to the class of those $\mathcal{F}_\Omega$-QFMs which satisfy (Ω-4), and can therefore be expressed as $\mathcal{F}_\omega$-QFMs.

9.6 Characterisation of the $\mathcal{F}_\omega$-Models

It is then convenient to switch from (Ω-1)–(Ω-5) to corresponding conditions on $\omega : \mathbb{L} \longrightarrow \mathbf{I}$. To accomplish this, we will first define a preorder $\sqsubseteq \subseteq \mathbb{L} \times \mathbb{L}$, which is needed to express a monotonicity condition.

Definition 9.26.
For all $s, s' \in \mathbb{L}$, $s \sqsubseteq s'$ if and only if the following two conditions hold:

a. for all $z \in \mathbf{I}$, $\inf\{s'(z') : z' \geq z\} \leq s(z)$;
b. for all $z' \in \mathbf{I}$, $\inf\{s(z) : z \leq z'\} \leq s'(z')$.

In the case of $\mathcal{F}_\omega$-QFMs, we can express the conditions on $\omega : \mathbb{L} \longrightarrow \mathbf{I}$ even more succinctly.

Definition 9.27. *We impose the following conditions on* $\omega : \mathbb{L} \longrightarrow \mathbf{I}$. *For all* $s \in \mathbb{L}$,

$$\text{If } s^{-1}([0,1)) = \{a\}, \text{ then } \omega(s) = a. \tag{ω-1}$$
$$\text{If } s'(z) = s(1-z) \text{ for all } z \in \mathbf{I}, \text{ then } \omega(s') = 1 - \omega(s). \tag{ω-2}$$
$$\text{If } s(1) = 0 \text{ and } s^{-1}([0,1)) \subseteq \{0,1\}, \text{ then } \omega(s) = \tfrac{1}{2} + \tfrac{1}{2}s(0). \tag{ω-3}$$
$$\text{If } s' \in \mathbb{L} \text{ with } s \sqsubseteq s', \text{ then } \omega(s) \leq \omega(s'). \tag{ω-4}$$

Theorem 9.28.
Let $\omega : \mathbb{L} \longrightarrow \mathbf{I}$ *be given and suppose that* $\Omega : \mathbb{K} \longrightarrow \mathbf{I}$ *is defined in terms of* ω *according to* (9.14). *Then the following relationships hold.*

a. *Ω satisfies (Ω-1) if and only if ω satisfies (ω-1);*
b. *Ω satisfies (Ω-2) if and only if ω satisfies (ω-2);*
c. *Ω satisfies (Ω-3) if and only if ω satisfies (ω-3);*
d. *Ω satisfies (Ω-4);*
e. *Ω satisfies (Ω-5) if and only if ω satisfies (ω-4).*

The following theorems are now obvious from the results for Ω.

Theorem 9.29.
The conditions (ω-1)–(ω-4) are sufficient for $\mathcal{F}_\omega$ to be a standard DFS.

Theorem 9.30.
The conditions (ω-1)–(ω-4) are necessary for $\mathcal{F}_\omega$ to be a DFS.

Theorem 9.31.
The conditions (ω-1)–(ω-4) are independent.

To sum up, $\mathcal{F}_\omega$-DFSes comprise all $\mathcal{F}_\Omega$-DFSes, they are usually easier to define, and simpler conditions (ω-1)–(ω-4) have to be checked. However, the monotonicity condition (ω-4) on ω is somewhat more complicated compared to the monotonicity condition (Ω-5) on Ω. In the following, let us therefore introduce a simpler preorder $\trianglelefteq$ for expressing monotonicity, which when combined with an additional condition can replace $\sqsubseteq$ and the corresponding monotonicity condition (ω-4). $\trianglelefteq$ is defined as follows.

Definition 9.32.
For all $s, s' \in \mathbb{L}$, $s \trianglelefteq s'$ if and only if the following two conditions hold:

a. *for all $z \in \mathbf{I}$, there exists $z' \geq z$ with $s'(z') \leq s(z)$;*
b. *for all $z' \in \mathbf{I}$, there exists $z \leq z$ with $s(z) \leq s'(z')$.*

In order to state the additional condition, it is necessary to introduce a construction on $s \in \mathbb{L}$ which corresponds to the gap-filling operation $S^\ddagger$ defined on $S \in \mathbb{K}$.

Definition 9.33.
For all $s \in \mathbb{L}$, $s^\ddagger : \mathbf{I} \longrightarrow \mathbf{I}$ *is defined by*

$$s^\ddagger(z) = \max(\inf\{s(z') : z' \leq z\}, \inf\{s(z'') : z'' \geq z\}),$$

for all $z \in \mathbf{I}$.

Some basic properties of ‡ are the following.

Theorem 9.34.
Let $s \in \mathbb{L}$ *be given. Then*

a. $s^\ddagger \leq s$;
b. $s^\ddagger \in \mathbb{L}$;
c. $s^\ddagger$ *is concave, i.e.* $s^\ddagger(z_2) \leq \max(s^\ddagger(z_1), s^\ddagger(z_3))$ *whenever* $z_1 \leq z_2 \leq z_3$.

This 'concavification' construction was needed for the proof that ω's satisfying (ω-4) entails that Ω defined by (9.14) satisfies (Ω-5). However, it will also play its role in defining examples of $\mathcal{F}_\omega$ models, see Def. 9.36 and Def. 9.43. The connection between ‡ and monotonic behaviour of ω becomes visible in the next theorem, which facilitates the proof that a given ω satisfies (ω-4), by reducing it to the ‡-invariance of ω, and its monotonicity with respect to the simplified preorder $\trianglelefteq$.

Theorem 9.35.

For all $\omega : \mathbb{L} \longrightarrow \mathbf{I}$, *the monotonicity condition* (ω-4) *is equivalent to the conjunction of the following two conditions:*

a. for all $s, s' \in \mathbb{L}$ *with* $s \trianglelefteq s'$, *it holds that* $\omega(s) \leq \omega(s')$;
b. for all $s \in \mathbb{L}$, $\omega(s^\ddagger) = \omega(s)$.

9.7 Examples of $\mathcal{F}_\Omega$-Models

We will now present four examples of 'genuine' $\mathcal{F}_\omega$-models, i.e. of $\mathcal{F}_\omega$-DFSes which do not belong to the class of $\mathcal{F}_\xi$-DFSes. To this end, it is necessary to introduce some coefficients defined in terms of a given $s \in \mathbb{L}$.

Definition 9.36.
For all $s \in \mathbb{L}$, *the coefficients* $s_*^{\top,0}, s_*^{\bot,0}, s_1^{\top,*}, s_1^{\bot,*}, s_*^{\leq\frac{1}{2}}, s_*^{\geq\frac{1}{2}} \in \mathbf{I}$ *are defined by*

$$s_*^{\top,0} = \sup s^{\ddagger^{-1}}(0) \tag{9.18}$$

$$s_*^{\perp,0} = \inf s^{\ddagger^{-1}}(0) \tag{9.19}$$

$$s_1^{\top,*} = \sup s^{-1}([0,1)) \tag{9.20}$$

$$s_1^{\perp,*} = \inf s^{-1}([0,1)) \tag{9.21}$$

$$s_*^{\leq\frac{1}{2}} = \inf\{s(z) : z \leq \tfrac{1}{2}\} \tag{9.22}$$

$$s_*^{\geq\frac{1}{2}} = \inf\{s(z) : z \geq \tfrac{1}{2}\}\,. \tag{9.23}$$

Based on these coefficients, we can now define the examples of $\mathcal{F}_\omega$-models.

Definition 9.37.
By $\omega_M : \mathbb{L} \longrightarrow \mathbf{I}$ we denote the following mapping,

$$\omega_M(s) = \begin{cases} \min(s_*^{\perp,0}, \frac{1}{2} + \frac{1}{2} s_*^{\leq\frac{1}{2}}) & : \quad s_*^{\perp,0} > \frac{1}{2} \\ \max(s_*^{\top,0}, \frac{1}{2} - \frac{1}{2} s_*^{\geq\frac{1}{2}}) & : \quad s_*^{\top,0} < \frac{1}{2} \\ \frac{1}{2} & : \quad \text{else} \end{cases}$$

for all $s \in \mathbb{L}$. The QFM $\mathcal{F}_M$ is defined in terms of ω_M according to Def. 9.24, i.e. $\mathcal{F}_M = \mathcal{F}_{\omega_M}$.

Let us first notice that the QFM $\mathcal{F}_M$ so defined is indeed a DFS.

Theorem 9.38.
$\mathcal{F}_M$ is a standard DFS.

Let us remark that $\mathcal{F}_M$ is indeed a 'genuine' $\mathcal{F}_\omega$-DFS:

Theorem 9.39.
$\mathcal{F}_M$ is not an $\mathcal{F}_\xi$-DFS, i.e. there exists no $\xi : \mathbb{T} \longrightarrow \mathbf{I}$ with $\mathcal{F}_M = \mathcal{F}_\xi$.

In particular, this proves that the $\mathcal{F}_\omega$-DFSes are really more general than $\mathcal{F}_\xi$-DFSes, i.e. the $\mathcal{F}_\xi$-DFSes form a proper subclass of the $\mathcal{F}_\omega$-DFSes.

Definition 9.40.
By $\omega_P : \mathbb{L} \longrightarrow \mathbf{I}$ we denote the mapping defined by

$$\omega_P(s) = \begin{cases} \min(s_1^{\top,*}, \frac{1}{2} + \frac{1}{2} s_*^{\leq\frac{1}{2}}) & : \quad s_*^{\perp,0} > \frac{1}{2} \\ \max(s_1^{\perp,*}, \frac{1}{2} - \frac{1}{2} s_*^{\geq\frac{1}{2}}) & : \quad s_*^{\top,0} < \frac{1}{2} \\ \frac{1}{2} & : \quad \text{else} \end{cases}$$

for all $s \in \mathbb{L}$. We define the QFM $\mathcal{F}_P$ in terms of ω_P according to Def. 9.24, i.e. $\mathcal{F}_P = \mathcal{F}_{\omega_P}$.

Theorem 9.41.
$\mathcal{F}_P$ is a standard DFS.

Let us also observe that $\mathcal{F}_P$ is a genuine $\mathcal{F}_\omega$-DFS.

Theorem 9.42.
$\mathcal{F}_P$ is not an $\mathcal{F}_\xi$-DFS, i.e. there exists no $\xi : \mathbb{T} \longrightarrow \mathbf{I}$ such that $\mathcal{F}_P = \mathcal{F}_\xi$.

It is possible to obtain an even more specific DFS by slightly changing the definition of $\mathcal{F}_P$.

Definition 9.43.
By $\omega_Z : \mathbb{L} \longrightarrow \mathbf{I}$ we denote the mapping defined by

$$\omega_Z(s) = \begin{cases} \min(s_1^{\top,*}, \frac{1}{2} + \frac{1}{2} s_*^{\le \frac{1}{2}}) & : \quad {s^{\ddagger}}^{-1}(0) \subseteq [\frac{1}{2}, 1] \\ \max(s_1^{\perp,*}, \frac{1}{2} - \frac{1}{2} s_*^{\ge \frac{1}{2}}) & : \quad {s^{\ddagger}}^{-1}(0) \subseteq [0, \frac{1}{2}] \\ \frac{1}{2} & : \quad else \end{cases}$$

for all $s \in \mathbb{L}$. We define the QFM $\mathcal{F}_Z$ in terms of ω_Z according to Def. 9.24, i.e. $\mathcal{F}_Z = \mathcal{F}_{\omega_Z}$.

Note 9.44. It has been shown in [58, p. 43+] that ω_Z is well-defined.

Theorem 9.45.
$\mathcal{F}_Z$ is a standard DFS.

Again, it is easily shown that $\mathcal{F}_Z$ is a genuine $\mathcal{F}_\omega$-DFS.

Theorem 9.46.
$\mathcal{F}_Z$ is not an $\mathcal{F}_\xi$-DFS, i.e. there exists no $\xi : \mathbb{T} \longrightarrow \mathbf{I}$ such that $\mathcal{F}_Z = \mathcal{F}_\xi$.

Definition 9.47.
By $\omega_K : \mathbb{L} \longrightarrow \mathbf{I}$ we denote the mapping defined by

$$\omega_K(s) = \begin{cases} \min(s_*^{\perp,0}, \frac{1}{2} + \frac{1}{2} s(0)) & : \quad s_*^{\perp,0} > \frac{1}{2} \\ \max(s_*^{\top,0}, \frac{1}{2} - \frac{1}{2} s(1)) & : \quad s_*^{\top,0} < \frac{1}{2} \\ \frac{1}{2} & : \quad else \end{cases}$$

for all $s \in \mathbb{L}$. We define the QFM $\mathcal{F}_K$ in terms of ω_K according to Def. 9.24, i.e. $\mathcal{F}_K = \mathcal{F}_{\omega_K}$.

Theorem 9.48.
$\mathcal{F}_K$ is a standard DFS.

Again, it can be asserted that $\mathcal{F}_K$ is a genuine $\mathcal{F}_\omega$-DFS.

Theorem 9.49.
$\mathcal{F}_K$ is not an $\mathcal{F}_\xi$-DFS, i.e. there exists no $\xi : \mathbb{T} \longrightarrow \mathbf{I}$ with $\mathcal{F}_K = \mathcal{F}_\xi$.

9.8 Properties of $\mathcal{F}_\Omega$-Models

Now that the defining conditions of $\mathcal{F}_\Omega$-DFSes and $\mathcal{F}_\omega$-DFS have been established and examples of the new classes of models have been given, we will turn to additional properties like propagation of fuzziness. Usually the corresponding conditions will be stated both for the representation in terms of $\mathcal{F}_\Omega$ and in terms of $\mathcal{F}_\omega$. This provides maximum flexibility in later proofs whether a model at hand does or does not possess these properties.

Definition 9.50.
For all $S, S' \in \mathbb{K}$, we say that S is fuzzier (less crisp) than S', in symbols: $S \preceq_c S'$, if and only if the following conditions are satisfied for all $\gamma \in \mathbf{I}$.

$$\text{for all } z' \in S'(\gamma), \text{ there exists } z \in S(\gamma) \text{ such that } z \preceq_c z'; \tag{9.24}$$

$$\text{for all } z \in S(\gamma), \text{ there exists } z' \in S'(\gamma) \text{ such that } z \preceq_c z'. \tag{9.25}$$

Definition 9.51.
Let $\Omega : \mathbb{K} \longrightarrow \mathbf{I}$ be given. We say that Ω propagates fuzziness *if and only if*

$$\Omega(S) \preceq_c \Omega(S')$$

whenever $S, S' \in \mathbb{K}$ satisfy $S \preceq_c S'$.

Theorem 9.52.
For all $\Omega : \mathbb{K} \longrightarrow \mathbf{I}$, $\mathcal{F}_\Omega$ propagates fuzziness in quantifiers if and only if Ω propagates fuzziness.

The following condition permits a simplified check if a given Ω propagates fuzziness.

Theorem 9.53.
Suppose $\Omega : \mathbb{K} \longrightarrow \mathbf{I}$ satisfies (Ω-1)–(Ω-5). Then Ω propagates fuzziness if and only if

$$\Omega(S) = \Omega(S^\ddagger \cap [\tfrac{1}{2}, 1])\,,$$

for all $S \in \mathbb{K}$ with $S(0) \subseteq [\frac{1}{2}, 1]$.

Definition 9.54.
Let $S, S' \in \mathbb{K}$ be given. We say that S is less specific than S', in symbols $S \Subset S'$, if and only if

$$S(\gamma) \supseteq S'(\gamma)$$

for all $\gamma \in \mathbf{I}$.

that these conditions are independent. This analysis also reveals that all $\mathcal{F}_\Omega$-models are in fact standard DFSes.
We then focused on an apparent subclass of $\mathcal{F}_\Omega$-QFMs, the class of $\mathcal{F}_\omega$-QFMs. These are obtained by defining coefficients $s_{Q,X_1,\ldots,X_n}(z) = \inf\{\gamma : z \in S_{Q,X_1,\ldots,X_n}(\gamma)\}$ which extract an important characteristic of the result sets $S_{Q,X_1,\ldots,X_n}(\gamma)$. This construction offers the advantage that we no longer need to work with *sets* of results, as it was the case with the $S_{Q,X_1,\ldots,X_n}(\gamma)$'s, which are subsets of the unit interval. By contrast, we can now focus on scalars $s_{Q,X_1,\ldots,X_n}$ in the unit range, and a subsequent aggregation by applying the chosen $\omega : \mathbb{L} \longrightarrow \mathbf{I}$. Among other things, this greatly simplifies the definition of models, and therefore all examples of $\mathcal{F}_\Omega$-DFSes were presented in this succinct format.
Noticing that the new coefficients $s_{Q,X_1,\ldots,X_n}$ are functions of $S_{Q,X_1,\ldots,X_n}$ which suppress some of the original information, the question then arises if some of the $\mathcal{F}_\Omega$-models are lost under the new construction. To resolve this issue and gain some insight into the structure of $\mathcal{F}_\omega$-DFSes, a set of independent conditions that precisely describe the $\mathcal{F}_\omega$-DFSes in terms of necessary and sufficient criteria on ω was developed. In addition, the $\mathcal{F}_\omega$-QFMs were related to their superclass of $\mathcal{F}_\Omega$-QFMs. This analysis revealed that the move from $\mathcal{F}_\Omega$-QFMs to $\mathcal{F}_\omega$-QFMs does not result in any loss of intended models, i.e. the classes of $\mathcal{F}_\Omega$-DFSes and $\mathcal{F}_\omega$-DFSes coincide.
Turning to examples of $\mathcal{F}_\Omega$-DFSes (or synonymously, $\mathcal{F}_\omega$-DFSes), the simplified format was utilized to define the four $\mathcal{F}_\omega$-models $\mathcal{F}_M$, $\mathcal{F}_P$, $\mathcal{F}_Z$ and $\mathcal{F}_K$, all of which were shown to be 'genuine' members which go beyond the class of $\mathcal{F}_\xi$-QFMs. We then formulated conditions on Ω and ω which permit an investigation of all characteristic properties of the corresponding models.

To this end, we first extended the specificity order to the case of $S \preceq_c S'$ and $s \preceq_c s'$. This allowed us to reduce $\mathcal{F}_\Omega$'s propagating fuzziness in quantifiers to the requirement that Ω propagate fuzziness, i.e. $S \preceq_c S'$ entails $\Omega(S) \preceq_c \Omega(S')$. Based on a different relation $S \Subset S'$ defined on $\mathbb{K}$, it was further possible to define a condition of propagating unspecificity on Ω, and to prove that $\mathcal{F}_\Omega$ propagates fuzziness in arguments if and only if Ω propagates unspecificity. In addition, it was shown that both the condition of propagating fuzziness and the condition of propagating unspecificity can be simplified if the considered Ω is well-behaved (in particular if $\mathcal{F}_\Omega$ is a DFS). In this common case, a very elementary test on Ω is sufficient for detecting or rejecting these properties. All of these results have also been transferred to $\mathcal{F}_\omega$-QFMs, and thus turned into corresponding conditions on ω. After developing the formal apparatus required to investigate propagation of fuzziness in $\mathcal{F}_\Omega$- and $\mathcal{F}_\omega$-QFMs, the issue of most specific and least specific models was discussed to some depth. Acknowledging its relevance to the existence of most specific models, we first extended the notion of specificity consistency to collections $\boldsymbol{\Omega}^*$ of aggregation mappings Ω and proved that the resulting criterion on $\boldsymbol{\Omega}^*$ precisely captures specificity consistency of the class of QFMs $\mathbb{F} = \{\mathcal{F}_\Omega : \Omega \in \boldsymbol{\Omega}^*\}$. Hence the question whether $\mathbb{F}$ has a least upper specificity bound can be decided by looking at the aggregation mappings in $\boldsymbol{\Omega}^*$. It was also shown how the criterion can be simplified in common situations. Following this, the question was addressed how a specificity comparison $\mathcal{F}_\Omega \preceq_c \mathcal{F}_{\Omega'}$ can be reformulated as a condition $\Omega \preceq_c \Omega'$ imposed on the aggregation mappings. Again, the condition for $\Omega \preceq_c \Omega'$ can be reduced to a very simple check in many typical situations. All of the above concepts and theorems were then adapted to $\mathcal{F}_\omega$-QFMs, in order to provide similar support for specificity comparison in those cases where the models of interest are defined in terms of an aggregation mapping ω.

Based on these preparations, it was easy to prove some results concerning propagation of fuzziness that elucidate the structure of the class of $\mathcal{F}_\omega$-DFSes, and that relate the examples of $\mathcal{F}_\omega$-DFSes to the class as a whole. First of all, the full class of $\mathcal{F}_\omega$-DFSes is not specificity consistent (because its subclass of $\mathcal{F}_\xi$-DFSes is known to violate specificity consistency), which means that a 'most specific $\mathcal{F}_\omega$-DFS' does not exist. However, the class of models that propagate fuzziness in quantifiers was shown to be specificity consistent, and the most specific $\mathcal{F}_\omega$-DFS with this property was also identified, which turned out to be $\mathcal{F}_Z$. Recalling that $\mathcal{F}_Z$ is not an $\mathcal{F}_\xi$-QFM, this demonstrates that the class of $\mathcal{F}_\omega$-DFSes which propagate fuzziness in quantifiers is an extension proper of the class of $\mathcal{F}_\xi$-DFSes with the same behaviour. Turning to propagation of fuzziness in quantifiers, it was possible to prove a similar result. The corresponding class of $\mathcal{F}_\omega$-models was shown to be specificity consistent, and $\mathcal{F}_K$ was established to be the most specific $\mathcal{F}_\omega$-DFS with this property. Again, we can conclude from the fact that $\mathcal{F}_K$ is a 'genuine' $\mathcal{F}_\omega$-DFS that the $\mathcal{F}_\omega$-DFSes contain models which propagate fuzziness in arguments beyond those already known from the study of $\mathcal{F}_\xi$-DFSes. We then investigated those standard models that propagate fuzziness both in quantifiers and arguments. The model $\mathcal{F}_M$ was shown to be the most specific $\mathcal{F}_\omega$-DFS with these properties. The class of $\mathcal{F}_\xi$-DFSes that fulfill both conditions is known to coincide with the class of $\mathcal{M}_\mathcal{B}$-DFSes. Because $\mathcal{F}_M$ is not an $\mathcal{F}_\xi$-DFS, this proves that there are standard models beyond the $\mathcal{M}_\mathcal{B}$-type which propagate fuzziness both in quantifiers and arguments.

The problem of identifying the greatest lower specificity bound has also been addressed. In fact, the least specific $\mathcal{F}_\omega$-DFS was proven to coincide with one of the $\mathcal{M}_\mathcal{B}$-models, namely $\mathcal{M}_U$, which was already known to be the least specific $\mathcal{M}_\mathcal{B}$- and $\mathcal{F}_\xi$-DFS.

Finally, we also addressed the continuity issue. It is indispensible for applications that the chosen QFM be robust against slight variations in the chosen quantifier and in its arguments, which might e.g. result from noise. In addition, both continuity conditions are desirable in order to account for imperfect knowledge of the precise interpretation of the involved NL quantifier and NL concepts in terms of numeric membership grades. Based on an auxiliary filling construction $S^\square$, it was shown that every $\mathcal{F}_\Omega$-QFM which is continuous in quantifiers is in fact an $\mathcal{F}_\xi$-QFM. The class of $\mathcal{F}_\omega$-DFSes which are Q-continuous therefore collapses into the class of Q-continuous $\mathcal{F}_\xi$-DFSes, and those Q-continuous $\mathcal{F}_\omega$-DFSes which propagate fuzziness in quantifiers and arguments collapse into the class of $\mathcal{M}_\mathcal{B}$-DFSes. This proves that all *practical* models are already contained in the class of $\mathcal{F}_\xi$-DFSes, and those practical models which propagate fuzziness both in quantifiers and arguments are contained in the class of $\mathcal{M}_\mathcal{B}$-DFSes. This justifies the development and thorough analysis of these simpler classes in the previous chapters, because every model of practical interest will belong to one of them. It can thus be expressed through constructions simpler than those used to define $\mathcal{F}_\Omega$- and $\mathcal{F}_\omega$-QFMs, which in turn permit a simpler check of the relevant formal properties, like being a DFS, propagation of fuzziness, specificity comparisons, and continuity, and which suggest simple algorithms for implementing quantifiers in the model.

10

The Class of Models Based on the Extension Principle

10.1 Motivation and Chapter Overview

In this chapter, an attempt is made to establish a new class of fuzzification mechanisms not constructed from three-valued cuts. Starting from a straightforward definition of argument similarity, we will first introduce the class of QFMs defined in terms of the similarity measure, the class of $\mathcal{F}_\psi$-QFMs. It encloses the interesting subclass of $\mathcal{F}_\varphi$-QFMs, i.e. the class of models defined in terms of the standard extension principle (which serves to aggregate similarity grades). We will then formalize the necessary and sufficient conditions on the aggregation mappings which make the fuzzification mechanism a DFS. This analysis will demonstrate that the classes of $\mathcal{F}_\psi$ and $\mathcal{F}_\xi$-models are identical. Moreover the classes of $\mathcal{F}_\Omega$-DFSes and $\mathcal{F}_\psi$-DFSes/$\mathcal{F}_\varphi$-DFSes coincide as well, which is the main result of this chapter. Because the same class of models is obtained from independent considerations, this provides evidence that it indeed represents a natural class of standard models of fuzzy quantification.

10.2 A Similarity Measure on Fuzzy Arguments

To begin with, the similarity grade $\Xi_{Y_1,\ldots,Y_n}(X_1, \ldots, X_n)$ of the fuzzy arguments $(X_1, \ldots, X_n)$ to a choice of crisp arguments $(Y_1, \ldots, Y_n)$ can be defined as follows.

Definition 10.1.
Let $E \neq \varnothing$ be some base set and $Y \in \mathcal{P}(E)$. The mapping $\Xi_Y : \widetilde{\mathcal{P}}(E) \longrightarrow \mathbf{I}$ is defined by

$$\Xi_Y(X) = \min(\inf\{\mu_X(e) : e \in Y\}, \inf\{1 - \mu_X(e) : e \notin Y\})$$

for all $X \in \widetilde{\mathcal{P}}(E)$. For n-tuples of arguments $Y_1, \ldots, Y_n \in \mathcal{P}(E)$, we define $\Xi^{(n)}_{Y_1,\ldots,Y_n} : \widetilde{\mathcal{P}}(E)^n \longrightarrow \mathbf{I}$ by

I. Glöckner: *Fuzzy Quantifiers*, StudFuzz **193**, 259–276 (2006)
www.springerlink.com © Springer-Verlag Berlin Heidelberg 2006

$$\Xi^{(n)}_{Y_1,\dots,Y_n}(X_1,\dots,X_n) = \bigwedge_{i=1}^{n} \Xi_{Y_i}(X_i)$$

for all $X_1,\dots,X_n \in \widetilde{\mathcal{P}}(E)$.
Whenever n *is clear from the context, we shall omit the superscript and write* $\Xi_{Y_1,\dots,Y_n}(X_1,\dots,X_n)$ *rather than* $\Xi^{(n)}_{Y_1,\dots,Y_n}(X_1,\dots,X_n)$.

At times, it will be convenient to use the following abbreviation. Let us recall the fuzzy equivalence connective $\leftrightarrow : \mathbf{I} \times \mathbf{I} \longrightarrow \mathbf{I}$, defined by

$$x \leftrightarrow y = (x \wedge y) \vee (\neg x \wedge \neg y)$$

for all $x, y \in \mathbf{I}$. In the case that $y \in \{0,1\}$, this apparently becomes

$$x \leftrightarrow y = \begin{cases} x & : \quad y = 1 \\ \neg x & : \quad y = 0 \end{cases}$$

Now consider a base set $E \neq \varnothing$ and let $X \in \widetilde{\mathcal{P}}(E)$, $Y \in \mathcal{P}(E)$. Making use of the $\leftrightarrow$-connective we can define $\delta_{X,Y} : E \longrightarrow \mathbf{I}$ by

$$\delta_{X,Y}(e) = (\mu_X(e) \leftrightarrow \chi_Y(e)) = \begin{cases} \mu_X(e) & : \quad e \in Y \\ 1 - \mu_X(e) & : \quad e \notin Y \end{cases} \tag{10.1}$$

for all $e \in E$. In terms of $\delta_{Y,E}$, we can now conveniently reformulate $\Xi_Y(X)$. In particular, we can express $\Xi_{Y_1,\dots,Y_n}(X_1,\dots,X_n)$, where $X_1,\dots,X_n \in \widetilde{\mathcal{P}}(E)$ and $Y_1,\dots,Y_n \in \mathcal{P}(E)$, by

$$\Xi_{Y_1,\dots,Y_n}(X_1,\dots,X_n) = \inf\{\delta_{X_i,Y_i}(e) : e \in E, i = 1,\dots,n\}\,. \tag{10.2}$$

In order to illustrate the proposed definition of argument similarity, let us consider an example, which also profits from the more succinct alternative notation. Hence let a two-element base set $E = \{a, b\}$ be given, and suppose that $X \in \widetilde{\mathcal{P}}(E)$ is defined by $\mu_X(a) = \frac{1}{3}$, $\mu_X(b) = \frac{3}{4}$. Then

$$\begin{aligned}
\Xi_{\varnothing}(X) &= \min\{\delta_{X,\varnothing}(a),\ \delta_{X,\varnothing}(b)\} \\
&= \min\{1 - \mu_X(a), 1 - \mu_X(b)\} \\
&= \min\{\tfrac{2}{3}, \tfrac{1}{4}\} \\
&= \tfrac{1}{4} \\
\Xi_{\{a\}}(X) &= \min\{\delta_{X,\{a\}}(a),\ \delta_{X,\{a\}}(b)\} \\
&= \min\{\mu_X(a), 1 - \mu_X(b)\} \\
&= \min\{\tfrac{1}{3}, \tfrac{1}{4}\} \\
&= \tfrac{1}{4} \\
\Xi_{\{b\}}(X) &= \min\{\delta_{X,\{b\}}(a),\ \delta_{X,\{b\}}(b)\} \\
&= \min\{1 - \mu_X(a), \mu_X(b)\} \\
&= \min\{\tfrac{2}{3}, \tfrac{3}{4}\} \\
&= \tfrac{2}{3}
\end{aligned}$$

$$\begin{aligned}\Xi_{\{a,b\}}(X) &= \min\{\delta_{X,\{a,b\}}(a), \delta_{X,\{a,b\}}(b)\} \\ &= \min\{\mu_X(a), \mu_X(b)\} \\ &= \min\{\tfrac{1}{3}, \tfrac{3}{4}\} \\ &= \tfrac{1}{3}\,.\end{aligned}$$

Next we define the set of all compatibility grades which corresponds to a given choice of fuzzy arguments $X_1, \ldots, X_n$.

Definition 10.2.
Let $E \neq \varnothing$ be a given base set and $X_1, \ldots, X_n \in \widetilde{\mathcal{P}}(E)$, $n \geq 0$. Then $D^{(n)}_{X_1,\ldots,X_n} \in \mathcal{P}(\mathbf{I})$ is defined by

$$D^{(n)}_{X_1,\ldots,X_n} = \{\Xi_{Y_1,\ldots,Y_n}(X_1, \ldots, X_n) : Y_1, \ldots, Y_n \in \mathcal{P}(E)\}\,.$$

Whenever this causes no ambiguity, the superscript (n) will be omitted, thus abbreviating $D_{X_1,\ldots,X_n} = D^{(n)}_{X_1,\ldots,X_n}$.

Note 10.3. The superscript is only needed to discern $D^{(0)}_\varnothing$ (which corresponds to the empty tuple and thus, no arguments) from $D^{(1)}_\varnothing$ (which corresponds to the empty set supplied as a single argument).

Definition 10.4.
By $\mathbb{D} \in \mathcal{P}(\mathcal{P}(\mathbf{I}))$ we denote the set of all $D \in \mathcal{P}(\mathbf{I})$ with the following properties:

1. *$D \cap [\frac{1}{2}, 1] = \{r_+\}$ for some $r_+ = r_+(D) \in [\frac{1}{2}, 1]$;*
2. *for all $D' \subseteq D$ with $D' \neq \varnothing$, $\inf D' \in D$;*
3. *if $r_+ > \frac{1}{2}$, then $\sup D \setminus \{r_+\} = 1 - r_+$.*

Theorem 10.5.
Suppose $E \neq \varnothing$ is some base set and $X_1, \ldots, X_n \in \widetilde{\mathcal{P}}(E)$ are fuzzy subsets of E. Then $D_{X_1,\ldots,X_n} \in \mathbb{D}$.

Hence $\mathbb{D}$ is large enough to contain all $D_{X_1,\ldots,X_n}$. As we shall see later in Th. 10.11, $\mathbb{D}$ is indeed the smallest possible subset of $\mathcal{P}(\mathbf{I})$ which contains every $D_{X_1,\ldots,X_n}$. (The theorem has been delayed because it then becomes a corollary).

10.3 The Unrestricted Class of $\mathcal{F}_\psi$-QFMs

In order to define the new class of fuzzification mechanisms, we will now relate the similarity information expressed by $\Xi_{Y_1,\ldots,Y_n}(X_1, \ldots, X_n)$ to the behaviour of a quantifier on its arguments.

Definition 10.6.
Let $Q : \mathcal{P}(E)^n \longrightarrow \mathbf{I}$ *be a given semi-fuzzy quantifier and* $X_1, \ldots, X_n \in \widetilde{\mathcal{P}}(E)$. *Then* $A^{(n)}_{Q,X_1,\ldots,X_n} : \mathbf{I} \longrightarrow \mathcal{P}(\mathbf{I})$ *is defined by*

$$A^{(n)}_{Q,X_1,\ldots,X_n}(z) = \{\Xi_{Y_1,\ldots,Y_n}(X_1, \ldots, X_n) : (Y_1, \ldots, Y_n) \in Q^{-1}(z)\}$$

for all $z \in \mathbf{I}$. *When* n *is clear from the context, we will omit the superscript* (n), *thus abbreviating* $A_{Q,X_1,\ldots,X_n} = A^{(n)}_{Q,X_1,\ldots,X_n}$.

Note 10.7. Again, the superscript is only needed to eliminate the ambiguity between $A^{(0)}_{Q,\varnothing}$, where Q is a nullary quantifier and $\varnothing$ the empty tuple, and $A^{(1)}_{Q,\varnothing}$, where Q is a one-place quantifier and $\varnothing$ is the empty argument set.

Next let us describe the range of all possible $A_{Q,X_1,\ldots,X_n}$.

Definition 10.8.
By $\mathbb{A}$ *we denote the set of all mappings* $A : \mathbf{I} \longrightarrow \mathcal{P}(\mathbf{I})$ *with the following properties:*

a. $\cup\{A(z) : z \in \mathbf{I}\} \in \mathbb{D}$;
b. *for all* $z, z' \in \mathbf{I}$, $\sup A(z) > \frac{1}{2}$ *and* $\sup A(z') > \frac{1}{2}$ *entails that* $z = z'$.

In the following, $D(A)$ denotes the set

$$D(A) = \cup\{A(z) : z \in \mathbf{I}\}\,. \tag{10.3}$$

In addition, r_+ abbreviates $r_+(A) = r_+(D(A))$. It is then apparent from Def. 10.8.a and Def. 10.4 that there exists $z_+ = z_+(A) \in \mathbf{I}$ with

$$r_+ \in A(z_+)\,. \tag{10.4}$$

In the following, z_+ is assumed to be an arbitrary but fixed choice of $z_+ \in \mathbf{I}$ which satisfies (10.4) for a considered $A \in \mathbb{A}$.

Theorem 10.9.
Suppose $Q : \mathcal{P}(E)^n \longrightarrow \mathbf{I}$ *is a semi-fuzzy quantifier and* $X_1, \ldots, X_n \in \widetilde{\mathcal{P}}(E)$. *Then* $A_{Q,X_1,\ldots,X_n} \in \mathbb{A}$.

Hence $\mathbb{A}$ contains every $A_{Q,X_1,\ldots,X_n}$.

Theorem 10.10.
Let $A \in \mathbb{A}$ *be given and* $D(A) = \cup\{A(z) : z \in \mathbf{I}\}$.

a. *If* $D(A) = \{1\}$, *then* $A = A^{(0)}_{Q,\varnothing}$, *where* $Q : \mathcal{P}(\{*\})^0 \longrightarrow \mathbf{I}$ *is the constant quantifier* $Q(\varnothing) = z_+$.

b. If $D(A) \neq \{1\}$, then we can choose some mapping $\zeta : D(A) \longrightarrow \mathbf{I}$ with

$$r \in A(\zeta(r)) \tag{10.5}$$

for all $r \in D(A)$. If $r_+ = r_+(A)$ equals $\frac{1}{2}$, then $r_+ \in D(A) \cap [0, 1 - r_+]$. If $r_+ > \frac{1}{2}$, then we recall from Def. 10.8 that $\sup D(A) \setminus \{r_+\} = 1 - r_+$. Because $D(A) \neq \{1\}$ by assumption, we know that there exists

$$r_- \in D(A) \cap [0, 1 - r_+] \tag{10.6}$$

and we shall assume an arbitrary choice of r_- with this property. Based on r_-, we define $X \in \widetilde{\mathcal{P}}(\mathbf{I} \times \mathbf{I})$ by

$$\mu_X(z, r) = \begin{cases} r & : \ r \in A(z) \setminus \{r_+\} \\ r_- & : \ r \notin A(z) \vee r = r_+ > \frac{1}{2} \\ \frac{1}{2} & : \ r = r_+ = \frac{1}{2} \end{cases} \tag{10.7}$$

for all $z, r \in \mathbf{I}$. For all $Y \in \mathcal{P}(\mathbf{I} \times \mathbf{I})$, we abbreviate

$$r' = r'(Y) = \Xi_Y(X) \tag{10.8}$$

$$z' = z'(Y) = \inf\{z \in \mathbf{I} : (z, r') \in Y \text{ and } r' = r'(Y) \in A(z)\} . \tag{10.9}$$

Based on ζ, we define $Q : \mathcal{P}(\mathbf{I} \times \mathbf{I}) \longrightarrow \mathbf{I}$ by

$$Q(Y) = \begin{cases} z' & : \ r' \in A(z') \\ \zeta(r') & : \ r' \notin A(z') \end{cases} \tag{10.10}$$

for all $Y \in \mathcal{P}(\mathbf{I} \times \mathbf{I})$.
Then $A = A_{Q,X}$.

We also obtain the following corollary concerning $\mathbb{D}$:

Theorem 10.11.
For all $D \in \mathbb{D}$,

a. If $D = \{1\}$, then $D = D^{(0)}_{\varnothing}$, where $\varnothing$ is the empty tuple $\varnothing \in \mathcal{P}(\{*\})^0$.
b. If $D \neq \{1\}$, then there exists $X \in \widetilde{\mathcal{P}}(\mathbf{I} \times \mathbf{I})$ such that $D = D_X$.

Hence $\mathbb{D}$ is indeed the smallest subset of $\mathcal{P}(\mathbf{I})$ which contains every $D_{X_1,\ldots,X_n}$.

In order to carry out the desired aggregation, which will turn the compatibility grades into a fuzzification mechanism, we will use aggregation mappings $\psi : \mathbb{A} \longrightarrow \mathbf{I}$. These can be used to define a QFM in the apparent way, by composing with the $A_{Q,X_1,\ldots,X_n}$'s:

Definition 10.12.
Let $\psi : \mathbb{A} \longrightarrow \mathbf{I}$ be given. The QFM $\mathcal{F}_\psi$ is defined by

$$\mathcal{F}_\psi(Q)(X_1, \ldots, X_n) = \psi(A_{Q,X_1,\ldots,X_n})$$

for all semi-fuzzy quantifiers $Q : \mathcal{P}(E)^n \longrightarrow \mathbf{I}$ and all $X_1, \ldots, X_n \in \widetilde{\mathcal{P}}(E)$.

This definition spans the full class of QFMs definable in terms of argument similarity, and we will now investigate its well-behaved models, i.e. the class of $\mathcal{F}_\psi$-DFSes.

10.4 Characterisation of the $\mathcal{F}_\psi$-Models

Let us now tackle the goal of characterising the class of $\mathcal{F}_\psi$-DFSes, by making explicit the structure of all plausible models. To this end, we need some more notation, for expressing the properties expected from legal choices of ψ. As usual, the goal is that of characterising the new class of models in terms of the necessary and sufficient conditions on the aggregation mapping.

In order to describe the desired monotonicity properties, we first define a suitable preorder on $\mathbb{A}$.

Definition 10.13.
For all $A, A' \in \mathbb{A}$, we say that $A \sqsubseteq A'$ if and only if the following conditions are satisfied by A, A'.

a. *for all $z \in \mathbf{I}$ and all $r \in A(z)$, there exists $z' \geq z$ with $r \in A'(z')$;*
b. *for all $z' \in \mathbf{I}$ and all $r \in A'(z')$, there exists $z \leq z'$ with $r \in A(z)$.*

Next we introduce a 'cut/fill operator' $\boxplus$ on $\mathbb{A}$. The invariance of ψ with respect to $\boxplus$ turns out to be essential for ψ to satisfy (Z-4).

Definition 10.14.
For all $A \in \mathbb{A}$, $\boxplus A \in \mathbb{A}$ is defined by

$$\boxplus A(z) = [0, \widehat{\boxplus} A(z)]\,, \tag{10.11}$$

where

$$\widehat{\boxplus} A(z) = \min(\sup A(z), \tfrac{1}{2}) \tag{10.12}$$

for all $z \in \mathbf{I}$.

Notes 10.15.

- It is immediate from the definition of $\boxplus A$ that $\boxplus A \in \mathbb{A}$, see Def. 10.8.
- For every semi-fuzzy quantifier $Q : \mathcal{P}(E)^n \longrightarrow \mathbf{I}$ and all $X_1, \ldots, X_n \in \widetilde{\mathcal{P}}(E)$, we abbreviate

$$\boxplus_{Q,X_1,\ldots,X_n} = \boxplus A_{Q,X_1,\ldots,X_n} \tag{10.13}$$
$$\widehat{\boxplus}_{Q,X_1,\ldots,X_n} = \widehat{\boxplus} A_{Q,X_1,\ldots,X_n}\,. \tag{10.14}$$

- In the above definition, $\widehat{\boxplus}$ has been used to express $\boxplus$. In fact, both are definable in terms of each other, because conversely

$$\widehat{\boxplus}A(z) = \sup \boxplus A(z)\,, \tag{10.15}$$

for all $A \in \mathbb{A}$ and $z \in \mathbf{I}$.

The cut/fill operator $\boxplus$ is of special relevance to the characterisation of $\mathcal{F}_\psi$-DFSes because $\boxplus$-invariance ensures that (Z-6) be valid.
In order to define the conditions on ψ succinctly, it is useful to stipulate the following abbreviation. For all $A \in \mathbb{A}$,

$$\mathrm{NV}(A) = \{z \in \mathbf{I} : A(z) \neq \varnothing\} \tag{10.16}$$

We have now introduced all notation required to express the conditions on admissible choices of ψ.

Definition 10.16. *Let $\psi : \mathbb{A} \longrightarrow \mathbf{I}$ be given. The conditions (ψ-1)–(ψ-5) are defined as follows. For all $A, A' \in \mathbb{A}$,*

If $D(A) = \{1\}$, then $\psi(A) = z_+$. (ψ-1)

If $A(z) = A'(1-z)$ for all $z \in \mathbf{I}$, then $\psi(A) = 1 - \psi(A')$. (ψ-2)

If $\mathrm{NV}(A) \subseteq \{0,1\}$ and $r_+ \in A(1)$,
then $\psi(A) = 1 - \sup A(0)$. (ψ-3)

If $A \sqsubseteq A'$, then $\psi(A) \leq \psi(A')$. (ψ-4)

$\psi(\boxplus A) = \psi(A)$. ($\psi$-5)

Let us first notice that the above set of conditions indeed captures the essential requirements on ψ which ensure that $\mathcal{F}_\psi$ be a DFS. This is straightforward from the following observation on the behaviour of ψ for two-valued quantifiers.

Theorem 10.17.
If $\psi : \mathbb{A} \longrightarrow \mathbf{I}$ satisfies (ψ-2) and (ψ-3), then $\mathcal{F}_\psi$ coincides with all standard DFSes on two-valued quantifiers, i.e. for every standard DFS $\mathcal{F}$ and two-valued quantifier $Q : \mathcal{P}(E)^n \longrightarrow \{0,1\}$, it holds that $\mathcal{F}_\psi(Q) = \mathcal{F}(Q)$.

Based on this theorem, it is then easy to prove the following result.

Theorem 10.18.
The conditions (ψ-1)–(ψ-5) on $\psi : \mathbb{A} \longrightarrow \mathbf{I}$ are sufficient for $\mathcal{F}_\psi$ to be a standard DFS.

The converse claim is also true, i.e. it can be shown that the underlying mapping $\psi : \mathbb{A} \longrightarrow \mathbf{I}$ satisfies (ψ-1)–(ψ-5) whenever $\mathcal{F}_\psi$ is a DFS.

Theorem 10.19.
The conditions (ψ-1)–(ψ-5) on $\psi : \mathbb{A} \longrightarrow \mathbf{I}$ are necessary for $\mathcal{F}_\psi$ to be a DFS.

$\hat{\hat{Q}} : \widetilde{\mathcal{P}}(\mathcal{P}(E)^n) \longrightarrow \widetilde{\mathcal{P}}(\mathbf{I})$. Hence for a given $V \in \widetilde{\mathcal{P}}(\mathcal{P}(E)^n)$, $\hat{\hat{Q}}(V) \in \widetilde{\mathcal{P}}(\mathbf{I})$ is the fuzzy subset defined by

$$\mu_{\hat{\hat{Q}}(V)}(z) = \sup\{\mu_V(Y_1, \dots, Y_n) : (Y_1, \dots, Y_n) \in Q^{-1}(z)\}$$

for all $z \in \mathbf{I}$, see Def. 3.23. Given a choice of fuzzy arguments $X_1, \dots, X_n \in \widetilde{\mathcal{P}}(E)$, we now express $V = V_{X_1,\dots,X_n}$ in terms of argument similarity, viz

$$\mu_V(Y_1, \dots, Y_n) = \Xi_{Y_1,\dots,Y_n}(X_1, \dots, X_n)\,,$$

for all $Y_1, \dots, Y_n \in \mathcal{P}(E)$. It is then apparent from Def. 10.30 that

$$f_{Q,X_1,\dots,X_n}(z) = \mu_{\hat{\hat{Q}}(V)}(z)$$

for all $z \in \mathbf{I}$. Because $V = V_{X_1,\dots,X_n}$ represents argument similarity, $\hat{\hat{Q}}$ is obtained from Q by applying the standard extension principle, $f_{Q,X_1,\dots,X_n}$ is defined by composing $\hat{\hat{Q}}$ and $V_{X_1,\dots,X_n}$, and $\mathcal{F}_\varphi(Q)(X_1, \dots, X_n) = \varphi(f_{Q,X_1,\dots,X_n})$ is a function of $f_{Q,X_1,\dots,X_n}$, this proves the claim that every $\mathcal{F}_\varphi$ is defined from the argument similarity grades by applying the extension principle. Noticing that no additional assumptions were made in defining $f_{Q,X_1,\dots,X_n}$, which merely composes similarity assessment and the extended $\hat{\hat{Q}}$, this demonstrates that the $\mathcal{F}_\varphi$-QFMs are precisely the QFMs definable in terms of argument similarity and the standard extension principle. The $\mathcal{F}_\varphi$-QFMs thus constitute an interesting class of fuzzification mechanisms.

10.7 The Classes of $\mathcal{F}_\varphi$-Models and $\mathcal{F}_\psi$-Models Coincide

Before going into the details of the new class of models, and disclosing the structure of its well-behaved models, let us first make two observations, which establish the precise relationship between $\mathcal{F}_\varphi$-QFMs and their apparent superclass of $\mathcal{F}_\psi$-QFMs.

Theorem 10.36.
All $\mathcal{F}_\varphi$-QFMs are $\mathcal{F}_\psi$-QFMs, i.e. $\mathcal{F}_\varphi = \mathcal{F}_\psi$, provided that $\psi : \mathbb{A} \longrightarrow \mathbf{I}$ is defined in dependence on $\varphi : \mathbb{X} \longrightarrow \mathbf{I}$ by

$$\psi(A) = \varphi(f_A)\,, \tag{10.24}$$

for all $A \in \mathbb{A}$.

Conversely, one can assert the following.

Theorem 10.37.
Suppose that $\psi : \mathbb{A} \longrightarrow \mathbf{I}$ satisfies (ψ-5). Then $\mathcal{F}_\psi = \mathcal{F}_\varphi$, where $\varphi : \mathbb{X} \longrightarrow \mathbf{I}$ is defined by

$$\varphi(f) = \psi(A_f)\,, \tag{10.25}$$

for all $f \in \mathbb{X}$, *and*

$$A_f(z) = \begin{cases} [0, f(z)] & : \quad f(z) \leq \frac{1}{2} \\ [0, 1 - f(z)] \cup \{f(z)\} & : \quad f(z) > \frac{1}{2} \end{cases} \tag{10.26}$$

for all $z \in \mathbf{I}$.

These relationships are straightforward from the structure of the involved base constructions, and the earlier remark on equation (10.23). Recalling from Th. 10.19 that the condition (ψ-5) is necessary for $\mathcal{F}_\psi$ to be a DFS, this substantiates that the two classes of $\mathcal{F}_\varphi$-DFSes and $\mathcal{F}_\psi$-DFSes indeed coincide, and merely provide different views on the models definable in terms of the extension principle.

10.8 Characterisation of the $\mathcal{F}_\varphi$-Models

Let us now impose a number of conditions on admissable choices of φ. Firstly we define a pre-eorder on $\mathbb{X}$, again needed to express a monotonicity condition.

Definition 10.38.
For all $f, f' \in \mathbb{X}$, *we write* $f \sqsubseteq f'$ *if and only if the following conditions are satisfied for* f, f'.

a. for all $z \in \mathbf{I}$, $\sup\{f'(z') : z' \geq z\} \geq f(z)$;
b. for all $z' \in \mathbf{I}$, $\sup\{f(z) : z \leq z'\} \geq f'(z')$.

We can now state the conditions that must be obeyed by φ in order to make $\mathcal{F}_\varphi$ a DFS.

Definition 10.39. *Let* $\varphi : \mathbb{X} \longrightarrow \mathbf{I}$ *be given. The conditions* (φ-1)–(φ-5) *are defined as follows. For all* $f, f' \in \mathbb{X}$,

If $f^{-1}((0,1]) = \{z_+\}$ *and* $f(z_+) = 1$, *then* $\varphi(f) = z_+$. (φ-1)

If $f'(z) = f(1 - z)$ *for all* $z \in \mathbf{I}$, *then* $\varphi(f') = 1 - \varphi(f)$. ($\varphi$-2)

If $f^{-1}((0,1]) \subseteq \{0, 1\}$ *and* $f(1) \geq \frac{1}{2}$, *then* $\varphi(f) = 1 - f(0)$. (φ-3)

If $f \sqsubseteq f'$, *then* $\varphi(f) \leq \varphi(f')$. ($\varphi$-4)

If $f'(z) = \min(f(z), \frac{1}{2})$ *for all* $z \in \mathbf{I}$, *then* $\varphi(f') = \varphi(f)$. (φ-5)

The proof that these conditions describe precisely the intended class of models becomes feasible once we notice the close relationship between the 'φ-conditions' and corresponding 'ψ-conditions'.

the definition of $\mathcal{F}_\varphi$-QFMs. Due to the close relationship between $A_{Q,X_1,\ldots,X_n}$ and the derived $f_{Q,X_1,\ldots,X_n}$, the precise conditions on φ which make $\mathcal{F}_\varphi$ a DFS are apparent from the corresponding conditions (ψ-1)–(ψ-5) imposed on ψ. By adapting these conditions, it was easy to obtain a set of necessary and sufficient conditions (φ-1)–(φ-5) imposed on φ, and to prove that these conditions are independent.

Finally we reviewed the fuzzification mechanism proposed by Gaines [49] and its reformulation as a base construction for QFMs proposed in [53]. In the course of this investigation, it was shown that all of the resulting QFMs are $\mathcal{F}_\varphi$-QFMs and thus definable in terms of argument similarity and the extension principle. Conversely, all 'reasonable' choices of $\mathcal{F}_\varphi$ where φ satisfies at least (φ-5) can be expressed as 'Gainesian' QFMs, and thus be reduced to a mechanism claimed to provide a 'foundation of fuzzy reasoning' [49].

11

Implementation of Quantifiers in the Models

11.1 Motivation and Chapter Overview

By abstracting from computational considerations in the first part of this book, it became possible to develop a theory of fuzzy quantification which rests on an axiomatic foundation. These theoretical advances will only be significant in practice, however, if the theory is also *computational* and thus permits an efficient implementation of quantifiers in the new models.

A model of potential relevance to applications must show sufficient robustness against small changes in the data. We will therefore focus on practical models which are both Q-continuous and arg-continuous (see Def. 6.1 and Def. 6.2). Recalling from Th. 9.96 that the class of $\mathcal{F}_\xi$-DFSes contains all known models with these properties, we will thus be concerned with the implementation of quantifiers in models of the $\mathcal{F}_\xi$-type only. Due to the diversity of these models, one cannot expect a 'generic' implementation which fits all examples. However, we can try to develop techniques which will prove useful as building-blocks for implementing arbitrary $\mathcal{F}_\xi$-models. These techniques will be concerned with common patterns found in all $\mathcal{F}_\xi$-DFSes. For example, consider the joint constructive principle underlying these models, i.e. the quantification result is always computed by applying an aggregation mapping ξ to the upper and lower bounds $\top_{Q,X_1,\dots,X_n}$ and $\bot_{Q,X_1,\dots,X_n}$. The development of algorithms for computing these upper and lower bounds, then, is of obvious utility to the implementation of quantifiers in arbitrary $\mathcal{F}_\xi$-models.

The chapter on implementation is organized as follows. We will start with a discussion of simple two-valued quantifiers which share the same interpretation across all standard models. These quantifiers can be expressed in terms of the membership grades of the arguments and the connectives min, max and $\neg$. We can use such quantifiers to compare two fuzzy sets by cardinality, to assess their degree of being equal, etc. Next we turn to the implementation of unary quantifiers on finite base sets which are no longer required to be two-valued. This investigation will reveal an important relationship between quantitativity and cardinality assessments. Based on this analysis, we

I. Glöckner: *Fuzzy Quantifiers*, StudFuzz **193**, 277–353 (2006)
www.springerlink.com
© Springer-Verlag Berlin Heidelberg 2006

will then show that for the considered quantifiers, the model $\mathcal{M}_{CX}$ can be expressed in terms of a measure of fuzzy cardinality. This reformulation will be useful, among other things, to establish the interpretation of quantifiers like "between ten and twenty" or "exactly five" in the standard models.
Having considered these basic examples, we will focus on the development of algorithms for implementing more complex quantifiers. To this end, we first explain how the mappings $\top_{Q,X_1,\ldots,X_n}$ and $\bot_{Q,X_1,\ldots,X_n}$ on which every $\mathcal{F}_\xi$-DFS is based, can be given a finite representation. The problem is that in practice, one can consider only a limited number of cutting levels, rather than the continuous range of all $\gamma \in \mathbf{I}$. It will be shown that the reduction to a finite sample is always possible if the base set is finite. This means that every $\mathcal{F}_\xi$-DFS can be expressed in terms of a finite sample of the cut levels. The prototypical models $\mathcal{M}$, $\mathcal{M}_{CX}$ and $\mathcal{F}_{\text{owa}}$ will serve to demonstrate this reformulation. In this way, we obtain the basic computational procedures for implementing quantifiers in these models.

Some efficiency improvements are necessary to make these methods useful in practice. Specifically, we will show that $\top_{Q,X_1,\ldots,X_n}$ and $\bot_{Q,X_1,\ldots,X_n}$ can be computed from cardinality coefficients sampled from $X_1, \ldots, X_n$ and their Boolean combinations. We will also explain how these coefficients can be efficiently computed from histogram information. By combining these techniques, we obtain practical algorithms for implementing quantifiers in the chosen models. These algorithms are surprisingly simple, but sufficiently powerful to cover absolute quantifiers like "at least ten", quantifiers of exception like "all except ten", proportional quantifiers like "most" and cardinal comparatives like "more than". Some practical examples will also be presented along with measured processing times, which demonstrate that these algorithms are pretty fast, and ready for use in practical applications.

11.2 'Simple' Quantifiers

Before turning to more complex cases, let us first consider the standard quantifiers. The following results are straightforward from the representation of the universal and existential quantifiers developed in Th. 4.55 and Th. 4.61. It is sufficient to notice that the considered quantifiers are constructed from the logical ones in terms of negation, antonym, or unions/intersections of arguments, to which every model is known to conform by Th. 4.20, Th. 4.19, (Z-4) and Th. 4.24, respectively.

Theorem 11.1.
In every standard DFS $\mathcal{F}$ and for all $E \neq \varnothing$,

$$\begin{aligned}
\mathcal{F}(\exists)(X) &= \sup\{\mu_X(e) : e \in E\} \\
\mathcal{F}(\forall)(X) &= \inf\{\mu_X(e) : e \in E\} \\
\mathcal{F}(\mathbf{all})(X_1, X_2) &= \inf\{\max(1 - \mu_{X_1}(e), \mu_{X_2}(e)) : e \in E\} \\
\mathcal{F}(\mathbf{some})(X_1, X_2) &= \sup\{\min(\mu_{X_1}(e), \mu_{X_2}(e)) : e \in E\} \\
\mathcal{F}(\mathbf{no})(X_1, X_2) &= \inf\{\max(1 - \mu_{X_1}(e), 1 - \mu_{X_2}(e)) : e \in E\}
\end{aligned}$$

for all $X, X_1, X_2 \in \widetilde{\mathcal{P}}(E)$.

These interpretations of the elementary quantifiers are quite satisfactory. Next we shall discuss the NL quantifier "at least k" and its variants. It is convenient to start with the unrestricted version of the quantifier, and to transfer these results in a subsequent step to the case of restricted quantification which involves two arguments. Hence let us focus on this type of quantifiers:

Definition 11.2.
Suppose $E \neq \varnothing$ *is a base set and* $k \in \mathbb{N}$. *The quantifier* $[\geq k] : \mathcal{P}(E) \longrightarrow \mathbf{2}$ *is defined by*

$$[\geq k](Y) = \begin{cases} 1 & : \ |Y| \geq k \\ 0 & : \ \textit{else} \end{cases}$$

for all $Y \in \mathcal{P}(E)$.

$[\geq k]$ is a standard quantifier because in the crisp case, it can be expressed as a Boolean combination involving the existential and universal quantifiers, see introduction p. 6. As to the interpretation of $[\geq k]$ in the models, let us first establish generic bounds on the possible quantification results, which are valid in arbitrary DFSes, i.e. not limited in scope to the standard models. Based on the coefficients $\mu_{[j]}(X)$ stipulated in Def. 7.91, which denote the j-th largest membership grade of the fuzzy subset X, these upper and lower bounds on the interpretation of $[\geq k]$ can be expressed as follows.

Theorem 11.3.
Let $E \neq \varnothing$ *be a given finite base set and* $k \in \mathbb{N}$. *Then in every DFS* $\mathcal{F}$,

$$\mu_{[1]}(X) \,\widetilde{\wedge} \ldots \widetilde{\wedge}\, \mu_{[k]}(X) \leq \mathcal{F}([\geq k])(X) \leq \mu_{[k]}(X) \,\widetilde{\vee} \ldots \widetilde{\vee}\, \mu_{[m]}(X)\,,$$

for all $X \in \widetilde{\mathcal{P}}(E)$, *where* $m = |E|$.

In the case of a standard model, the upper and lower bounds coincide, because $\widetilde{\wedge}$ and $\widetilde{\vee}$ then become min and max, respectively. Consequently, the above theorem uniquely determines the interpretation of $[\geq k]$ in the common models (this is apparent from the fact that the $\mu_{[j]}(X)$ form a non-increasing sequence). In addition, it is a rather straightforward task to extend this analysis to base sets of infinite cardinality as well. Summing up, the following result can then be proven for the regular models.

Theorem 11.4.
Suppose that $\mathcal{F}$ is a standard DFS, $E \neq \varnothing$ is a nonempty base set and $k \in \mathbb{N}$. Then

$$\mathcal{F}([\geq k])(X) = \sup\{\alpha \in \mathbf{I} : |X_{\geq\alpha}| \geq k\}\,,$$

for all $X \in \widetilde{\mathcal{P}}(E)$. In particular, if E is finite, then

$$\mathcal{F}([\geq k])(X) = \mu_{[k]}(X)\,.$$

Note 11.5. Recalling the notion of FG-count [197], the theorem thus asserts that

$$\mathcal{F}([\geq k])(X) = \mu_{\text{FG-count}(X)}(k)\,.$$

The analysis of $[\geq k]$ also reveals how some derived NL quantifiers are interpreted in the models. By decomposing the two-place quantifier "at least k" into **at least k** $= [\geq k]\cap$, and by further utilizing that all models are compatible with intersections of argument sets, we can now deduce that

$$\mathcal{F}(\textbf{at least k})(X_1, X_2) = \sup\{\alpha \in \mathbf{I} : |(X_1 \cap X_2)_{\geq\alpha}| \geq k\}\,,$$

for all $X_1, X_2 \in \widetilde{\mathcal{P}}(E)$, which in the finite case again becomes

$$\mathcal{F}(\textbf{at least k})(X_1, X_2) = \mu_{[k]}(X_1 \cap X_2)\,.$$

Next we consider some apparent derivations of "at least k": the quantifiers "more than k", "less than k" and "at most k". In terms of the known construction of *external negation*, all of these quantifiers can be reduced to "at least k", i.e.

$$\begin{aligned}\textbf{more than k} &= \textbf{at least k+1}\\ \textbf{less than k} &= 1 - \textbf{at least k}\\ \textbf{at most k} &= 1 - \textbf{more than k}\end{aligned}$$

(this is immediate from the definition of these quantifiers, see Def. 2.3). Owing to Th. 4.20, these quantifiers can thus be computed from the above interpretation of "at least k" in the apparent ways.

Finally let us consider some examples of 'simple' quantifiers which are not directly available in NL, but still useful to express relationships between fuzzy subsets and to compare these by cardinality. The following definition introduces these quantifiers and also stipulates the corresponding notation.

Definition 11.6.
Let $E \neq \varnothing$ be some base set. For finite E, we define quantifiers $[\textbf{card}\geq]$, $[\textbf{card}>]$, $[\textbf{card}=] : \mathcal{P}(E)^2 \longrightarrow \mathbf{2}$ by

$$[\mathbf{card} \geq](Y_1, Y_2) = \begin{cases} 1 & : \quad |Y_1| \geq |Y_2| \\ 0 & : \quad \text{else} \end{cases} \tag{11.1}$$

$$[\mathbf{card} >](Y_1, Y_2) = \begin{cases} 1 & : \quad |Y_1| > |Y_2| \\ 0 & : \quad \text{else} \end{cases} \tag{11.2}$$

$$[\mathbf{card} =](Y_1, Y_2) = \begin{cases} 1 & : \quad |Y_1| = |Y_2| \\ 0 & : \quad \text{else} \end{cases} \tag{11.3}$$

for all $Y_1, Y_2 \in \mathcal{P}(E)$. *In addition, we define the quantifier* $\mathbf{eq} : \mathcal{P}(E)^2 \longrightarrow \mathbf{I}$ *by*

$$\mathbf{eq}(Y_1, Y_2) = \begin{cases} 1 & : \quad Y_1 = Y_2 \\ 0 & : \quad Y_1 \neq Y_2 \end{cases} \tag{11.4}$$

for all $Y_1, Y_2 \in \mathcal{P}(E)$. *(In this case, no assumptions on the finiteness of* E *are necessary).*

Notes 11.7.

- These quantifiers serve the following purposes: $[\mathbf{card} \geq](Y_1, Y_2)$ checks if the cardinality of Y_1 is at least as large as the cardinality of Y_2; $[\mathbf{card} >](Y_1, Y_2)$ checks if the cardinality of Y_1 exceeds that of Y_2; the quantifier $[\mathbf{card} =](Y_1, Y_2)$ checks if Y_1 and Y_2 have the same cardinality; and $\mathbf{eq}(Y_1, Y_2)$ checks if Y_1 and Y_2 are identical.
- Some further quantifiers which capture set-theoretic notions have already been introduced without mentioning their set-theoretic counterparts. In particular, the quantifier $\mathbf{all}(Y_1, Y_2)$ checks 'subsethood' (inclusion) of Y_1 in Y_2; $\mathbf{no}(Y_1, Y_2)$ checks Y_1 and Y_2 for being disjoint; and $\pi_e(Y)$ checks membership of the element e in a set Y. As pointed out by W.V. Quine, this close relationship between the universal quantifier and subsethood has long been known in logic, and "figured already in Peirce's 1870 algebra of absolute and relative terms, thus even antedating any coherent logic of the variable itself" [71, p. 355].

It is interesting to consider the corresponding fuzzy quantifiers, which generalize these kinds of comparisons to fuzzy sets:

Theorem 11.8.
Let $E \neq \varnothing$ *be some finite base set and* $X_1, X_2 \in \widetilde{\mathcal{P}}(E)$. *Then in every standard DFS* $\mathcal{F}$,

a. $\mathcal{F}([\mathbf{card} \geq])(X_1, X_2) = \max\{\min(\mu_{[k]}(X_1), 1 - \mu_{[k+1]}(X_2)) : 0 \leq k \leq |E|\}$;
b. $\mathcal{F}([\mathbf{card} >])(X_1, X_2) = \max\{\min(\mu_k(X_1), 1 - \mu_k(X_2)) : 1 \leq k \leq |E|\}$;
c. $\mathcal{F}([\mathbf{card} =])(X_1, X_2) = \max\{\min\{\mu_{[k]}(X_1), 1 - \mu_{[k+1]}(X_1), \mu_{[k]}(X_2), 1 - \mu_{[k+1]}(X_2)\} : 0 \leq k \leq |E|\}$.

Further results on the interpretation of two-valued quantifiers in standard models are easily proven from this representation of $\mathcal{M}_{CX}$, recalling the earlier Th. 5.22. For example, consider the unary convex quantifier $[\geq r : \leq s]$ defined as follows.

Definition 11.20.
Let $E \neq \varnothing$ be some base set and $r, s \in \mathbb{N}$, $r \leq s$. The quantifier $[\geq r : \leq s] : \mathcal{P}(E) \longrightarrow \mathbf{2}$ is defined by

$$[\geq r : \leq s](Y) = \begin{cases} 1 & : \quad r \leq |Y| \leq s \\ 0 & : \quad \text{else} \end{cases}$$

for all $Y \in \mathcal{P}(E)$.

The quantifier is apparently useful for interpreting statements like *"Between ten and twenty of the married persons have children"*. Let us now make explicit the concrete interpretation of $[\geq r : \leq s]$ in the standard models. The following theorem is rather straightforward from the achieved representation of $\mathcal{M}_{CX}$, and the known fact that all standard models coincide on two-valued quantifiers:

Theorem 11.21.

Let $E \neq \varnothing$ be a finite base set and $r, s \in \mathbb{N}$, where $r \leq s$. Further suppose that $\mathcal{F}$ is a given standard DFS. Then $\mathcal{F}([\geq r : \leq s]) : \widetilde{\mathcal{P}}(E) \longrightarrow \mathbf{I}$ is the fuzzy quantifier defined by

$$\mathcal{F}([\geq r : \leq s])(X) = \mu_{\|X\|_{\text{iv}}}(r, s) = \min(\mu_{[r]}(X), 1 - \mu_{[s+1]}(X))$$

for all $X \in \widetilde{\mathcal{P}}(E)$.

This analysis of $\mathcal{F}([\geq r : \leq s])$ is readily extended to the convex natural language quantifiers "between r and s" that were introduced in Def. 2.3. Recalling the operation of intersecting arguments defined in Def. 4.23, the two-place quantifier "between r and s" becomes **between r and s** $= [\geq r : \leq s]\cap$, i.e.

$$\textbf{between r and s}(Y_1, Y_2) = [\geq r : \leq s](Y_1 \cap Y_2)$$

for all $Y_1, Y_2 \in \mathcal{P}(E)$. Now applying Th. 4.24, we directly obtain from the above Th. 11.21 that in every standard DFS,

$$\begin{aligned}\textbf{between r and s}(X_1, X_2) &= \mu_{\|X_1 \cap X_2\|_{\text{iv}}}(r, s) \\ &= \min(\mu_{[r]}(X_1 \cap X_2), 1 - \mu_{[s+1]}(X_1 \cap X_2)),\end{aligned}$$

for all $X_1, X_2 \in \widetilde{\mathcal{P}}(E)$, where $E \neq \varnothing$ is again assumed to be finite. The latter result also covers the quantifier "exactly k", which can be expressed as **exactly k** = **between k and k**. Therefore

$$\begin{aligned}\mathcal{F}(\mathbf{exactly\ k})(X_1, X_2) &= \mu_{\|X_1 \cap X_2\|_{\mathrm{iv}}}(k, k) \\ &= \min(\mu_{[k]}(X_1 \cap X_2), 1 - \mu_{[k+1]}(X_1 \cap X_2)) \\ &= \mu_{\text{FE-count}(X_1 \cap X_2)}(k)\end{aligned}$$

in all standard models.

11.4 The Core Algorithms for General Quantifiers

In the previous section, it was shown that $\mathcal{M}_{CX}(Q)(X)$ can be reduced to a fuzzy propositional formula built from constants $q^{\min}(\ell, u), q^{\max}(\ell, u) \in \mathbf{I}$ sampled from the quantifier, and from cardinality coefficients $\mu_{\|X\|_{\mathrm{iv}}}(\ell, u) \in \mathbf{I}$ obtained from the argument.

For the important proportional kind and other two-place quantifiers, however, there is no known reduction of $\mathcal{M}_{CX}(Q)(X_1, X_2)$ to a closed-form expression involving some notion of (relative) fuzzy cardinality. This makes it necessary to develop a more general, iterative procedure for computing quantification results. The need for such a general procedure is also obvious in the case of the other models. For example, there is no apparent method of directly computing $\mathcal{M}(Q)(X)$ or $\mathcal{F}_{\mathrm{owa}}(Q)(X)$ from a formula of fixed structure like that presented in Th. 11.18, even in the simplest case of a quantitative unary quantifier on a finite domain.

When trying to implement quantifiers in such general $\mathcal{F}_\xi$-DFSes, the first hindrance that we face is this. According to Def. 8.5, these models are defined by

$$\mathcal{F}_\xi(Q)(X_1, \ldots, X_n) = \xi(\top, \bot)\,,$$

where $\top = \top_{Q,X_1,\ldots,X_n}$ and $\bot = \bot_{Q,X_1,\ldots,X_n}$ are obtained from the three-valued cuts of the argument sets at *all* cutting levels $\gamma \in \mathbf{I}$. The point at issue is that the cutting parameter γ takes its values in a *continuous range*. An implementation on digital computers, however, can only consider a finite sample of relevant cut levels $\Gamma = \{\gamma_0, \ldots, \gamma_m\}$, along with the corresponding results of $\top$ and $\bot$ at these levels. In order to solve this problem, we will now take a closer look at the specific shape of $\top$ and $\bot$ for a certain type of fuzzy sets, which includes all fuzzy arguments on finite base sets and certain classes of fuzzy sets defined on base sets of transfinite cardinality. We will show that in this case, the desired reduction to a finite sample of $\top(\gamma)$ and $\bot(\gamma)$ is always possible. Moreover, we will describe the fundamental computational procedures for $\mathcal{F}_{\mathrm{owa}}$, $\mathcal{M}$ and $\mathcal{M}_{CX}$ by reformulating these prototypical models in such a way that they operate on the chosen finite sample only. In this way, it becomes possible to compute quantification results of arbitrary quantifiers on finite base sets in the prototype models.

In order to develop the desired finite representation of $\top$ and $\bot$, we need some further notation. Hence let some base set $E \neq \varnothing$ be given (not required to be finite) and further let $X_1, \ldots, X_n \in \widetilde{\mathcal{P}}(E)$. We shall denote by

$A(X_1, \ldots, X_n) \in \mathcal{P}(\mathbf{I})$ the set of membership grades assumed by one of the X_i's. Hence

$$A(X_1, \ldots, X_n) = \cup\{\operatorname{Im} \mu_{X_i} : i \in \{1, \ldots, n\}\}, \tag{11.13}$$

i.e. $A(X_1, \ldots, X_n) = \{\mu_{X_i}(e) : e \in E, i \in \{1, \ldots, n\}\}$. In dependence on $A(X_1, \ldots, X_n)$, we can further define the corresponding set of three-valued cut levels $\Gamma(X_1, \ldots, X_n) \in \mathcal{P}(\mathbf{I})$ according to

$$\begin{aligned} \Gamma(X_1, \ldots, X_n) = &\{2\alpha - 1 : \alpha \in A(X_1, \ldots, X_n) \cap [\tfrac{1}{2}, 1]\} \\ &\cup\{1 - 2\alpha : \alpha \in A(X_1, \ldots, X_n) \cap [0, \tfrac{1}{2})\} \\ &\cup\{0, 1\}. \end{aligned} \tag{11.14}$$

Notes 11.22.

- It is obvious that for finite base sets, $\Gamma(X_1, \ldots, X_n)$ is always finite as well.
- It should be pointed out that $\Gamma(X_1, \ldots, X_n)$ always includes the boundary cases $\gamma = 0$ and $\gamma = 1$, which we enforced by an explicit union with $\{0, 1\}$ in the defining equation for $\Gamma(X_1, \ldots, X_n)$. These boundary cases have been included into $\Gamma(X_1, \ldots, X_n)$ because knowing that $0 \in \Gamma(X_1, \ldots, X_n)$ and $1 \in \Gamma(X_1, \ldots, X_n)$ will considerably simplify the presentation of the later algorithms which operate on $\Gamma(X_1, \ldots, X_n)$.
- Obviously, $\Gamma(X_1, \ldots, X_n)$ can always be decomposed into a union of the components $\Gamma(X_i)$, i.e.

$$\Gamma(X_1, \ldots, X_n) = \Gamma(X_1) \cup \cdots \cup \Gamma(X_n). \tag{11.15}$$

The following observation on the behaviour of $\Gamma(\bullet)$ for Boolean combinations will later prove useful when implementing proportional quantifiers.

Theorem 11.23.

Let $E \neq \varnothing$ be a given base set. Then

a. *for all $X \in \widetilde{\mathcal{P}}(E)$, $\Gamma(\neg X) = \Gamma(X)$;*
b. *for all $X_1, X_2 \in \widetilde{\mathcal{P}}(E)$,*

$$\Gamma(X_1 \cap X_2) \subseteq \Gamma(X_1, X_2)$$

and

$$\Gamma(X_1 \cup X_2) \subseteq \Gamma(X_1, X_2).$$

Now suppose that $\Gamma(X_1, \ldots, X_n)$ is finite and that $\Gamma \supseteq \Gamma(X_1, \ldots, X_n)$ is a finite superset of $\Gamma(X_1, \ldots, X_n)$.[1] Knowing that $\{0, 1\} \subseteq \Gamma(X_1, \ldots, X_n)$, Γ

[1] The use of the finite superset $\Gamma \supseteq \Gamma(X_1, \ldots, X_n)$, rather than $\Gamma(X_1, \ldots, X_n)$, has shown itself more convenient for describing the algorithms based on integer-arithmetics.

can then be written as $\Gamma = \{\gamma_0, \ldots, \gamma_m\}$ where $0 = \gamma_0 < \gamma_1 < \cdots < \gamma_{m-1} < \gamma_m = 1$. In dependence on $\gamma_0, \ldots, \gamma_m$, we shall define derived coefficients $\overline{\gamma}_0, \ldots, \overline{\gamma}_{m-1}$ according to

$$\overline{\gamma}_j = \tfrac{\gamma_j + \gamma_{j+1}}{2} \tag{11.16}$$

for all $j \in \{0, \ldots, m-1\}$. In the following, it will be convenient to introduce a succinct notation for the results of $\top_{Q,X_1,\ldots,X_n}(\overline{\gamma}_j)$ and $\bot_{Q,X_1,\ldots,X_n}(\overline{\gamma}_j)$ that are observed at each $\overline{\gamma}_j$. Hence let us stipulate that

$$\top_j = \top_{Q,X_1,\ldots,X_n}(\overline{\gamma}_j) \tag{11.17}$$

$$\bot_j = \bot_{Q,X_1,\ldots,X_n}(\overline{\gamma}_j) \tag{11.18}$$

for $j \in \{0, \ldots, m-1\}$. When discussing $\mathcal{M}_\mathcal{B}$-models, it will also be useful to have a shorthand notation for $Q_{\overline{\gamma}_j}(X_1, \ldots, X_n)$. We therefore abbreviate

$$C_j = Q_{\overline{\gamma}_j}(X_1, \ldots, X_n) = \mathrm{med}_{\frac{1}{2}}(\top_j, \bot_j) \tag{11.19}$$

for all $j \in \{0, \ldots, m-1\}$, where the equalities are immediate from Th. 8.9, (11.17) and (11.18).

Theorem 11.24.

Let $Q : \mathcal{P}(E)^n \longrightarrow \mathbf{I}$ be a semi-fuzzy quantifier and $X_1, \ldots, X_n \in \widetilde{\mathcal{P}}(E)$ a choice of fuzzy arguments such that $\Gamma(X_1, \ldots, X_n)$ is finite. Further let $0 = \gamma_0 < \gamma_1 < \cdots < \gamma_{m-1} < \gamma_m = 1$ be given such that $\Gamma = \{\gamma_0, \ldots, \gamma_m\} \supseteq \Gamma(X_1, \ldots, X_n)$. Then for all $j \in \{0, \ldots, m-1\}$ and all $\gamma \in (\gamma_j, \gamma_{j+1})$,

$$\top_{Q,X_1,\ldots,X_n}(\gamma) = \top_j$$
$$\bot_{Q,X_1,\ldots,X_n}(\gamma) = \bot_j$$
$$Q_\gamma(X_1, \ldots, X_n) = C_j\,.$$

In other words, $\top_{Q,X_1,\ldots,X_n}, \bot_{Q,X_1,\ldots,X_n} : \mathbf{I} \longrightarrow \mathbf{I}$ reduce to simple step functions (with a finite number of steps), which are locally constant in the open intervals (γ_j, γ_{j+1}), $j = 0, \ldots, m-1$. In turn, the mapping $Q_\gamma(X_1, \ldots, X_n)$ which underlies the construction of $\mathcal{M}_\mathcal{B}$-QFMs also reduces to a simple step function. The relevance of these observations manifests itself in the next theorem.

Theorem 11.25.

Let $Q : \mathcal{P}(E)^n \longrightarrow \mathbf{I}$ be a semi-fuzzy quantifier and $(\gamma_j)_{j\in\{0,\ldots,m\}}$ a finite $\mathbf{I}$-valued sequence such that

$$0 = \gamma_0 < \gamma_1 < \cdots < \gamma_{m-1} < \gamma_m = 1\,.$$

a. *If* $\bot_0 > \frac{1}{2}$, *then*

$$C_j = \begin{cases} \bot_j & : \ j < j^* \\ \frac{1}{2} & : \ j \geq j^* \end{cases}$$

for all $j \in \{0, \ldots, m-1\}$.

b. *If* $\top_0 < \frac{1}{2}$, *then*

$$C_j = \begin{cases} \top_j & : \ j < j^* \\ \frac{1}{2} & : \ j \geq j^* \end{cases}$$

for all $j \in \{0, \ldots, m-1\}$.

Based on these preparations, it is now easy to decompose $\mathcal{M}(Q)(X_1, \ldots, X_n)$ into a simple weighted summation:

Theorem 11.30.

Let a semi-fuzzy quantifier $Q : \mathcal{P}(E)^n \longrightarrow \mathbf{I}$ *and fuzzy arguments* $X_1, \ldots, X_n \in \widetilde{\mathcal{P}}(E)$ *be given, and suppose that* $\Gamma(X_1, \ldots, X_n)$ *is finite. Further assume a choice of* $\Gamma = \{\gamma_0, \ldots, \gamma_m\} \supseteq \Gamma(X_1, \ldots, X_n)$ *with* $0 = \gamma_0 < \gamma_1 < \cdots < \gamma_{m-1} < \gamma_m = 1$. *Then*

$$\mathcal{M}(Q)(X_1, \ldots, X_n) = \begin{cases} \left(\sum\limits_{j=0}^{j^*-1} (\gamma_{j+1} - \gamma_j) \bot_j \right) + \frac{1}{2}(1 - \gamma_{j^*}) & : \ \bot_0 > \frac{1}{2} \\ \frac{1}{2} & : \ \bot_0 \leq \frac{1}{2} \leq \top_0 \\ \left(\sum\limits_{j=0}^{j^*-1} (\gamma_{j+1} - \gamma_j) \top_j \right) + \frac{1}{2}(1 - \gamma_{j^*}) & : \ \top_0 < \frac{1}{2} \end{cases}$$

The computation of $\mathcal{M}_{CX}$ is even simpler compared to $\mathcal{F}_{\mathrm{owa}}$ and $\mathcal{M}$, which require a summation in order to determine the final outcome of quantification from the results obtained at the individual j's.

In fact, it is sufficient for calculating $\mathcal{M}_{CX}(Q)(X_1, \ldots, X_n)$ to determine the minimal choice of $j \in \{0, \ldots, m-1\}$ which satisfies a certain inequality.

Theorem 11.31.

Let $Q : \mathcal{P}(E)^n \longrightarrow \mathbf{I}$ *be a semi-fuzzy quantifier and* $X_1, \ldots, X_n \in \widetilde{\mathcal{P}}(E)$ *a choice of fuzzy arguments such that* $\Gamma(X_1, \ldots, X_n)$ *is finite. Further let* $\Gamma = \{\gamma_0, \ldots, \gamma_m\} \supseteq \Gamma(X_1, \ldots, X_n)$ *be a subset of* $\mathbf{I}$ *with* $0 = \gamma_0 < \gamma_1 < \cdots < \gamma_{m-1} < \gamma_m = 1$. *In the case that* $C_0 > \frac{1}{2}$, *let us abbreviate*

$$B_j = 2\bot_j - 1\,, \tag{11.24}$$

while in the case that $C_0 < \frac{1}{2}$, *we stipulate*

$$B_j = 1 - 2\top_j \tag{11.25}$$

for all $j \in \{0, \ldots, m-1\}$*. Let us further abbreviate*

$$\widehat{J} = \{j \in \{0, \ldots, m-1\} : B_j \leq \gamma_{j+1}\} \tag{11.26}$$

$$\widehat{\jmath} = \min \widehat{J}\,. \tag{11.27}$$

Then

$$\mathcal{M}_{CX}(Q)(X_1, \ldots, X_n) = \begin{cases} \frac{1}{2} + \frac{1}{2}\max(\gamma_{\widehat{\jmath}}, B_{\widehat{\jmath}}) & : \bot_0 > \frac{1}{2} \\ \frac{1}{2} & : \bot_0 \leq \frac{1}{2} \leq \top_0 \\ \frac{1}{2} - \frac{1}{2}\max(\gamma_{\widehat{\jmath}}, B_{\widehat{\jmath}}) & : \top_0 < \frac{1}{2}\,. \end{cases}$$

To sum up, $\bot_{Q,X_1,\ldots,X_n}$, $\top_{Q,X_1,\ldots,X_n}$ and $Q_\gamma(X_1, \ldots, X_n)$ can be specified with sufficient detail by a finite number of γ_j, $\top_j$, $\bot_j$, from which the quantification results $\mathcal{F}_\xi(Q)(X_1, \ldots, X_n)$ of arbitrary $\mathcal{F}_\xi$-DFSes can be computed. For discussing the $\mathcal{M}_\mathcal{B}$-type, it proved convenient to utilize a derived coefficient $C_j = \operatorname{med}_{\frac{1}{2}}(\top_j, \bot_j)$ instead of $\top_j$ and $\bot_j$. The representation of these C_j's was further simplified based on the known monotonicity properties of $(Q_\gamma(X_1, \ldots, X_n))_{\gamma \in \mathbf{I}} \in \mathbb{B}$. The analysis of $\top_{Q,X_1,\ldots,X_n}$, $\bot_{Q,X_1,\ldots,X_n}$ and $Q_\gamma(X_1, \ldots, X_n)$ made it possible to develop more explicit formulations of the prototypical models $\mathcal{F}_{\text{owa}}$, $\mathcal{M}$ and $\mathcal{M}_{CX}$, which lend themselves directly to implementation.

In the remainder of the chapter, we will be concerned with further refinements which improve efficiency compared to the analysis achieved so far. These efficiency improvements will not require any changes to the computation procedures themselves. By contrast, they rest on the observation that the analysis achieved in theorems Th. 11.27, Th. 11.30 and Th. 11.31 still depends on $\top_j$, $\bot_j$ and/or C_j, and it is these coefficients which now receive more attention. This will also provide the necessary support for implementing further models of the $\mathcal{F}_\xi$- and $\mathcal{M}_\mathcal{B}$-type, which are all known to be expressible in terms of $\top_j$ and $\bot_j$ by Th. 11.25.

11.5 Refinement for Quantitative Quantifiers

In this section, we will be concerned with the issue of computing $\top_j = \top_{Q,X_1,\ldots,X_n}(\overline{\gamma}_j)$ and $\bot_j = \bot_{Q,X_1,\ldots,X_n}(\overline{\gamma}_j)$ efficiently, where $Q : \mathcal{P}(E)^n \longrightarrow \mathbf{I}$ is assumed to have a finite domain $E \neq \varnothing$.

In principle, Def. 8.1 permits the direct computation of $\top_j$ and $\bot_j$ based on an exhaustive search of the maximum and minimum[3] of $\mathcal{S}_{Q,X_1,\ldots,X_n}(\overline{\gamma}_j)$, i.e. $\top_j = \max S_j$ and $\bot_j = \min S_j$, where

[3] The use of min and max rather than inf and sup is possible because $\mathcal{S}_{Q,X_1,\ldots,X_n}(\overline{\gamma}_j)$ is finite. This is immediate from Def. 9.1 and the finiteness of the base set.

$$S_j = S_{Q,X_1,\dots,X_n}(\overline{\gamma}_j) = \{Q(Y_1,\dots,Y_n) : (Y_1,\dots,Y_n) \in \mathcal{T}_{\overline{\gamma}_j}(X_1,\dots,X_n)\}\,.$$

However, the exhaustive search of the minimum and maximum becomes impractical for domains of realistic size. In the worst case, the number of elements in S_j can be exponential in the size of the domain. Therefore the effort for directly computing the minimum and maximum soon becomes prohibitive.
In some cases, it is possible to shortcut the exhaustive search. For example, consider a quantifier $Q : \mathcal{P}(E)^n \longrightarrow \mathbf{I}$ which is nondecreasing in all arguments. The coefficients $\top_j$ and $\bot_j$ can then be expressed as $\top_j = Q((X_1)^{\max}_{\overline{\gamma}_j},\dots,(X_n)^{\max}_{\overline{\gamma}_j})$ and $\bot_j = Q((X_1)^{\min}_{\overline{\gamma}_j},\dots,(X_n)^{\min}_{\overline{\gamma}_j})$. It is thus sufficient to consider a single choice of $Y_i \in \mathcal{T}_{\overline{\gamma}_j}(X_i)$, and all other choices of Y_i are known to be irrelevant. In particular, this approach will work for $n = 1$, i.e. in the case of nondecreasing unary quantifiers $Q : \mathcal{P}(E) \longrightarrow \mathbf{I}$. It therefore renders possible the efficient evaluation of *fuzzy measures* in the models of interest. In the remaining cases where this a priori simplification of $\top_j$ and $\bot_j$ is not applicable, a careful analysis of other regularities of NL quantifiers is necessary in order to identify possible simplifications, from which an efficient implementation of $\top_j$ and $\bot_j$ can be derived. The particular regularity that will be assumed in the following is that of *quantitativity*, see Def. 4.41. Consequently, we will restrict attention to those semi-fuzzy quantifiers $Q : \mathcal{P}(E)^n \longrightarrow \mathbf{I}$ on a finite domain $E \neq \varnothing$ which also exhibit automorphism-invariance, i.e.

$$Q(Y_1,\dots,Y_n) = Q(\widehat{\beta}(Y_1),\dots,\widehat{\beta}(Y_n))$$

for all automorphisms β of E and $Y_1,\dots,Y_n \in \mathcal{P}(E)$. How can we exploit this regularity in order to simplify the computation of $\top_j$ and $\bot_j$? To this end, let us observe that $\top_j$ and $\bot_j$ are computed from S_j, which in turn is computed from the set $\mathcal{T}_{\overline{\gamma}_j}(X_1,\dots,X_n) \subseteq \mathcal{P}(E)^n$ defined by Def. 7.15. This suggests that we can spare unnecessary work, if we define an equivalence relation on $\mathcal{P}(E)^n$ which identifies argument tuples $(Y_1,\dots,Y_n), (Y'_1,\dots,Y'_n) \in \mathcal{T}_{\overline{\gamma}_j}(X_1,\dots,X_n)$ with

$$Q(Y_1,\dots,Y_n) = Q(Y'_1,\dots,Y'_n)\,.$$

Hence let $\sim\, \subseteq \mathcal{P}(E)^n \times \mathcal{P}(E)^n$ denote the following relation,

$$(Y_1,\dots,Y_n) \sim (Y'_1,\dots,Y'_n) \quad \Leftrightarrow$$
$$\text{there exists an automorphism } \beta \text{ of } E \text{ s.th. } (Y'_1,\dots,Y'_n) = (\widehat{\beta}(Y_1),\dots,\widehat{\beta}(Y_n))\,.$$

It is apparent that $\sim$ is indeed an equivalence relation on $\mathcal{P}(E)^n$. In addition, we can conclude from $(Y_1,\dots,Y_n) \sim (Y'_1,\dots,Y'_n)$ that $Q(Y_1,\dots,Y_n) = Q(Y'_1,\dots,Y'_n)$, because Q is assumed to be automorphism-invariant. This means that we can avoid redundant effort in computing $\top_j$ and $\bot_j$ in the following way. For each $(Y_1,\dots,Y_n) \in \mathcal{P}(E)^n$, let $(Y_1,\dots,Y_n)^*$ denote a representative under $\sim$, which is linked to the given $(Y_1,\dots,Y_n)$ by $(Y_1,\dots,Y_n) \sim (Y_1,\dots,Y_n)^*$. Due to the fact that Q is automorphism invariant, we then know that $Q(Y_1, \ldots, Y_n) = Q(Y_1, \ldots, Y_n)^*$. Now let $\mathcal{T}^*$ denote the set of all representatives for argument tuples in $\mathcal{T}_{\overline{\gamma}_j}(X_1, \ldots, X_n)$, i.e.

$$\mathcal{T}^* = \{(Y_1, \ldots, Y_n)^* : (Y_1, \ldots, Y_n) \in \mathcal{T}_{\overline{\gamma}_j}(X_1, \ldots, X_n)\}.$$

Clearly

$$\begin{aligned} S_j &= \{Q(Y_1, \ldots, Y_n) : (Y_1, \ldots, Y_n) \in \mathcal{T}_{\overline{\gamma}_j}(X_1, \ldots, X_n)\} \\ &= \{Q(Y_1, \ldots, Y_n)^* : (Y_1, \ldots, Y_n) \in \mathcal{T}_{\overline{\gamma}_j}(X_1, \ldots, X_n)\} \\ &= \{Q(Z_1, \ldots, Z_n) : (Z_1, \ldots, Z_n) \in \mathcal{T}^*\} \end{aligned} \tag{11.28}$$

and

$$|\mathcal{T}^*| \leq |\mathcal{T}_{\overline{\gamma}_j}(X_1, \ldots, X_n)|.$$

Consequently, the search of the maximum and minimum can now be restricted to the small(er) number of representatives in $\mathcal{T}^*$, and need not exhaust the total collection of alternatives in $\mathcal{T}_{\overline{\gamma}_j}(X_1, \ldots, X_n)$.

In practice, we will proceed another step and eliminate these representatives as well, in favour of a purely numerical scheme which rests on cardinality information only. The promise of such a scheme is that of avoiding any reference to set-based information (like $\mathcal{T}_{\overline{\gamma}_j}(X_1, \ldots, X_n)$ or $\mathcal{T}^*$), which might be hard to represent in a uniform way and awkward for processing purposes. By contrast, the cardinal numbers on which the numerical scheme will operate are easy to represent and permit very fast processing on digital computers. For the development of such a scheme, we can utilize a special regularity observed with all quantitative quantifiers, which establishes a link between quantitativity (automorphism-invariance) and definability of the quantifier in terms of cardinalities sampled from the arguments. For example, consider the quantitative unary quantifiers anticipated in Sect. 11.3. It was shown there that every quantifier of the considered type can be expressed in terms of cardinality information, i.e. there is a choice of $q : \{0, \ldots, |E|\} \longrightarrow \mathbf{I}$ such that $Q(Y) = q(|Y|)$ for all $Y \in \mathcal{P}(E)$. Based on this representation, we developed efficient reformulations of $\top_j = \top_{Q,X}(\overline{\gamma}_j)$ and $\bot_j = \bot_{Q,X}(\overline{\gamma}_j)$, which reduce to simple coefficients $q^{\max}(\ell, u)$ and $q^{\min}(\ell, u)$, respectively.

The goal of the present section is to abstract from this special case and develop a similar representation for arbitrary quantifiers, which are only assumed to be quantitative and declared on a finite base set. We will show that the previous analysis of unary quantifiers can be broadened to a fully general result concerning quantitative quantifiers on finite base sets. In particular, the computation of $\top_j$, $\bot_j$ and C_j from cardinality information is always possible if Q belongs to this type of quantifier, which eliminates the need to operate on the original set-based data in $\mathcal{T}_{\overline{\gamma}_j}(X_1, \ldots, X_n)$. For the main class of NL quantifiers which are two-place and conservative (like the absolute and proportional kinds), this basic approach will then be further refined in order to achieve optimal performance for all quantifiers of relevance to applications.

To begin with, it is well-known from TGQ that the quantitative two-valued quantifiers on finite base sets are exactly those quantifiers which can be computed from the cardinalities of the arguments and their Boolean combinations [9, p. 471]. As will be shown now, this characterisation of quantitative quantifiers on finite base sets generalizes to semi-fuzzy quantifiers, i.e. the quantitative semi-fuzzy quantifiers on finite domains are precisely those semi-fuzzy quantifiers which solely depend on the cardinality of their arguments (or Boolean combinations thereof). In order to state the theorem, we need some more notation. Hence let a finite base set $E \neq \varnothing$ and $n \in \mathbb{N}$ be given. For every choice of $\ell_1, \dots, \ell_n \in \{0, 1\}$, we define the Boolean combination $\Phi_{\ell_1,\dots,\ell_n}$ by

$$\Phi_{\ell_1,\dots,\ell_n}(Y_1, \dots, Y_n) = Y_1^{(\ell_1)} \cap \dots \cap Y_n^{(\ell_n)} \tag{11.29}$$

for all $Y_1, \dots, Y_n \in \mathcal{P}(E)$, where

$$Y^{(\ell)} = \begin{cases} Y & : \ \ell = 1 \\ \neg Y & : \ \ell = 0 \end{cases} \tag{11.30}$$

for $Y \in \mathcal{P}(E)$ and $\ell \in \{0, 1\}$. Based on these abbreviations, we can now assert the following.

Theorem 11.32.

A semi-fuzzy quantifier $Q : \mathcal{P}(E)^n \longrightarrow \mathbf{I}$ on a finite base set $E \neq \varnothing$ is quantitative if and only if Q can be computed from the cardinalities of its arguments and their Boolean combinations, i.e. there exist Boolean expressions $\Phi_1(Y_1, \dots, Y_n), \dots, \Phi_K(Y_1, \dots, Y_n)$ for some $K \in \mathbb{N}$, and a mapping $q : \{0, \dots, |E|\}^K \longrightarrow \mathbf{I}$ such that

$$Q(Y_1, \dots, Y_n) = q(|\Phi_1(Y_1, \dots, Y_n)|, \dots, |\Phi_K(Y_1, \dots, Y_n)|)\,, \tag{11.31}$$

for all $Y_1, \dots, Y_n \in \mathcal{P}(E)$. In particular, Q can be expressed as $Q(Y_1, \dots, Y_n) = q(c)$, where $c : \{0, 1\}^n \longrightarrow \{0, \dots, |E|\}$ is defined by

$$c_{\ell_1,\dots,\ell_n} = |\Phi_{\ell_1,\dots,\ell_n}(Y_1, \dots, Y_n)|$$

for all $\ell_1, \dots, \ell_n \in \{0, 1\}$, see (11.29).

Notes 11.33.

- For convenience, we will assume that the Boolean expressions $\Phi_i(Y_1, \dots, Y_n)$ are constructed from the arguments $Y_1, \dots, Y_n$ by forming unions $\cup$, intersections $\cap$ and complementation $\neg$; of course, a minimal set of operations op $\in \{\cap, \neg\}$ would be sufficient as well. It is not required that all variables $Y_1, \dots, Y_n$ actually participate in every considered Φ_i. In particular, trivial combinations like $\Phi_1(Y_1, Y_2, Y_3) = Y_2$ and more complex ones like $\Phi_2(Y_1, Y_2, Y_3) = (Y_1 \cap \neg Y_2) \cup Y_3$ are equally admissible.

- The second part of the theorem establishes a worst-case analysis in terms of complexity. It asserts that for unrestricted quantifiers, the number of Boolean combinations required to express Q in terms of cardinalities is bounded by 2^n, where n is the arity of the quantifier.

It is instructive to compare the general result of Th. 11.32 to the earlier analysis of quantitative one-place quantifiers that was given in Sect. 11.3. In the new notation, the special case covered there can now be represented in terms of $K = 1$, $\Phi(Y) = Y$, $c = |\Phi(Y)| = |Y|$ and $q : \{0, \dots, |E|\} \longrightarrow \mathbf{I}$. It is then guaranteed by Th. 11.10 that indeed $Q(Y) = q(c) = q(|Y|)$, provided that q is defined according to (11.5).

Obviously, things will not remain that simple once we consider a more general type of quantifiers. However, most NL quantifiers will not require the worst-case analysis presented in Th. 11.32. These quantifiers are far from being 'unrestricted', and they often permit considerable simplifications due to their regular structure. In particular, most NL quantifiers are known to be two-place and conservative, so it is worthwhile studying this type of quantifiers. Hence let us start by discussing general two-place quantifiers, and subsequently taylor this analysis to the conservative examples. It is well-known from TGQ that every quantitative two-valued quantifier $Q : \mathcal{P}(E)^2 \longrightarrow \mathbf{2}$ on a finite base set can be computed from $a = |Y_1 \setminus Y_2|$, $b = |Y_2 \setminus Y_1|$, $c = |Y_1 \cap Y_2|$ and $d = |E \setminus (Y_1 \cup Y_2)|$, where $Y_1, Y_2 \in \mathcal{P}(E)$ are the arguments of Q, c.f. [10, p. 457]. This result obviously transfers to the more general situation of semi-fuzzy quantifiers:

Theorem 11.34.
Let $Q : \mathcal{P}(E)^n \longrightarrow \mathbf{I}$ be a two-place quantifier on a finite base set $E \neq \varnothing$. If Q is quantitative, then $Q(Y_1, Y_2)$ can be expressed in terms of $a = |Y_1 \setminus Y_2|$, $b = |Y_2 \setminus Y_1|$, $c = |Y_1 \cap Y_2|$ and $d = |E \setminus (Y_1 \cup Y_2)|$, for any choice of crisp arguments $Y_1, Y_2 \in \mathcal{P}(E)$.

Note 11.35. The theorem merely instantiates the general framework introduced above for $n = 2$ (actually, it is an apparent corollary to Th. 11.32). Its sole purpose is that of stipulating symbols a, b, c, d which refer to the cardinality coefficients sampled from $\Phi_1(Y_1, Y_2) = Y_1 \setminus Y_2$, $\Phi_2(Y_1, Y_2) = Y_2 \setminus Y_1$, $\Phi_3(Y_1, Y_2) = Y_1 \cap Y_2$, and $\Phi_4(Y_1, Y_2) = E \setminus (Y_1 \cup Y_2)$.

Things become more interesting once additional assumptions are imposed on Q. In fact, it has been shown by van Benthem [9, p. 446], [10, p. 458] that in the case of a conservative two-valued quantifier, the coefficients a and c are sufficient for determining the quantification result $Q(Y_1, Y_2)$, which then becomes independent of b and d. Again, this result generalizes to semi-fuzzy quantifiers.

Theorem 11.36.
Let $Q : \mathcal{P}(E)^n \longrightarrow \mathbf{I}$ be a quantitative two-place quantifier on a finite base set

In the following, we will develop a more compact notation for these expressions. Hence let $\mathbb{V}$ denote the set of mappings,

$$\mathbb{V} = \{V : A \longrightarrow \{0,1\} | A \subseteq \{1,\dots,n\}\} . \tag{11.39}$$

Abusing notation, we will identify each $V \in \mathbb{V}$ and its graph, i.e. V will be written as $V = \{(i_1, v_1), \dots, (i_w, v_w)\}$, where $1 \leq i_1 < i_2 < \cdots < i_w \leq n$ and $v_t = V(i_t) \in \{0,1\}$ for all $t \in \{1,\dots,w\}$. Based on this representation of V, we can define a corresponding Boolean combination of the fuzzy subsets $X_1, \dots, X_n \in \widetilde{\mathcal{P}}(E)$ by

$$Z_V = \Psi_V(X_1, \dots, X_n) = X_{i_1}^{\langle v_1 \rangle} \cap \dots \cap X_{i_w}^{\langle v_w \rangle} . \tag{11.40}$$

In dependence on V, we further abbreviate

$$\ell_V = |Z_V|_\gamma^{\min} \tag{11.41}$$

$$u_V = |Z_V|_\gamma^{\max} \tag{11.42}$$

for all $V \in \mathbb{V}$, where $\gamma \in \mathbf{I}$ is a given choice of the cutting parameter. It is these Boolean combinations $Z_V = \Psi_V(X_1, \dots, X_n)$ from which the cardinality information expressed by the upper and lower cardinality bounds u_V and ℓ_V will be sampled. However, it is not a trivial task to express the relation $R_\gamma(X_1, \dots, X_n)$ in terms of the cardinality coefficients ℓ_V and u_V, $V \in \mathbb{V}$. we therefore take an intermediate step, and first show how the cardinalities of arbitrary Boolean combinations of the $(X_i)_\gamma^{\min}$ and $(X_i)_\gamma^{\max}$ can be computed from the known coefficients ℓ_V and u_V, $V \in \mathbb{V}$.
Hence let $E \neq \varnothing$ be some base set and $X \in \widetilde{\mathcal{P}}(E)$ be a fuzzy subset of E. For $p \in \{0, *, 1, +, -\}$, we define

$$X^{[p]} = \begin{cases} \neg(X_\gamma^{\max}) & : \ p = 0 \\ X_\gamma^{\max} \cap \neg X_\gamma^{\min} & : \ p = * \\ X_\gamma^{\min} & : \ p = 1 \\ X_\gamma^{\max} & : \ p = + \\ \neg(X_\gamma^{\min}) & : \ p = - \end{cases} \tag{11.43}$$

In order to express the relation $R_\gamma(X_1, \dots, X_n)$, it is sufficient to know the cardinality of all Boolean combinations ${X_1}^{[p_1]} \cap \dots \cap {X_n}^{[p_n]}$ where $p \in \{0, *, 1\}$. This will first be ascertained for the most fine-grained choice of $R_\gamma(X_1, \dots, X_n)$, which is sampled from all min-terms of the variables $Y_1, \dots, Y_n$:

Theorem 11.40.

Let $E \neq \varnothing$ be a finite base set and $X_1, \dots, X_n \in \widetilde{\mathcal{P}}(E)$. Further let $\gamma \in \mathbf{I}$ be given. For all $p = (p_1, \dots, p_n) \in \{0, *, 1\}^n$, we abbreviate

$$D(p) = \{d \in \{0,1\}^n : (d_1, p_1), \ldots, (d_n, p_n) \in \{(0,0), (1,1), (0,*), (1,*)\}\} \tag{11.44}$$

$$S(p) = X_1^{[p_1]} \cap \cdots \cap X_n^{[p_n]} \tag{11.45}$$

$$c(p) = |S(p)| = |X_1^{[p_1]} \cap \cdots \cap X_n^{[p_n]}| \tag{11.46}$$

We further define

$$\begin{aligned} \Lambda = \{(\lambda_p)_{p \in \{0,*,1\}^n} \mid & \text{ for all } p \in \{0,*,1\}^n,\ \lambda_p : \{0,1\}^n \longrightarrow \{0, \ldots, |E|\} \\ & \text{satisfies} \sum_{d \in \{0,1\}^n} \lambda_p(d) = c(p) \text{ and } \lambda_p(d) = 0 \text{ for all } d \notin D(p)\}\,. \end{aligned} \tag{11.47}$$

The relation $R = R_\gamma^{\Phi_{0,\ldots,0}, \ldots, \Phi_{1,\ldots,1}}(X_1, \ldots, X_n)$ *generated by all min-terms*

$$\Phi_{d_1,\ldots,d_n}(Y_1, \ldots, Y_n) = Y_1^{(d_1)} \cap \cdots \cap Y_n^{(d_n)}$$

and corresponding cardinality coefficients $c_{d_1,\ldots,d_n} = |\Phi_{d_1,\ldots,d_n}(Y_1, \ldots, Y_n)|$ *according to* (11.34) *can then be expressed as*

$$\begin{aligned} R = \{c : \{0,1\}^n \longrightarrow \{0, \ldots, |E|\} \mid & \text{ there exists } (\lambda_p)_{p \in \{0,*,1\}^n} \in \Lambda \\ & \text{such that for all } d \in \{0,1\}^n,\ c_d = \sum_{p \in \{0,*,1\}^n} \lambda_p(d)\}\,. \end{aligned} \tag{11.48}$$

In other words, R can be computed from the cardinalities of the sets $X_1^{[p_1]} \cap \cdots \cap X_n^{[p_n]}$, where $p = (p_1, \ldots, p_n) \in \{0,*,1\}^n$. This result is important because R is the 'universal' relation assumed in the worst-case analysis of Th. 11.32 which can express every quantitative Q. As shown in the next theorem, the above reduction to the universal relation R is sufficient to compute every other relation $R_\gamma(X_1, \ldots, X_n) = R_\gamma^{\Phi_1, \ldots, \Phi_K}(X_1, \ldots, X_n)$ as well, which results from a choice of Boolean combinations $\Phi_1(Y_1, \ldots, Y_n), \ldots, \Phi_K(Y_1, \ldots, Y_n)$:

Theorem 11.41.

Let $E \neq \varnothing$ *be a finite base set,* $X_1, \ldots, X_n \in \widetilde{\mathcal{P}}(E)$*, and suppose that* $\Phi'_1(Y_1, \ldots, Y_n)$*, ...,* $\Phi'_K(Y_1, \ldots, Y_n)$ *are Boolean combinations of the crisp variables* $Y_1, \ldots, Y_n \in \mathcal{P}(E)$*. We shall further abbreviate*

$$R = R_\gamma^{\Phi_{0,\ldots,0}, \ldots, \Phi_{1,\ldots,1}}(X_1, \ldots, X_n) \tag{11.49}$$

$$R' = R_\gamma^{\Phi'_1, \ldots, \Phi'_K}(X_1, \ldots, X_n) \tag{11.50}$$

where $\Phi_{d_1,\ldots,d_n}(Y_1, \ldots, Y_n) = Y_1^{(d_1)} \cap \cdots \cap Y_n^{(d_n)}$*, see* (11.30)*. In addition, we will use cardinality coefficients* $c_{d_1,\ldots,d_n} = |\Phi_{d_1,\ldots,d_n}(Y_1, \ldots, Y_n)|$.
Then

$$R' = \{(c'_1, \dots, c'_K) : (c_{0,\dots,0}, \dots, c_{1,\dots,1}) \in R, c'_t = \sum_{d \in D_t} c_d, \ t = 1, \dots, K\}, \tag{11.51}$$

where

$$D_t = \{d = (d_1, \dots, d_n) \in \{0,1\}^n : \Phi'_t(Y_1, \dots, Y_n) \cap \Phi_d(Y_1, \dots, Y_n) \neq \varnothing\} \tag{11.52}$$

for all $t \in \{1, \dots, K\}$.

Hence $\top_j$ and $\bot_j$ can be computed from some choice of R', and R' can be computed from the most fine-grained, or 'universal' relation R, which in turn can be computed from the cardinalities $|X_1^{[p_1]} \cap \cdots \cap X_n^{[p_n]}|$, $p_i \in \{0, *, 1\}$. The next goal is that of computing the cardinality of $X_1{}^{[p_1]} \cap \cdots \cap X_n{}^{[p_n]}$ from the cardinality coefficients u_V and ℓ_V, $V \in \mathbb{V}$. In order to accomplish this, we need additional intermediate representations. Rather than fixed-length Boolean combinations of the above type, i.e. $X_1{}^{[p_1]} \cap \cdots \cap X_n{}^{[p_n]}$, we therefore introduce more flexible Boolean combinations which take the following form,

$$X_{i_1}^{[p_1]} \cap \cdots \cap X_{i_w}^{[p_w]}$$

where $1 \leq i_1 < i_2 < \cdots < i_w \leq n$ and $p_t \in \{0, 1, *, +, -\}$ for all $t \in \{1, \dots, w\}$. The relevant argument positions i_t and corresponding p_t can again be viewed as the graph $P = \{(i_1, p_1), \dots, (i_w, p_w)\}$ of a mapping $P : A \longrightarrow \{0, 1, *, +, -\}$, where $A \subseteq \{1, \dots, n\} = \{i_t : t = 1, \dots, w\}$ and $P(i_t) = p_t$. We can therefore capture all intended combinations of the arguments by letting P range over the following collection,

$$\mathbb{P}^* = \{P : A \longrightarrow \{0, 1, *, +, -\} | A \subseteq \{1, \ \dots, \ n\}\}. \tag{11.53}$$

Based on a given $P \in \mathbb{P}^*$ and its graph representation $P = \{(i_1, p_1), \dots, (i_w, p_w)\}$, we can now define the associated crisp set $S(P) \in \mathcal{P}(E)$ and its cardinality $c(P)$ by

$$S(P) = X_{i_1}^{[p_1]} \cap \cdots \cap X_{i_w}^{[p_w]} \tag{11.54}$$

$$c(P) = |S(P)| = |X_{i_1}^{[p_1]} \cap \cdots \cap X_{i_w}^{[p_w]}|. \tag{11.55}$$

For reasons that will soon become clear, we also define the following restricted range $\mathbb{P}$, in which P is allowed to assume values in $\{0, 1, +, -\}$ only:

$$\mathbb{P} = \{P : A \longrightarrow \{0, 1, +, -\} | A \subseteq \{1, \ \dots, \ n\}\}. \tag{11.56}$$

Compared to $\mathbb{P}^*$, $\mathbb{P}$ thus comprises those mappings $P \in \mathbb{P}^*$ which satisfy $P(i) \neq *$ for all $i \in \operatorname{Dom} P$. The following theorem asserts that the restricted range $\mathbb{P}$ is sufficient for computing $c(P_*)$ for all $P_* \in \mathbb{P}^*$, by presenting a suitable reduction:

Theorem 11.42.

Let $E \neq \varnothing$ be a finite base set and $X_1, \ldots, X_n \in \widetilde{\mathcal{P}}(E)$. Then for all $P_* \in \mathbb{P}^*$,

$$c(P_*) = \sum_{P \in \mathbb{P}} \sigma(P) \cdot c(P)\,, \tag{11.57}$$

where the integer-valued mapping $\sigma = \sigma_{P_*} : \mathbb{P} \longrightarrow \mathbb{Z}$ is inductively defined as follows. Given $P_* \in \mathbb{P}$, let

$$i_* = \max\{i \in \operatorname{Dom} P_* : P_*(i) = *\}\,. \tag{11.58}$$

In dependence on i_*, we stipulate

a. If $i_* = 0$, then

$$\sigma_{P_*}(P) = \begin{cases} 1 & : \quad P = P_* \\ 0 & : \quad P \neq P_* \end{cases} \tag{11.59}$$

for all $P \in \mathbb{P}$.

b. If $i_* > 0$, then

$$\sigma_{P_*}(P) = \sigma_{P'}(P) - \sigma_{P''}(P) \tag{11.60}$$

for all $P \in \mathbb{P}$, where

$$P' = (P \setminus \{(i_*, *)\}) \cup \{(i_*, +)\} \tag{11.61}$$

$$P'' = (P \setminus \{(i_*, *)\}) \cup \{(i_*, 1)\}\,. \tag{11.62}$$

Hence $c(P_*)$ can indeed be computed from the cardinalities $c(P)$, $P \in \mathbb{P}$. Having shown that the case $p = *$ can be eliminated, we shall assume in the following that $p \in \{0, 1, +, -\}$ only, thus focussing on the case that $P \in \mathbb{P}$. It remains to be shown that the cardinalities $c(P) = |X_{i_1}^{[p_1]} \cap \cdots \cap X_{i_w}^{[p_w]}|$ can be computed from the quantities u_V and ℓ_V, $V \in \mathbb{V}$ sampled from $X_1, \ldots, X_n$. In the following, we will introduce some new concepts which simplify the description of the necessary calculations. For $p \in \{0, 1, +, -\}$, we define its 'polarity' $\operatorname{pol}(p) \in \{0, 1\}$ and its 'type', $\operatorname{type}(p) \in \{\max, \min\}$, by

$$\operatorname{pol}(p) = \begin{cases} 1 & : \quad p \in \{1, +\} \\ 0 & : \quad p \in \{0, -\} \end{cases} \tag{11.63}$$

$$\operatorname{type}(p) = \begin{cases} \max & : \quad p \in \{+, -\} \\ \min & : \quad p \in \{0, 1\} \end{cases} \tag{11.64}$$

The polarity of p is 0 if $X^{[p]}$ is negated, i.e. in the cases $X^{[0]} = \neg(X_\gamma^{\max})$ and $X^{[-]} = \neg(X_\gamma^{\min})$, while the polarity is one in the non-negated cases $X^{[1]} = X_\gamma^{\min}$ and $X^{[+]} = X_\gamma^{\max}$. The type of p is 'max' if $X^{[p]} = (X^{\langle \operatorname{pol}(p) \rangle})_\gamma^{\max}$ i.e. in the cases $X^{[+]} = X_\gamma^{\max}$ and $X^{[-]} = \neg(X_\gamma^{\min}) = (\neg X)_\gamma^{\max}$, see

Th. 7.17. The type of p is 'min' if $X^{[p]} = (X^{\langle \mathrm{pol}(p)\rangle})_\gamma^{\min}$ and thus in the cases $X^{[0]} = \neg(X_\gamma^{\max}) = (\neg X)_\gamma^{\min}$ and $X^{[1]} = X_\gamma^{\min}$. Combining the results for the individual cases, we obtain that

$$X^{[p]} = (X^{\langle \mathrm{pol}(p)\rangle})_\gamma^{\mathrm{type}(p)} \tag{11.65}$$

for all $p \in \{0, 1, +, -\}$. The following theorem completes the intended reduction, by showing how $c(P)$, $P \in \mathbb{P}$ can be computed from the cardinality coefficients ℓ_V and u_V, $V \in \mathbb{V}$.

Theorem 11.43.

Let $E \neq \varnothing$ be a finite base set, $X_1, \ldots, X_n \in \widetilde{\mathcal{P}}(E)$ and $\gamma \in \mathbf{I}$. Then for all $P \in \mathbb{P}$,

$$c(P) = \sum_{V \in \mathbb{V}} \zeta_P(V, \min) \cdot \ell_V + \sum_{V \in \mathbb{V}} \zeta_P(V, \max) \cdot u_V\,, \tag{11.66}$$

where the integer-valued mapping $\zeta = \zeta_P : \mathbb{V} \times \{\min, \max\} \longrightarrow \mathbb{Z}$ is inductively defined as follows. Recalling the graph representation $P = \{(i_1, p_1), \ldots, (i_w, p_w)\}$ of P, where $p_t = P(i_t)$, let

$$i_* = \max\{i \in \mathrm{Dom}\, P : \mathrm{type}(P(i)) \neq \mathrm{type}(p_w)\}\,. \tag{11.67}$$

In dependence on i_*, we stipulate

a. If $i_* = 0$, then

$$\zeta_P(V, y) = \begin{cases} 1 & : \quad V = \{(i_1, \mathrm{pol}(p_1)), \ldots, (i_w, \mathrm{pol}(p_w))\}, y = \mathrm{type}(p_w) \\ 0 & : \quad \text{else} \end{cases} \tag{11.68}$$

for all $V \in \mathbb{V}$ and $y \in \{\min, \max\}$.

b. If $i_* > 0$, then

$$\zeta_P(V, y) = \zeta_{P'}(V, y) - \zeta_{P''}(V, y) \tag{11.69}$$

for all $V \in \mathbb{V}$ and $y \in \{\min, \max\}$, where

$$P' = P \setminus \{(i_*, p_*)\} \tag{11.70}$$
$$P'' = P' \cup \{(i_*, p')\} \tag{11.71}$$
$$p_* = P(i_*) \tag{11.72}$$

and

$$p' = \begin{cases} 0 & : \quad p_* = + \\ 1 & : \quad p_* = - \\ + & : \quad p_* = 0 \\ - & : \quad p_* = 1\,. \end{cases} \tag{11.73}$$

Subsequently applying Th. 11.40 to Th. 11.43, every relation $R_\gamma(X_1,\dots,X_n)$ can be computed from a choice of cardinality coefficients $\ell_r = |Z_r|_\gamma^{\min}$ and $u_r = |Z_r|_\gamma^{\max}$, which are sampled from suitable Boolean combinations $Z_r = \Psi_r(X_1,\dots,X_n)$ of the arguments. All of the theorems are constructive in style, i.e. they also present a specific choice of Boolean combinations and show how the relation $R_\gamma(X_1,\dots,X_n)$ can be computed from the resulting quantities ℓ_r and u_r.

In order to spell out a procedure which covers all possible $R_\gamma(X_1,\dots,X_n)$ in full generality, it was of course necessary to make worst-case assumptions in terms of complexity of the quantifiers to be modelled. In fact, the number of Boolean combinations required for the fully general analysis amounts to $|V| = 3^n$. Although the arity n of NL quantifiers is very small in practice (usually $n = 2$ or $n = 3$), this indicates that the generic solution might not be optimal from the performance point of view, and should rather be viewed as a proof of concept, and as a point of departure for developing more efficient, dedicated solutions for typical classes of quantifiers.

The potential for simplification can be demonstrated nicely for conservative quantifiers, which we will now analyse in some more depth. Our goal is to develop a simple and efficient way of computing $R_\gamma(X_1, X_2)$. Hence let $Q : \mathcal{P}(E)^2 \longrightarrow \mathbf{I}$ be a two-place quantifier on a finite base set which is both quantitative and conservative. As has been shown in Th. 11.37 above, we can describe the quantifier in terms of a mapping $q : \{0,\dots,|E|\}^2 \longrightarrow \mathbf{I}$, based on an analysis in terms of $K = 2$, $\Phi_1(Y_1,Y_2) = Y_1$, $\Phi_2(Y_1,Y_2) = Y_1 \cap Y_2$, $c_1 = |Y_1|$ and $c_2 = |Y_1 \cap Y_2|$ for all $Y_1, Y_2 \in \mathcal{P}(E)$. The corresponding relation $R_\gamma(X_1,X_2)$, $X_1, X_2 \in \widetilde{\mathcal{P}}(E)$, can then be determined from $Z_1 = \Psi_1(X_1,X_2) = X_1$, $Z_2 = \Psi_2(X_1,X_2) = X_1 \cap X_2$ and $Z_3 = \Psi_3(X_1,X_2) = X_1 \cap \neg X_2$ in the following way.

Theorem 11.44.

Let $E \neq \varnothing$ be a finite base set and suppose that $X_1, X_2 \in \widetilde{\mathcal{P}}(E)$ are fuzzy subsets of E. Further let $\gamma \in \mathbf{I}$, and suppose that $R_\gamma(X_1,X_2) \subseteq \{0,\dots,|E|\} \times \{0,\dots,|E|\}$, is the relation defined by

$$R_\gamma(X_1,X_2) = \{(|Y_1|, |Y_1 \cap Y_2|) : Y_1 \in \mathcal{T}_\gamma(X_1), Y_2 \in \mathcal{T}_\gamma(X_2)\}\,.$$

Then $R_\gamma(X_1,X_2)$ can be computed from $\ell_1 = |X_1|_\gamma^{\min}$, $\ell_2 = |X_1 \cap X_2|_\gamma^{\min}$, $\ell_3 = |X_1 \cap \neg X_2|_\gamma^{\min}$, $u_1 = |X_1|_\gamma^{\max}$, $u_2 = |X_1 \cap X_2|_\gamma^{\max}$, and $u_3 = |X_1 \cap \neg X_2|_\gamma^{\max}$, viz

$$R_\gamma(X_1,X_2) = \{(c_1,c_2) : \ell_1 \leq c_1 \leq u_1, \max(\ell_2, c_1 - u_3) \leq c_2 \leq \min(u_2, c_1 - \ell_3)\}\,.$$

Note 11.45. In this case, the 'worst-case analysis' for $n = 2$ results in $3^2 = 9$ combinations of X_1, X_2 that must possibly be considered for computing R, i.e. $X_1 \cap X_2$, $X_1 \cap \neg X_2$, $\neg X_1 \cap X_2$, $\neg X_1 \cap \neg X_2$, X_1, $\neg X_1$, X_2, $\neg X_2$, and E. Of these, $\neg X_1$ and $\neg X_2$ are obviously redundant, because their cardinality bounds can be computed from the cardinality bounds of X_1 and X_2. The full domain E can also be eliminated, because it bears no information about X_1 and X_2. This leaves a total of 6 combinations which are potentially necessary for computing $R_\gamma(X_1, X_2)$. Hence the conservativity of Q shrinks down this set of required combinations by one half, i.e. only the combinations X_1, $X_1 \cap X_2$ and $X_1 \cap \neg X_2$ must actually be considered.

Let us now put this theorem into context. The earlier Th. 11.37 states that a conservative quantifier can be computed from the cardinality of $\Phi_1(Y_1, Y_2) = Y_1$ and $\Phi_2(Y_1, Y_2) = Y_1 \cap Y_2$. Hence $\top_{Q,X_1,X_2}$ and $\bot_{Q,X_1,X_2}$ can be computed from that choice of $R_\gamma(X_1, X_2)$ assumed in the above theorem. In other words, the theorem ensures that if Q is conservative, then there exist mappings $q^{\min}, q^{\max} : \{0, \ldots, |E|\}^4 \longrightarrow \mathbf{I}$ such that

$$\top_{Q,X_1,X_2}(\gamma) = q^{\max}(\ell_1, \ell_2, \ell_3, u_1, u_2, u_3) \tag{11.74}$$

$$\bot_{Q,X_1,X_2}(\gamma) = q^{\min}(\ell_1, \ell_2, \ell_3, u_1, u_2, u_3) \tag{11.75}$$

for all $X_1, X_2 \in \widetilde{\mathcal{P}}(E)$ and $\gamma \in \mathbf{I}$, where the coefficients $\ell_1, \ell_2, \ell_3, u_1, u_2$ and u_3 are defined as above. This is obvious from the Th. 11.44 if we define these mappings as follows,

$$q^{\max}(\ell_1, \ell_2, \ell_3, u_1, u_2, u_3) = \max\{q(c_1, c_2) : (c_1, c_2) \in R\}$$
$$q^{\min}(\ell_1, \ell_2, \ell_3, u_1, u_2, u_3) = \min\{q(c_1, c_2) : (c_1, c_2) \in R\}\,,$$

where $R \subseteq \{0, \ldots, |E|\} \times \{0, \ldots, |E|\}$ is the relation

$$R = \{(c_1, c_2) : \ell_1 \le c_1 \le u_1, \max(\ell_2, c_1 - u_3) \le c_2 \le \min(u_2, c_1 - \ell_3)\}\,.$$

We shall see below in Sect. 11.9 how this analysis of conservative quantifiers can be developed into an efficient implementation of proportional quantifiers in the models. Other classes of quantifiers can be analysed in a similar way. This will be demonstrated below in Sect. 11.10, where the method is extended to cardinal comparatives like "more... than". The proposed scheme of analysing quantitative quantifiers on finite base sets therefore enables us to implement the relevant NL quantifiers in all models of interest. As witnessed by Th. 11.44, for example, the proposed solution of computing $\top_j$ and $\bot_j$ from a relation $R_j = R_{\overline{\gamma}_j}$ on cardinals achieves a very simple description which is also computationally straightforward. In the next section, we will be concerned with the remaining issue of computing the quantities $\ell_r = |Z_r|_{\overline{\gamma}_j}^{\min}$ and $u_r = |Z_r|_{\overline{\gamma}_j}^{\max}$ efficiently, to which the computation of R_j has been reduced. Upon solving this problem, all necessary ingredients will then be available for implementing quantifiers in the models. We will finally put all the pieces together and restate the complete algorithms in pseudo-code. In particular, this self-contained presentation is intended as a point of departure for concrete implementations.

11.6 Computation of Cardinality Bounds

Given Q and $X_1, \ldots, X_n$, we are now able to compute $\top_{Q,X_1,\ldots,X_n}$, $\bot_{Q,X_1,\ldots,X_n}$ and $Q_\gamma(X_1, \ldots, X_n)$ from the upper cardinality bounds $u_r = |Z_r|_\gamma^{\max}$ and lower cardinality bounds $\ell_r = |Z_r|_\gamma^{\min}$ of Boolean combinations $Z_1, \ldots, Z_L$ of the arguments. We have therefore succeeded in reformulating the computation of $\top_{Q,X_1,\ldots,X_n}$, $\bot_{Q,X_1,\ldots,X_n}$ and $Q_\gamma(X_1, \ldots, X_n)$ in such a way that it depends only on cardinality information expressed by the tuple of coefficients $(\ell_1, \ldots, \ell_L, u_1, \ldots, u_L)$, see Th. 11.40 to Th. 11.43 and equations (11.35), (11.36), (11.37), respectively. Once the cardinality bounds $\ell_1, \ldots, \ell_L, u_1, \ldots, u_L$ are known, the subsequent computations can dispense with any direct reference to the fuzzy sets $X_1, \ldots, X_n$, which might be too large and awkward to operate upon. The definition of the cardinality coefficients ℓ_r and u_r, however, still depends on $X_1, \ldots, X_n$, which means that up to this point, we have merely shifted the burden of dealing with the $X_1, \ldots, X_n$ into the computation of these coefficients.

When computing quantification results, it is necessary to calculate the u_r and ℓ_r for every considered choice of the cutting parameter γ that controls the three-valued cuts. Due to the fact that these coefficients must be computed repeatedly, their computation should be implemented as efficiently as possible. Consequently, the above strategy will now be re-iterated in order to ban any direct reference to $X_1, \ldots, X_n$ from the computation of the cardinality bounds ℓ_r and u_r. Basically, a pre-processing step will be applied which extracts the relevant information from the given fuzzy subsets, and the extracted cardinality information will then be represented in a way which lends itself to efficient processing of the ℓ_r and u_r.

In order to discuss the intermediate representation extracted from the fuzzy sets, and in order to construct a computation procedure which operates on this representation, it is sufficient to consider each Z_r in isolation. We will thus focus on one fuzzy set at a time, which for simplicity we denote by X, which is supposed to be chosen from $\{Z_1, \ldots, Z_L\}$. In addition, we will assume that a sample of cutting levels $\Gamma = \{\gamma_0, \ldots, \gamma_m\}$ be given, equipped with the usual properties, i.e. $0 = \gamma_0 < \gamma_1 < \cdots < \gamma_m = 1$ and $\Gamma(X) \subseteq \Gamma$. It is then convenient to index the results of the cardinality bounds for each $\overline{\gamma}_j$ by j. We will thus denote by $\ell(j)$ and $u(j)$ the lower and upper cardinality bounds of the considered fuzzy set $X \in \widetilde{\mathcal{P}}(E)$ in the j-th iteration, i.e.

$$\ell(j) = |X|_{\overline{\gamma}_j}^{\min} \tag{11.76}$$

$$u(j) = |X|_{\overline{\gamma}_j}^{\max} \tag{11.77}$$

for $j \in \{0, \ldots, m-1\}$. Rather than computing the coefficients $\ell(j)$ and $u(j)$ in each iteration $j = 0, \ldots, m-1$ from scratch, it would be advantageous to devise a procedure which computes the values of these coefficients in the current iteration, i.e. $\ell(j)$ and $u(j)$, from their respective values in the previous iteration, i.e. from $\ell(j-1)$ and $u(j-1)$. This re-use of earlier computations promises a cut in processing times due to the elimination of redundant effort. For the sake of efficiency, this procedure should not refer directly to the fuzzy set X from which the coefficients $\ell(j)$ and $u(j)$ are sampled. By contrast, all computations should rest on cardinality information only. A suitable mechanism which extracts exactly those aspects of a fuzzy subset that characterize its cardinality, is provided by the histogram.[4]

Definition 11.46. *By the* histogram *of a fuzzy subset $X \in \widetilde{\mathcal{P}}(E)$ (where E is some finite set), we denote the mapping $Hist_X : \mathbf{I} \longrightarrow \mathbb{N}$ defined by*

$$Hist_X(z) = |\mu_X^{-1}(z)| = |\{e \in E : \mu_X(e) = z\}| \tag{11.78}$$

for all $z \in \mathbf{I}$.

Notes 11.47.

- It is apparent from the assumed finiteness of E that Hist_X is well-defined, i.e. the term $|\mu_X^{-1}(z)|$ always refers to a finite cardinality.
- The proposed histogram representation can be likened to Rescher's *truth status statistics* ξ_z, which it generalizes from the m-valued to the continuous valued case, see [137, p. 201].

Let us now recall the notation

$$A(X) = \{\mu_X(e) : e \in E\}$$

introduced in (11.13). The finiteness of E obviously entails that $A(X)$ be a finite set as well. In addition, it is quite clear that $A(X)$ is the support of Hist_X, i.e.

$$A(X) = \{z \in \mathbf{I} : \mathrm{Hist}_X(z) \neq 0\}\,.$$

In particular, Hist_X has finite support. This is important because having finite support, Hist_X can be uniquely determined from a finite representation. To see this, let us consider a finite set $A \supseteq A(X)$; because $A(X)$ is finite, such sets are known to exist. We can then restrict Hist_X to the finite sample of $z \in A$, by defining a corresponding mapping $H_A : A \longrightarrow \mathbb{N}$ according to

$$H_A(z) = \mathrm{Hist}_X(z)$$

[4] This is apparent because two fuzzy subsets $X \in \mathcal{P}(E)$, $X' \in \mathcal{P}(E')$ of finite base sets $E, E' \neq \varnothing$ have the same histograms, i.e. $\mathrm{Hist}_X = \mathrm{Hist}_{X'}$, if and only if there exists a bijection $\beta : E \longrightarrow E'$ such that $X' = \hat{\hat{\beta}}(X)$ (and thus also $X = \hat{\hat{\beta^{-1}}}(X)$). Consequently, histograms capture exactly the cardinality aspect of fuzzy sets, because two finite sets have the same cardinality if and only if there exists a bijection between them.

for all $z \in A$. It is obvious that Hist_X can be recovered from H_A, because $A \supseteq A(X)$ grants that the support of Hist_X is contained in A. Therefore $z \in \mathbf{I} \setminus A$ entails that $\mathrm{Hist}_X(z) = 0$. This proves that knowing H_A is sufficient to determine $\mathrm{Hist}_X(z)$ in the full range of all $z \in \mathbf{I}$, because $\mathrm{Hist}_X(z)$ can be expressed in terms of H_A according to the obvious rule

$$\mathrm{Hist}_X(z) = \begin{cases} H_A(z) & : \quad z \in A \\ 0 & : \quad \text{else} \end{cases} \tag{11.79}$$

for all $z \in \mathbf{I}$. The benefit gained by this representation is that the 'generating' mapping $H_A : A \longrightarrow \mathbb{N}$ is declared on a finite domain. Therefore H_A can be specified by listing the finite number of pairs in its graph, i.e. as a finite list which comprises all pairs in $\{(z, \mathrm{Hist}_X(z)) : z \in A\}$. Obviously, this demonstrates that the histogram of X has a finite specification, and is thus suitable for representation on digital computers (modulo the finite precision of floating point numbers, of course).

Having introduced the histogram-based representation of the considered fuzzy subset X, we will now connect this representation with the given set of three-valued cut levels $\Gamma \supseteq \Gamma(X)$, in order to achieve an incremental computation of the cardinality coefficients. Hence suppose that $\Gamma = \{\gamma_0, \dots, \gamma_m\} \supseteq \Gamma(X_1, \dots, X_n)$ is given, where $0 = \gamma_0 < \gamma_1 < \dots < \gamma_{m-1} < \gamma_m = 1$. Let us define $A_\Gamma \in \mathcal{P}(\mathbf{I})$ by

$$A_\Gamma = \{\tfrac{1}{2} + \tfrac{1}{2}\gamma : \gamma \in \Gamma\} \cup \{\tfrac{1}{2} - \tfrac{1}{2}\gamma : \gamma \in \Gamma\} . \tag{11.80}$$

It is apparent from the finiteness of Γ that A_Γ has finite cardinality as well. Furthermore, A_Γ clearly satisfies the desired condition that

$$A_\Gamma \supseteq A_{\Gamma(X)} \supseteq A(X) . \tag{11.81}$$

This is obvious from $\Gamma \supseteq \Gamma(X)$, (11.13), (11.14) and (11.80). By the above reasoning, we then know that the support of Hist_X is contained in A_Γ, i.e. the histogram of X can be represented by $H_{A_\Gamma} : A_\Gamma \longrightarrow \mathbb{N}$, see (11.79). For notational convenience, we will not refer to H_{A_Γ} directly. It will be advantageous to split H_{A_Γ} into two mappings H^+ and H^-, which let us access the histogram information in terms of j rather than $\overline{\gamma}_j$. These mappings $H^+, H^- : \{0, \dots, m\} \longrightarrow \mathbf{I}$ are defined by

$$H^+(j) = \mathrm{Hist}_X(\tfrac{1}{2} + \tfrac{1}{2}\gamma_j) = |\{e \in E : \mu_X(e) = \tfrac{1}{2} + \tfrac{1}{2}\gamma_j\}| \tag{11.82}$$

$$H^-(j) = \mathrm{Hist}_X(\tfrac{1}{2} - \tfrac{1}{2}\gamma_j) = |\{e \in E : \mu_X(e) = \tfrac{1}{2} - \tfrac{1}{2}\gamma_j\}| \tag{11.83}$$

This is just a convenient representation of H_{A_Γ}, because H^+ and H^- are apparently sampled from H_{A_Γ}, and because the original mapping H_{A_Γ} can be recovered from H^+ and H^- as follows,

$$H_{A_\Gamma}(z) = \begin{cases} H^+(j) & : \quad z = \tfrac{1}{2} + \tfrac{1}{2}\gamma_j \\ H^-(j) & : \quad z = \tfrac{1}{2} - \tfrac{1}{2}\gamma_j \end{cases}$$

for all $z \in A_\Gamma$. Based on the mappings H^+ and H^-, which must be computed from X in a preprocessing step, it is now possible to formulate an incremental procedure which computes the current cardinality bounds $\ell(j)$ and $u(j)$ in the j-th iteration from the cardinality bounds $\ell(j-1)$ and $u(j-1)$ computed in the previous iteration and from the cardinality information stored in $H^+(j)$ and $H^-(j)$.

Theorem 11.48.

Let $E \neq \varnothing$ be a finite base set and $X \in \widetilde{\mathcal{P}}(E)$ a fuzzy subset of E. Further let $\Gamma = \{\gamma_0, \ldots, \gamma_m\} \supseteq \Gamma(X)$ be given such that $0 = \gamma_0 < \gamma_1 < \cdots < \gamma_{m-1} < \gamma_m = 1$. Then

$$\begin{aligned} \ell(0) &= \sum_{k=1}^{m} H^+(k) \\ u(0) &= \ell(0) + H^+(0)\,, \end{aligned}$$

and

$$\begin{aligned} \ell(j) &= \ell(j-1) - H^+(j) \\ u(j) &= u(j-1) + H^-(j) \end{aligned}$$

for all $j \in \{1, \ldots, m-1\}$.

Obviously, neither the formulas for initialisation nor those for updating refer to the original fuzzy set; it is only the histogram of X that must be known for computing the cardinality bounds. We have therefore succeeded in eliminating the fuzzy set X from the procedure for computing $\ell(j)$ and $u(j)$. Unlike the fuzzy sets themselves, their histograms all look alike. This greatly facilitates processing, because the histograms of arbitrary fuzzy sets can easily be represented in a uniform way. Due to the fact that the update rules described in Th. 11.48 only involve a single lookup operation, i.e. one access to the histogram data in $H^+(j)$ or $H^-(j)$, and one additional arithmetic operation (sum), the proposed method results in a very fast implementation. The presentation of a histogram-based update rule in Th. 11.48 completes the analysis of fuzzy quantification in $\mathcal{F}_\xi$-DFSes and the subsequent discussion of possible efficiency improvements.

From a more general perspective, we have shown that $\mathcal{F}_\xi(Q)(X_1, \ldots, X_n)$ can always be expressed as a function

$$\mathcal{F}_\xi(Q)(X_1, \ldots, X_n) = \Xi(\mathrm{Hist}_{Z_1}, \ldots, \mathrm{Hist}_{Z_L})$$

which depends only on the histograms of Boolean combinations $Z_1, \ldots, Z_L$ of the arguments. However, this should not lure us into *defining* fuzzy quantifiers as functions of histograms in the spirit of Rescher's treatment of quantifiers in many-valued logic, cf. [136] and [137, pp. 197-206]. This is because the histogram-based analysis rests on the assumption of automorphism-invariance, a property shared by most, but not all, NL quantifiers. When trying to define linguistic quantifiers directly in terms of $\Xi(\mathrm{Hist}_{Z_1}, \ldots, \mathrm{Hist}_{Z_L})$, the correspondence problem discussed in Chap. 2 returns with full force. Consequently, the analysis of certain quantifiers in terms of histograms is only possible in special cases, and it would not admit a comprehensive treatment of fuzzy NL quantification. In addition, it makes no provisions as to the systematicity of interpretation.

11.7 Implementation of Unary Quantifiers

This section finally presents the complete algorithms for quantitative one-place quantifiers on finite base sets. In principle, the necessary techniques for implementing these quantifiers in the prototypical models $\mathcal{F}_{\mathrm{owa}}$, $\mathcal{M}$ and $\mathcal{M}_{CX}$ have already been described. Nevertheless, it is presumably helpful to collect the results developed in the previous sections and show how the different parts of the implementation work in concert.

Hence let $E \neq \varnothing$ be a finite base set and suppose that $Q : \mathcal{P}(E) \longrightarrow \mathbf{I}$ is quantitative. Further let $X \in \widetilde{\mathcal{P}}(E)$ be a given fuzzy argument. We then know from Th. 11.27, Th. 11.30 and Th. 11.31 that $\mathcal{F}_{\mathrm{owa}}(Q)(X)$, $\mathcal{M}(Q)(X)$ and $\mathcal{M}_{CX}(Q)(X)$ can be computed from quantities $\top_j$, $\bot_j$ sampled from some choice of $\Gamma = \{\gamma_0, \ldots, \gamma_m\} \supseteq \Gamma(X)$, where $0 = \gamma_0 < \gamma_1 < \cdots < \gamma_m = 1$. For convenience, we will assume in the following that $\Gamma = \Gamma(X)$. Recalling (11.14), (11.82) and (11.83), it is then apparent how H^+, H^- as well as γ_j, $j = 0, \ldots, m$ can be computed from the fuzzy set X. We will write `H+[j]` and `H-[j]` to denote $H^+(j)$ and $H^-(j)$, respectively. An $(m+1)$-dimensional array `G` will be used to access the γ_j's, i.e. `G[j]` denotes γ_j, $j = 0, \ldots, m$. In particular, this means that `G[0]` $= 0$ and `G[m]` $= 1$, cf. (11.14). The resulting algorithms which compute $\mathcal{F}_{\mathrm{owa}}(Q)(X)$, $\mathcal{M}(Q)(X)$ and $\mathcal{M}_{CX}(Q)(X)$ from these data are depicted in Table 11.1, 11.2 and 11.3, respectively. Starting with $j = 0$, each of the algorithms enters a main loop which increments j in each iteration. This lets us compute the current value of the cardinality bounds $\ell = |X|_{\overline{\gamma}_j}^{\min}$ and $u = |X|_{\overline{\gamma}_j}^{\max}$ in the j-th iteration from their values in the previous iteration and from the histogram information stored in `H+` and `H-` as described in Th. 11.48. The updating of these coefficients takes place in the code fragment with the heading 'update clauses'. The quantities $\top_j = \top_{Q,X}(\overline{\gamma}_j)$, $\bot_j = \bot_{Q,X}(\overline{\gamma}_j)$ and $C_j = \mathrm{med}_{\frac{1}{2}}(\top_j, \bot_j)$ can then be computed from $q^{\min}(\ell, u)$ and $q^{\max}(\ell, u)$, as described in Th. 11.15. Apart from incorporating the update rule for the cardinality coefficients ℓ and u, the algorithms merely reflect the earlier analysis in Th. 11.27, Th. 11.30 and Th. 11.31.

In these algorithms, the fuzzy set $X \in \widetilde{\mathcal{P}}(E)$ was supplied as the input parameter, which directly reflects the expected input-output behaviour (i.e.

Table 11.1. Algorithm for evaluating quantitative one-place quantifiers in $\mathcal{F}_{\mathrm{owa}}$ (based on floating-point arithmetics)

Algorithm for computing $\mathcal{F}_{\mathrm{owa}}(Q)(X)$
(floating-point arithmetics)

```
INPUT: X
Compute G, H+, H- and m; // see text
// initialize ℓ, u
ℓ := Σ_{j=1}^{m} H+[j];
u := ℓ + H+[0];
sum = G[1] * (q^min(ℓ,u) + q^max(ℓ,u));
for( j := 1; j<m; j := j + 1 ) {
    // update clauses for ℓ and u
    ℓ := ℓ - H+[j];
    u := u + H-[j];
    sum := sum + (G[j+1] - G[j])
                  * (q^min(ℓ,u) + q^max(ℓ,u));
}
return sum / 2;
END
```

Table 11.2. Algorithm for evaluating quantitative one-place quantifiers in $\mathcal{M}$ (based on floating-point arithmetics)

Algorithm for computing $\mathcal{M}(Q)(X)$
(floating-point arithmetics)

```
INPUT: X
Compute G, H+, H- and m; // see text
// initialize ℓ, u
ℓ := Σ_{j=1}^{m} H+[j];
u := ℓ + H+[0];
⊤ := q^max(ℓ,u);
⊥ := q^min(ℓ,u);
if( ⊥ > 1/2 ) {
    sum := G[1] * ⊥;
    for( j := 1; j<m; j := j + 1 ) {
        // update clauses for ℓ and u
        ℓ := ℓ - H+[j];
        u := u + H-[j];
        ⊥ := q^min(ℓ,u);
        if( ⊥ ≤ 1/2 )
            break;
        sum := sum + (G[j+1] - G[j]) * ⊥;
        }
    }
else if( ⊤ < 1/2 ) {
    sum := G[1] * ⊤;
    for( j := 1; j<m; j := j + 1 ) {
        // update clauses for ℓ and u
        ℓ := ℓ - H+[j];
        u := u + H-[j];
        ⊤ := q^max(ℓ,u);
        if( ⊤ ≥ 1/2 )
            break;
        sum := sum + (G[j+1] - G[j]) * ⊤;
        }
    }
else
    { return 1/2; }
return sum + 1/2 * (1 - G[j]);
END
```

X is the input, and $\mathcal{F}(Q)(X)$ is the returned output). The histogram infor-

Table 11.3. Algorithm for evaluating quantitative one-place quantifiers in $\mathcal{M}_{CX}$ (based on floating-point arithmetics)

Algorithm for computing $\mathcal{M}_{CX}(Q)(X)$
(floating-point arithmetics)

```
INPUT: X
Compute G, H+, H- and m; // see text
// initialize ℓ, u
ℓ := Σ_{j=1}^{m} H+[j];
u := ℓ + H+[0];
⊤ := q^max(ℓ,u);
⊥ := q^min(ℓ,u);
j := 0;
if( ⊥ > 1/2 ) {
    B := 2 * ⊥ - 1;
    while( B > G[j+1] ) {
        j := j + 1;
        // update clauses for ℓ and u
        ℓ := ℓ - H+[j];
        u := u + H-[j];
        B := 2 * q^min(ℓ,u) - 1;
        }
    return 1/2 + 1/2 max(B, G[j]);
    }
else if( ⊤ < 1/2 ) {
    B := 1 - 2 * ⊤;
    while( B > G[j+1] ) {
        j := j + 1;
        // update clauses for ℓ and u
        ℓ := ℓ - H+[j];
        u := u + H-[j];
        B := 1 - 2 * q^max(ℓ,u);
        }
    return 1/2 - 1/2 max(B, G[j]);
    }
return 1/2;
END
```

mation in H^+ and H^- and the array G which collects the γ_j's, must then be computed from X in the obvious way. However, it also makes sense to consider a variant of these algorithms where the algorithms expect H^+, H^- and G as their inputs, rather than the fuzzy set X from which these coefficients have been sampled. This separation of the histogram extraction from the core algorithms will abstract from any specifics concerning the representation of the fuzzy set X (simply because the resulting histograms can be represented in a uniform way, which need not be the case for different types of fuzzy subsets and heterogeneous domains). By separating the histogram extraction from the main algorithms, it becomes possible to develop a generic solution which proves useful for implementing quantifiers in arbitrary domains.

The three algorithms presented so far resort to floating-point arithmetics, i.e. the membership grades $\mu_X(e)$ and consequently the γ_j's are allowed to assume arbitrary values in the unit interval. In some cases, however, it can be advantageous to use a reformulation in terms of integer arithmetics. Roughly speaking, the integer-based solution will be beneficial if the size of the domain is large compared to the number of supported membership grades. This will be the case e.g. in fuzzy image processing, where the size of the domain (number

Table 11.5. Algorithm for evaluating quantitative one-place quantifiers in $\mathcal{M}$ (based on integer arithmetics)

Algorithm for computing $\mathcal{M}(Q)(X)$
(integer arithmetics)

```
INPUT: X
// initialise H, ℓ, u
H := Hist_X;
ℓ := Σ_{j=1}^{m} H[m+j];
u := ℓ + H[m];
⊤ := q^max(ℓ,u);
⊥ := q^min(ℓ,u);
if( ⊥ > 1/2 ) {
    sum := ⊥;
    for( j := 1; j<m; j := j + 1 ) {
        nc:= true; // "no change"
        // update clauses for ℓ and u
        if( H[m+j] ≠ 0 )
            { ℓ := ℓ - H[m+j]; nc:= false; }
        if( H[m-j] ≠ 0 )
            { u := u + H[m-j]; nc:= false; }
        if( nc)
            { sum := sum + ⊥; continue; }
        // one of ℓ or u has changed
        ⊥ := q^min(ℓ,u);
        if( ⊥ ≤ 1/2 )
            break;
        sum := sum + ⊥;
        }
    }
else if( ⊤ < 1/2 ) {
    sum := ⊤;
    for( j := 1; j<m; j := j + 1 ) {
        nc:= true;
        if( H[m+j] ≠ 0 )
            { ℓ := ℓ - H[m+j]; nc:= false; }
        if( H[m-j] ≠ 0 )
            { u := u + H[m-j]; nc:= false; }
        if( nc)
            { sum := sum + ⊤; continue; }
        // one of ℓ or u has changed
        ⊤ := q^max(ℓ,u);
        if( ⊤ ≥ 1/2 )
            break;
        sum := sum + ⊤;
        }
    }
else
    { return 1/2; }
return (sum + 1/2*(m-j)) / m;
END
```

11.8 Implementation of Absolute Quantifiers and Quantifiers of Exception

Unary quantifiers, like those discussed in the previous section, usually do not express directly at the NL surface. Nevertheless, the modelling and implementation of these quantifiers proves invaluable for treating important classes of quantifiers in NL: Those which depend on an absolute count (i.e., *absolute quantifiers*), and those which specify the admissible exceptions to a general rule (i.e., *quantifiers of exception*). Here we shall formally define these classes of quantifiers and explain how quantifying expressions involving such quantifiers can be evaluated.

Table 11.6. Algorithm for evaluating quantitative one-place quantifiers in $\mathcal{M}_{CX}$ (based on integer arithmetics)

Algorithm for computing $\mathcal{M}_{CX}(Q)(X)$
(integer arithmetics)

```
INPUT: X
// initialise H, ℓ, u
H := Hist_X;
ℓ := Σ_{j=1}^{m} H[m+j];
u := ℓ + H[m];
⊤ := q^max(ℓ,u);
⊥ := q^min(ℓ,u);
if( ⊥ > 1/2 ) {
    for( j := 1; j<m; j := j + 1 ) {
        ch := false; // "change"
        // update clauses for ℓ and u
        if( H[m+j] ≠ 0 )
            { ℓ := ℓ - H[m+j]; ch := true; }
        if( H[m-j] ≠ 0 )
            { u := u + H[m-j]; ch := true; }
        if( ch )
            // one of ℓ or u has changed
            { ⊥ := q^min(ℓ,u); }
        if( ⊥ ≤ m+j )
            { return (m+j)/m'; }
        }
    return 1;
    }
else if( ⊤ < 1/2 ) {
    for( j := 1; j<m; j := j + 1 ) {
        ch := false;
        // update clauses for ℓ and u
        if( H[m+j] ≠ 0 )
            { ℓ := ℓ - H[m+j]; ch := true; }
        if( H[m-j] ≠ 0 )
            { u := u + H[m-j]; ch := true; }
        if( ch )
            // one of ℓ or u has changed
            { ⊤ := q^max(ℓ,u); }
        if( ⊤ ≥ m-j )
            { return (m-j)/m'; }
        }
    return 0;
    } return 1/2;
END
```

Definition 11.49. *For every two-place semi-fuzzy quantifier $Q : \mathcal{P}(E)^2 \longrightarrow \mathbf{I}$,*

- *Q is called* absolute *if there exists a quantitative unary quantifier $Q' : \mathcal{P}(E) \longrightarrow \mathbf{I}$ such that $Q = Q'\cap$, i.e. $Q(Y_1, Y_2) = Q'(Y_1 \cap Y_2)$ for all $Y_1, Y_2 \in \mathcal{P}(E)$. In particular,*

$$Q(Y_1, Y_2) = q(|Y_1 \cap Y_2|) \tag{11.87}$$

  for some $q : \{0, \dots, |E|\} \longrightarrow \mathbf{I}$.
- *Q is called a* quantifier of exception *if there exists an absolute quantifier $Q'' : \mathcal{P}(E)^2 \longrightarrow \mathbf{I}$ such that $Q = Q''\neg$, i.e. $Q(Y_1, Y_2) = Q''(Y_1, \neg Y_2)$ for all $Y_1, Y_2 \in \mathcal{P}(E)$. In particular*

$$Q(Y_1, Y_2) = q(|Y_1 \cap \neg Y_2|) \tag{11.88}$$

  for some $q : \{0, \dots, |E|\} \longrightarrow \mathbf{I}$.

Table 11.7. Examples of quantifiers of exception

Quantifier	Antonym (absolute)
all	no
all except exactly k	exactly k
all except about k	about k
all except at most k	at most k

Notes 11.50.

- Absolute quantifiers are of course well-known both in fuzzy set theory and in TGQ; the above definition only serves to describe more precisely the class of quantifiers handled by the algorithms to follow. As already mentioned in the introduction, absolute quantifiers are called 'quantifiers of the first kind' in Zadeh's publications on fuzzy quantifiers, e.g. [197, p. 149]. In TGQ, these quantifiers are normally called 'absolute', but Keenan and Moss [91, p. 98] also use the term 'cardinal determiner'.
- Quantifiers of exception are not mentioned in the literature on fuzzy quantifiers, but these quantifiers are well-known to TGQ, see e.g. [92]. The term 'exception determiner' is also common [91, p. 123].

To give one more example of the first type, the two-place quantifier "about 50" is an absolute quantifier. Some examples of quantifiers of exception are presented in Table 11.7. The DFS axioms ensure that

$$\mathcal{F}(Q)(X_1, X_2) = \mathcal{F}(Q')(X_1 \cap X_2)\,, \tag{11.89}$$

whenever $Q = Q'\cap$ is an absolute quantifier and $X_1, X_2 \in \widetilde{\mathcal{P}}(E)$. Similarly if $Q = Q'\cap\neg$ is a quantifier of exception, then

$$\mathcal{F}(Q)(X_1, X_2) = \mathcal{F}(Q')(X_1 \cap \neg X_2)\,, \tag{11.90}$$

for all $X_1, X_2 \in \widetilde{\mathcal{P}}(E)$, where $Q' : \mathcal{P}(E) \longrightarrow \mathbf{I}$ is the unary base quantifier from which Q is constructed. Owing to these reductions, we can therefore use the algorithms for computing $\mathcal{F}(Q')(X)$, $\mathcal{F} \in \{\mathcal{M}_{CX}, \mathcal{M}, \mathcal{F}_{\mathrm{owa}}\}$ to evaluate absolute quantifiers and quantifiers of exception.

For example, consider the absolute quantifier **about 50** $: \mathcal{P}(E)^2 \longrightarrow \mathbf{I}$. According to the above definition, we can then reduce the absolute quantifier to an intersection

$$\textbf{about 50} = [\sim\!50]\cap\,,$$

where $[\sim\!50] : \mathcal{P}(E) \longrightarrow \mathbf{I}$ is a suitable unary quantifier. A possible choice of $[\sim\!50]$ is the following.

$$[\sim\!50](Y) = \mathrm{SZ}(|Y|, 35, 45, 55, 85)$$

for all $Y \in \mathcal{P}(E)$, where 'SZ' is Zadeh's SZ-function [197, p. 184] (which has roughly the shape of a smoothened trapezoid), supposed to be nonzero in the range 35..85 and to reach full membership in the range 45..55. Owing to (11.89), we can then use the fuzzy quantifier $\mathcal{F}([\sim 50]) : \widetilde{\mathcal{P}}(E) \longrightarrow \mathbf{I}$ in order to interpret "about 50" in the chosen model, e.g. for $\mathcal{F} = \mathcal{M}_{CX}$,

$$\mathcal{M}_{CX}(\mathbf{about\ 50})(X_1, X_2) = \mathcal{M}_{CX}([\sim 50])(X_1 \cap X_2)\,,$$

for all $X_1, X_2 \in \widetilde{\mathcal{P}}(E)$. On the other hand, consider the quantifier of exception "all except k", which can be defined by

$$\mathbf{all\ except\ k}(Y_1, Y_2) = [\leq k](Y_1 \cap \neg Y_2)$$

for all $Y_1, Y_2 \in \mathcal{P}(E)$, i.e. **all except k** $= [\leq k]\widetilde{\cap}\neg$. Let us now see how $\mathcal{M}_{CX}([\leq k])$ can be used for implementing $\mathcal{M}_{CX}(\mathbf{all\ except\ k})$. Following the generic solution presented in (11.90), the computation of the quantifying expression $\mathcal{M}_{CX}(\mathbf{all\ except\ k})(X_1, X_2)$ then reduces to

$$\begin{aligned}\mathcal{M}_{CX}(\mathbf{all\,except\,k})(X_1, X_2) &= \mathcal{M}_{CX}([\leq k])(X_1 \cap \neg X_2) \\ &= 1 - \mu_{[k+1]}(X_1 \cap \neg X_2)\,,\end{aligned}$$

for all $X_1, X_2 \in \widetilde{\mathcal{P}}(E)$. Similar examples of practical relevance can be treated along the same lines. Noticing that the assumptions which this analysis makes on $\mathcal{F}$ are valid in every DFS, this reduction to the unary case is possible in arbitrary models.

11.9 Implementation of Proportional Quantifiers

Proportional quantifiers like "almost all", "most" or "many" are probably the most relevant from an application point of view: in many cases, it is not the absolute numbers, but rather some proportion or ratio of quantities that raises interest. Hence let us define this basic type of quantifiers formally, which lets us communicate about such proportions:

Definition 11.51.
A two-place semi-fuzzy quantifier $Q : \mathcal{P}(E)^2 \longrightarrow \mathbf{I}$ on a finite base set is called proportional *if there exist $f : \mathbf{I} \longrightarrow \mathbf{I}$ and $v_0 \in \mathbf{I}$ such that*

$$Q(Y_1, Y_2) = q(c_1, c_2) = \begin{cases} f(c_2/c_1) & : \quad c_1 \neq 0 \\ v_0 & : \quad \mathit{else} \end{cases}$$

for all $Y_1, Y_2 \in \mathcal{P}(E)$, where $c_1 = |Y_1|$ and $c_2 = |Y_1 \cap Y_2|$.

Notes 11.52.

- For example, the proposed definition of "almost all" in terms of equation (2.8) fits into this scheme, where $f(z) = S(x, 0.7, 0.9)$ and $v_0 = 1$.

Table 11.9. Algorithm for evaluating two-place proportional quantifiers in $\mathcal{M}$ (based on floating-point arithmetics)

Algorithm for computing $\mathcal{M}(Q)(X_1, X_2)$
(floating-point arithmetics)

```
INPUT: X₁, X₂
Compute G, m, H₁⁺, H₂⁺,H₃⁺,H₁⁻, H₂⁻, H₃⁻; // see text
// initialize ℓ, u:
for( r := 1; r ≤ 3; r := r+1 )
    { ℓ_r := Σ_{j=1}^{m} H_r⁺[j]; u_r := ℓ_r + H_r⁺[0]; }
⊤ := q^max(ℓ₁, ℓ₃, u₁, u₂);
⊥ := q^min(ℓ₁, ℓ₂, u₁, u₃);
if( ⊥ > 1/2 ) {
    sum := G[1] * ⊥;
    for( j := 1; j<m; j := j + 1 ) {
        // update clauses for ℓ₁, ℓ₂, u₁, u₃
        ℓ₁ := ℓ₁ - H₁⁺[j];
        ℓ₂ := ℓ₂ - H₂⁺[j];
        u₁ := u₁ + H₁⁻[j];
        u₃ := u₃ + H₃⁻[j];
        ⊥ := q^min(ℓ₁, ℓ₂, u₁, u₃);
        if( ⊥ ≤ 1/2 )
            break;
        sum := sum + (G[j+1] - G[j]) * ⊥;
        }
    }
else if( ⊤ < 1/2 ) {
    sum := G[1] * ⊤;
    for( j := 1; j<m; j := j + 1 ) {
        // update clauses for ℓ₁, ℓ₃, u₁, u₂
        ℓ₁ := ℓ₁ - H₁⁺[j];
        ℓ₃ := ℓ₃ - H₃⁺[j];
        u₁ := u₁ + H₁⁻[j];
        u₂ := u₂ + H₂⁻[j];
        ⊤ := q^max(ℓ₁, ℓ₃, u₁, u₂);
        if( ⊤ ≥ 1/2 )
            break;
        sum := sum + (G[j+1] - G[j]) * ⊤;
        }
    }
else
    { return 1/2; }
return sum + 1/2 * (1 - G[j]);
END
```

that if Q (and hence f) is nonincreasing, then $Q' = \neg Q$ is also a proportional quantifier, which can be described in terms of $f' = \neg f$ and $v_0' = \neg v_0$. Noticing that the resulting mapping f' is nondecreasing, the above algorithms can be used to compute $\mathcal{F}(Q')(X_1, X_2)$, where $\mathcal{F}$ is the model of interest. According to Th. 4.20, the computation of $\mathcal{F}(Q)(X_1, X_2)$ can therefore be reduced to $\mathcal{F}(Q)(X_1, X_2) = \neg\mathcal{F}(Q')(X_1, X_2)$.

We have stated the above algorithms for a special case often met in practice, in order to demonstrate the potential for optimization. However, these algorithms are easily adapted to general proportional quantifiers, which need not fulfill any monotonicity requirements. To this end, it is sufficient to replace all occurrences of $q^{\max}(\ell_1, \ell_3, u_1, u_2)$ and $q^{\min}(\ell_1, \ell_2, u_1, u_3)$ in the above algorithms with their generic counterparts, i.e. $q^{\max}(\ell_1, \ell_2, \ell_3, u_1, u_2, u_3)$ and $q^{\min}(\ell_1, \ell_2, \ell_3, u_1, u_2, u_3)$ defined by (11.91) and (11.92). Additional update

Table 11.10. Algorithm for evaluating two-place proportional quantifiers in $\mathcal{M}_{CX}$ (based on floating-point arithmetics)

Algorithm for computing $\mathcal{M}_{CX}(Q)(X_1, X_2)$
(floating-point arithmetics)

```
INPUT: X_1, X_2
Compute G, m, H_1^+, H_2^+, H_3^+, H_1^-, H_2^-, H_3^-; // see text
// initialize ℓ, u:
for( r := 1; r ≤ 3; r := r+1 )
   { ℓ_r := Σ_{j=1}^{m} H_r^+[j]; u_r := ℓ_r + H_r^+[0]; }
⊤ := q^max(ℓ_1, ℓ_3, u_1, u_2);
⊥ := q^min(ℓ_1, ℓ_2, u_1, u_3);
j := 0;
if( ⊥ > 1/2) {
   B := 2 * ⊥ - 1;
   while( B > G[j+1] ) {
      j := j + 1;
      // update clauses for ℓ_1, ℓ_2, u_1, u_3
      ℓ_1 := ℓ_1 - H_1^+[j];
      ℓ_2 := ℓ_2 - H_2^+[j];
      u_1 := u_1 + H_1^-[j];
      u_3 := u_3 + H_3^-[j];
      B := 2 * q^min(ℓ_1, ℓ_2, u_1, u_3) - 1;
   }
   return 1/2 + 1/2 max(B, G[j]);
}
else if( ⊤ < 1/2 ) {
   B := 1 - 2 * ⊤;
   while( B > G[j+1] ) {
      j := j + 1;
      // update clauses for ℓ_1, ℓ_3, u_1, u_2
      ℓ_1 := ℓ_1 - H_1^+[j];
      ℓ_3 := ℓ_3 - H_3^+[j];
      u_1 := u_1 + H_1^-[j];
      u_2 := u_2 + H_2^-[j];
      B := 1 - 2 * q^max(ℓ_1, ℓ_3, u_1, u_2);
   }
   return 1/2 - 1/2 max(B, G[j]);
}
return 1/2;
END
```

clauses must then be added, because it is now necessary to keep track of all coefficients ℓ_1, ℓ_2, ℓ_3, u_1, u_2 and u_3.

As in the case of unary quantifiers, it also makes sense to develop special versions of the algorithms, optimized for integer arithmetics. Hence let Q be a proportional quantifier and suppose that $X_1, X_2 \in \widetilde{\mathcal{P}}(E)$ are given fuzzy arguments. Further let $m' \in \mathbb{N} \setminus \{0\}$ be given, which specifies the available range of integers. Once m' is chosen, admissible fuzzy arguments are no longer allowed to assume arbitrary membership grades in $\mathbf{I} = [0, 1]$. By contrast, we again require that

$$\mu_{X_i}(e) \in U_{m'}$$

for all $e \in E$ and $i \in \{1, 2\}$, where $U_{m'}$ is defined as in the unary case, cf. (11.85). Under these assumptions on legal choices of fuzzy arguments, the above algorithms which implement proportional quantifiers in $\mathcal{F}_{\mathrm{owa}}$, $\mathcal{M}$ and $\mathcal{M}_{CX}$ can now be adapted in complete analogy to the earlier changes for

Table 11.11. Algorithm for evaluating two-place proportional quantifiers in $\mathcal{F}_{\mathrm{owa}}$ (based on integer arithmetics)

Algorithm for computing $\mathcal{F}_{\mathrm{owa}}(Q)(X_1, X_2)$
(integer arithmetics)

```
INPUT: X_1, X_2
// initialise H_r, ℓ, u
H_1 := Hist_{X_1};
H_2 := Hist_{X_1 ∩ X_2};
H_3 := Hist_{X_1 ∩ ¬X_2};
for( r := 1; r ≤ 3; r := r+1 )
    { ℓ_r := Σ_{j=1}^{m} H_r[m+j]; u_r := ℓ_r + H_r[m]; }
cq := q^min(ℓ_1, ℓ_2, u_1, u_3) + q^max(ℓ_1, ℓ_3, u_1, u_2);
sum := cq;
for( j := 1; j<m; j := j + 1 ) {
    ch := false; // "change"
    // update clauses for ℓ_1, ℓ_2, ℓ_3, u_1, u_2, u_3
    if( H_1[m+j] ≠ 0 )
        { ℓ_1 := ℓ_1 - H_1[m+j]; ch := true; }
    if( H_2[m+j] ≠ 0 )
        { ℓ_2 := ℓ_2 - H_2[m+j]; ch := true; }
    if( H_3[m+j] ≠ 0 )
        { ℓ_3 := ℓ_3 - H_3[m+j]; ch := true; }
    if( H_1[m-j] ≠ 0 )
        { u_1 := u_1 + H_1[m-j]; ch := true; }
    if( H_2[m-j] ≠ 0 )
        { u_2 := u_2 + H_2[m-j]; ch := true; }
    if( H_3[m-j] ≠ 0 )
        { u_3 := u_3 + H_3[m-j]; ch := true; }
    if( ch )
        // one of the ℓ_r or u_r has changed
        { cq := q^min(ℓ_1, ℓ_2, u_1, u_3) + q^max(ℓ_1, ℓ_3, u_1, u_2); }
    sum := sum + cq;
    }
return sum / m'; // where m' = 2*m
END
```

unary quantifiers, see pp. 315–317. The resulting algorithms for proportional quantification based on integer arithmetics are shown in Table 11.11 ($\mathcal{F}_{\mathrm{owa}}$), Table 11.12 ($\mathcal{M}$), and Table 11.13 (for $\mathcal{M}_{CX}$).

Specifically, the algorithms have again been augmented with flags `ch` ('change') or `nc` ('no change'), which keep track of any changes of the ℓ_r or u_r in the current iteration. Only if `ch` is set to true (or `nc` to false), a recomputation of $q^{\min}$ or $q^{\max}$ becomes necessary; otherwise, the algorithms will directly skip forward to the next iteration.
In practice, the use of these integer-based algorithms should again be considered when processing speed is more important than accuracy, i.e. the domain of quantification is very large, and the loss of precision due to a small choice of m' is acceptable. The final decision between the floating-point and integer-based variants should then be based on practical tests, which assess the processing times for typical instances of quantification.

The presented algorithms permit an efficient implementation which will be fast enough for most applications. Due to the inherent complexity of proportional quantification compared to the unary case, the algorithms are necessarily more complicated and computationally demanding than those for simple one-place quantifiers. In some situations, however, it is possible to short-cut the general solution and delegate proportional quantification to the fast

Table 11.12. Algorithm for evaluating two-place proportional quantifiers in $\mathcal{M}$ (based on integer arithmetics)

Algorithm for computing $\mathcal{M}(Q)(X_1, X_2)$
(integer arithmetics)

```
INPUT: X1, X2
// initialise Hr, ℓ, u
H1 := Hist_{X1};
H2 := Hist_{X1∩X2};
H3 := Hist_{X1∩¬X2};
for( r := 1; r ≤ 3; r := r+1 )
    { ℓr := Σ_{j=1}^{m} Hr[m+j]; ur := ℓr + Hr[m]; }
⊤ := q^max(ℓ1, ℓ3, u1, u2);
⊥ := q^min(ℓ1, ℓ2, u1, u3);
if( ⊥ > 1/2 ) {
    sum := ⊥;
    for( j := 1; j<m; j := j + 1 ) {
        nc:= true; // "no change"
        // update clauses for ℓ1, ℓ2, u1, u3
        if( H1[m+j] ≠ 0 )
            { ℓ1 := ℓ1 - H1[m+j]; nc:= false; }
        if( H2[m+j] ≠ 0 )
            { ℓ2 := ℓ2 - H2[m+j]; nc:= false; }
        if( H1[m-j] ≠ 0 )
            { u1 := u1 + H1[m-j]; nc:= false; }
        if( H3[m-j] ≠ 0 )
            { u3 := u3 + H3[m-j]; nc:= false; }
        if( nc)
            { sum := sum + ⊥; continue; }
        // one of ℓ1, ℓ2, u1, u3 has changed
        ⊥ := q^min(ℓ1, ℓ2, u1, u3);
        if( ⊥ ≤ 1/2 ) break;
        sum := sum + ⊥;
        }
    }
else if( ⊤ < 1/2 ) {
    sum := ⊤;
    for( j := 1; j<m; j := j + 1 ) {
        nc:= true;
        // update clauses for ℓ1, ℓ3, u1, u2
        if( H1[m+j] ≠ 0 )
            { ℓ1 := ℓ1 - H1[m+j]; nc:= false; }
        if( H3[m+j] ≠ 0 )
            { ℓ3 := ℓ3 - H3[m+j]; nc:= false; }
        if( H1[m-j] ≠ 0 )
            { u1 := u1 + H1[m-j]; nc:= false; }
        if( H2[m-j] ≠ 0 )
            { u2 := u2 + H2[m-j]; nc:= false; }
        if( nc)
            { sum := sum + ⊤; continue; }
        // one of ℓ1, ℓ3, u1, u2 has changed
        ⊤ := q^max(ℓ1, ℓ3, u1, u2);
        if( ⊤ ≥ 1/2 ) break;
        sum := sum + ⊤;
        }
    }
else
    { return 1/2; }
return (sum + 1/2*(m-j)) / m;
END
```

algorithms for unary quantifiers. Specifically, this simplification is possible when the first argument, i.e. the restriction of the quantifier, happens to be a crisp set, as in *"Most married men are bald"*. In this case, the restriction **married men** $\in \mathcal{P}(E)$ can serve as the domain of a derived unary quantifier, as shown by the following theorem.

where **bald**′ is the restriction of the fuzzy set **bald** to the new domain **married men**. Finally we utilize the earlier theorem Th. 11.4 and conclude that

$$\mathcal{F}(\mathbf{most})(\mathbf{married\ men}, \mathbf{bald}) = \mu_{[6]}(\mathbf{bald}')$$

in every standard model.

This example completes the discussion of proportional quantification. We are now prepared to evaluate proportional quantifiers efficiently, either based on floating-point or integer arithmetics. The proposed algorithms for proportional quantification in $\mathcal{F}_{\mathrm{owa}}$, $\mathcal{M}$ and $\mathcal{M}_{CX}$ cover the 'hard' case of proportional two-place quantification, i.e. the aggregation can be weighted by importances, which are specified by the first argument of the quantifier. In Th. 11.53, we further improved the earlier analysis of conservative quantifiers given in Th. 11.44, and thus achieved a more efficient computation in the frequent case of those proportional quantifiers which are also known to be monotonic. Finally, we considered the special case of unrestricted proportional quantification, i.e. unary proportional quantification, or proportional quantification restricted by crisp importances. In this case, we described an apparent reduction which renders applicable the algorithms for one-place quantifiers. In those applications which do not need importance qualification, this reduction will result in an additional boost of processing speed.

11.10 Implementation of Cardinal Comparatives

This section will discuss a natural class of quantifiers which has received little attention in the literature on fuzzy quantifiers: That of *cardinal comparatives* [92, p. 305] or *comparative determiners* [91, p. 123]. This class comprises all NL quantifiers that express a comparison of two cardinalities, which are sampled from two restriction arguments (linguistically realized by two noun phrases A, B) and a common scope argument C (linguistically realized by the verb phrase). For example, the quantifier "more than", which underlies the pattern "More A's than B's are C's", is a cardinal comparative. Formally, we define cardinal comparatives as follows.

Definition 11.56.
Let $E \neq \varnothing$ be a finite base set. A three-place quantifier $Q : \mathcal{P}(E)^3 \longrightarrow \mathbf{I}$ is called a cardinal comparative *if $Q(Y_1, Y_2, Y_3)$ depends on $|Y_1 \cap Y_3|$ and $|Y_2 \cap Y_3|$ only, i.e. there exists $q : \{0, \dots, |E|\}^2 \longrightarrow \mathbf{I}$ such that*

$$Q(Y_1, Y_2, Y_3) = q(c_1, c_2) \tag{11.96}$$

for all $Y_1, Y_2, Y_3 \in \mathcal{P}(E)$, where $c_1 = |Y_1 \cap Y_3|$ and $c_2 = |Y_2 \cap Y_3|$.

For example, "more than" can be defined as

$$\textbf{more than}(Y_1, Y_2, Y_3) = \begin{cases} 1 & : \quad |Y_1 \cap Y_3| - |Y_2 \cap Y_3| > 0 \\ 0 & : \quad \text{else} \end{cases} \tag{11.97}$$

for all $Y_1, Y_2, Y_3 \in \mathcal{P}(E)$. The quantifier can be used to interpret statements like "More men than women own a car", which becomes

$$\textbf{more than}(\textbf{men}, \textbf{women}, \textbf{own a car})\,.$$

We can also model quantifiers of the type "exactly k more than", i.e.

$$\textbf{exactly k more than}(Y_1, Y_2, Y_3) = \begin{cases} 1 & : \quad |Y_1 \cap Y_3| - |Y_2 \cap Y_3| = k \\ 0 & : \quad \text{else} \end{cases}$$

and "at least k more than", i.e.

$$\textbf{at least k more than}(Y_1, Y_2, Y_3) = \begin{cases} 1 & : \quad |Y_1 \cap Y_3| - |Y_2 \cap Y_3| \geq k \\ 0 & : \quad \text{else} \end{cases}$$

for all $Y_1, Y_2, Y_3 \in \mathcal{P}(E)$, where $k \in \mathbb{N}$. Special cases include "more than" and "same number of", which can be expressed as

$$\textbf{more than} = \textbf{at least 1 more than}$$
$$\textbf{same number of} = \textbf{exactly 0 more than}\,.$$

In particular, a statement like *"The same number of men and women are married"* now translates into the quantifying expression

$$\textbf{same number of}(\textbf{men}, \textbf{women}, \textbf{married})\,.$$

Further examples of cardinal comparatives comprise "less than", "twice as many", "at least twice as many", "many more than", etc. Resorting to (11.96), the comparatives "(exactly) twice as many" and "at least twice as many" can be expressed in terms of the following mappings q_1 and q_2, respectively:

$$q_1(c_1, c_2) = \begin{cases} 1 & : \quad c_1 = 2\,c_2 \\ 0 & : \quad \text{else} \end{cases} \qquad \text{for "exactly twice as many"}$$

$$q_2(c_1, c_2) = \begin{cases} 1 & : \quad c_1 \geq 2\,c_2 \\ 0 & : \quad \text{else} \end{cases} \qquad \text{for "at least twice as many"}$$

These quantifiers can be used for interpreting NL propositions like *"There are exactly twice as many male as female dogs in the kennel"* mentioned by [92, p. 305]. A conceivable definition of "many more than" is

$$\textbf{many more than}(Y_1, Y_2, Y_3) = q(c_1, c_2)$$

for all $Y_1, Y_2, Y_3 \in \mathcal{P}(E)$, where

$$q(c_1, c_2) = \begin{cases} 0 & : \quad c_1 \leq c_2 + \alpha \\ (c_1 - c_2 - \alpha) \cdot \beta & : \quad c_2 + \alpha < c_1 \leq c_2 + \alpha + 1/\beta \\ 1 & : \quad c_2 > c_2 + \alpha + 1/\beta \end{cases}$$

for $c_1, c_2 \in \{0, \ldots, |E|\}$. Obviously, suitable choices of $\alpha, \beta \in \mathbb{R}^+$ can only be made from the context. However, the rough shape of the resulting quantifier is easily grasped from q, so it should usually not pose problems to select a quantifier which suits the task at hand.
Now that we are able to define the cardinal comparatives of NL in terms of semi-fuzzy quantifiers, let us consider the relation $R_\gamma^{\Phi_1,\Phi_2}(X_1, X_2, X_3)$, where $\Phi_1(Y_1, Y_2, Y_3) = Y_1 \cap Y_3$ and $\Phi_2(Y_1, Y_2, Y_3) = Y_2 \cap Y_3$. Specifically, it must be shown how to express this relation in terms of quantities ℓ_r, u_r sampled from Boolean combinations of the fuzzy arguments $X_1, X_2, X_3 \in \widetilde{\mathcal{P}}(E)$. The simple description of this target relation in terms of four Boolean combinations that are required is set forth in the next theorem.

Theorem 11.57.

Let $E \neq \varnothing$ be a finite base set and $X_1, X_2, X_3 \in \widetilde{\mathcal{P}}(E)$. Further let Φ_1, Φ_2 denote the Boolean combinations $\Phi_1(Y_1, Y_2, Y_3) = Y_1 \cap Y_3$ and $\Phi_2(Y_1, Y_2, Y_3) = Y_2 \cap Y_3$ of the crisp variables $Y_1, Y_2, Y_3 \in \mathcal{P}(E)$. Then the relation $R = R_\gamma^{\Phi_1,\Phi_2}(X_1, X_2, X_3)$ defined by (11.34) can be computed from the cardinality coefficients $\ell_r = |Z_r|_\gamma^{\min}$ and $u_r = |Z_r|_\gamma^{\max}$ sampled from $Z_1 = \Psi_1(X_1, X_2, X_3) = X_1 \cap X_3$, $Z_2 = \Psi_2(X_1, X_2, X_3) = X_2 \cap X_3$, $Z_3 = \Psi_3(X_1, X_2, X_3) = X_1 \cap \neg X_2 \cap X_3$ and $Z_4 = \Psi_4(X_1, X_2, X_3) = \neg X_1 \cap X_2 \cap X_3$. In terms of $\ell_1, \ell_2, \ell_3, \ell_4, u_1, u_2, u_3, u_4 \in \{0, \ldots, |E|\}$, R then becomes

$$R = \{(c_1, c_2) : \ell_1 \leq c_1 \leq u_1, \tag{11.98}$$
$$\max(c_1 - u_3 + \ell_4, \ell_2) \leq c_2 \leq \min(c_1 - \ell_3 + u_4, u_2)\}\}\,. \tag{11.99}$$

In practice, it is not necessary to consider all pairs $(c_1, c_2) \in R$, because those cardinal comparatives actually found in NL conform to simple monotonicity patterns. Specifically, it is possible to simplify the computation of $\top_j$ and $\bot_j$ if the cardinal comparative Q is at least convex in its second argument. It is easily seen that Q is convex in the second argument if and only if q is convex in the second argument, i.e. $q(c_1, c_2') \geq \min(q(c_1, c_2), q(c_1, c_2''))$ given that $c_2 \leq c_2' \leq c_2''$. Consequently, there exists $c_{\mathrm{pk}} = c_{\mathrm{pk}}(c_1) \in \{0, \ldots, |E|\}$ such that $q(c_1, c_2)$ is nondecreasing in c_2 for $c_2 \leq c_{\mathrm{pk}}$, and nonincreasing for $c_2 \geq c_{\mathrm{pk}}$. In particular, c_{pk} maximizes $q(c_1, c_{\mathrm{pk}})$ given c_1. In the convex case, we therefore have

$$\top_j = \max\{q(c_1, c_2^*) : \ell_1 \leq c_1 \leq u_1\}$$

and

$$\bot_j = \min\{\min(\, q(c_1, \max(c_1 - u_3 + \ell_4, \ell_2)),\, q(c_1, \min(c_1 - \ell_3 + u_4, u_2))) \\ : \ell_1 \leq c_1 \leq u_1\}$$

where

$$c_2^* = c_2^*(c_1) = \begin{cases} \min(c_1 - \ell_3 + u_4, u_2) & : \quad c_{\mathrm{pk}} > \min(c_1 - \ell_3 + u_4, u_2) \\ \max(c_1 - u_3 + \ell_4, \ell_2) & : \quad c_{\mathrm{pk}} < \max(c_1 - u_3 + \ell_4, \ell_2) \\ c_{\mathrm{pk}} & : \quad \text{else} \end{cases}$$

for $c_{\mathrm{pk}} = c_{\mathrm{pk}}(c_1)$. This representation permits a simplified computation of $\top_j$ and $\bot_j$ for convex quantifiers like "about two times more than".

Most cardinal comparatives show even simpler monotonicity patterns. For example, *"More Y_1's than Y_2's are Y_3's"* is nondecreasing in the first and nonincreasing in the second argument, and the quantifier in *"Less Y_1's than Y_2's are Y_3's"* is nonincreasing in the first and nondecreasing in the second argument. Here we will restrict to the first of these monotonicity patterns, and study the corresponding quantifiers in some more depth; all results of this investigation are readily transfered to the second monotonicity pattern and we will omit the obvious modifications for the sake of brevity. Let us first link the monotonicity of a cardinal comparative Q to the monotonicity of the mapping q on cardinals.

Theorem 11.58.

Let $Q : \mathcal{P}(E)^3 \longrightarrow \mathbf{I}$ be a cardinal comparative on a finite base set $E \neq \varnothing$. Further let $q : \{0, \ldots, |E|\}^2 \longrightarrow \mathbf{I}$ be the corresponding mapping defined by (11.96), i.e. $Q(Y_1, Y_2, Y_3) = q(|Y_1 \cap Y_3|, |Y_2 \cap Y_3|)$ for all $Y_1, Y_2, Y_3 \in \mathcal{P}(E)$. Then the following are equivalent:

a. Q is nondecreasing in the first and nonincreasing in its second argument;
b. q is nondecreasing in its first and nonincreasing in its second argument.

Hence the considered monotonicity pattern of cardinal comparatives reflects in a very simple pattern of the mapping q. In turn, knowing these monotonicity properties permits a simplified description of $\top_{Q,X_1,X_2,X_3}$ and $\bot_{Q,X_1,X_2,X_3}$, because the minimum and maximum must then be assumed at the 'extreme poles'. Consequently, we obtain the following corollary to Th. 11.57 and Th. 11.58.

Theorem 11.59.

Let $Q : \mathcal{P}(E)^3 \longrightarrow \mathbf{I}$ be a cardinal comparative on a finite base set $E \neq \varnothing$, and let q be the corresponding mapping $q : \{0, \ldots, |E|\}^2 \longrightarrow \mathbf{I}$. If Q is nondecreasing in the first and nonincreasing in its second argument, then

$$\top_{Q,X_1,X_2,X_3}(\gamma) = \max\{q(c_1, \max(c_1 - u_3 + \ell_4, \ell_2)) : \ell_1 \leq c_1 \leq u_1\}$$
$$\bot_{Q,X_1,X_2,X_3}(\gamma) = \min\{q(c_1, \min(c_1 - \ell_3 + u_4, u_2)) : \ell_1 \leq c_1 \leq u_1\}$$

for all $X_1, X_2, X_3 \in \widetilde{\mathcal{P}}(E)$ and $\gamma \in \mathbf{I}$, referring to the cardinality coefficients ℓ_r and u_r, $r \in \{1, 2, 3, 4\}$ defined in Th. 11.57.

The particular algorithms which implement cardinal comparatives in $\mathcal{F}_{\mathrm{owa}}$, $\mathcal{M}$ and $\mathcal{M}_{CX}$ can easily be developed from this analysis of the considered variety of quantifiers, again utilizing the basic results that were proven for these models in Th. 11.27, Th. 11.30 and Th. 11.31, respectively. In fact, we can proceed in total analogy to the strategy used for absolute and proportional quantifiers. In the general case of unrestricted cardinal comparatives, we can express $\top_{Q,X_1,X_2,X_3}$ and $\bot_{Q,X_1,X_2,X_3}$, for a given choice of $\gamma \in \mathbf{I}$, as

$$\top_{Q,X_1,X_2,X_3}(\gamma) = q^{\max}(\ell_1, \ell_2, \ell_3, \ell_4, u_1, u_2, u_3, u_4)$$
$$\bot_{Q,X_1,X_2,X_3}(\gamma) = q^{\min}(\ell_1, \ell_2, \ell_3, \ell_4, u_1, u_2, u_3, u_4)\,,$$

where

$$q^{\max}(\ell_1, \ell_2, \ell_3, \ell_4, u_1, u_2, u_3, u_4) = \max\{q(c_1, c_2) : (c_1, c_2) \in R_\gamma(X_1, X_2, X_3)\}$$
$$q^{\min}(\ell_1, \ell_2, \ell_3, \ell_4, u_1, u_2, u_3, u_4) = \min\{q(c_1, c_2) : (c_1, c_2) \in R_\gamma(X_1, X_2, X_3)\}\,.$$

These coefficients can readily be computed, because we know the explicit representation of $R_\gamma(X_1, X_2, X_3)$, which was established in Th. 11.57. In the typical case that the quantifier of interest belongs to the monotonicity pattern assumed in Th. 11.59 (or the reverse pattern), a considerable cut in processing times becomes possible. In particular, $\top_{Q,X_1,X_2,X_3}$ and $\bot_{Q,X_1,X_2,X_3}$ can then be expressed in the following simplified way,

$$\top_{Q,X_1,X_2,X_3}(\gamma) = q^{\max}(\ell_1, \ell_2, \ell_4, u_1, u_3)$$
$$\bot_{Q,X_1,X_2,X_3}(\gamma) = q^{\min}(\ell_1, \ell_3, u_1, u_2, u_4)\,,$$

where $q^{\max}$ and $q^{\min}$ now read

$$q^{\max}(\ell_1, \ell_2, \ell_4, u_1, u_3) = \max\{q(c_1, \max(c_1 - u_3 + \ell_4, \ell_2)) : \ell_1 \le c_1 \le u_1\}$$
$$q^{\min}(\ell_1, \ell_3, u_1, u_2, u_4) = \min\{q(c_1, \min(c_1 - \ell_3 + u_4, u_2)) : \ell_1 \le c_1 \le u_1\}\,,$$

see Th. 11.59.

In the algorithms to be presented now, we will consider both the typical monotonic and the general quantifiers, in order to optimally account for both cases. Hence let us recall the analysis of the three prototypical models achieved in theorems Th. 11.27, Th. 11.30 and Th. 11.31, which will now be elaborated into a complete and practical implementation of cardinal comparatives. In order to instantiate the core formulas stated in these theorems, it is again necessary to fix the set of relevant cutting parameters, i.e. to choose some $\Gamma = \{\gamma_0, \dots, \gamma_m\}$ such that $0 = \gamma_0 < \gamma_1 < \dots < \gamma_{m-1} < \gamma_m = 1$ and $\Gamma \supseteq \Gamma(X_1, X_2, X_3)$.

When working with floating-point arithmetics, it is advantageous to minimize the set Γ, thus letting $\Gamma = \Gamma(X_1, X_2, X_3)$. Owing to Th. 11.23, we then know that

$$\begin{aligned}\Gamma(Z_1) &= \Gamma(X_1 \cap X_3) \subseteq \Gamma \\ \Gamma(Z_2) &= \Gamma(X_2 \cap X_3) \subseteq \Gamma \\ \Gamma(Z_3) &= \Gamma(X_1 \cap \neg X_2 \cap X_3) \subseteq \Gamma \\ \Gamma(Z_4) &= \Gamma(\neg X_1 \cap X_2 \cap X_3) \subseteq \Gamma\,,\end{aligned}$$

where

$$\begin{aligned}Z_1 &= \Phi_1(X_1, X_2, X_3) = X_1 \cap X_3 \\ Z_2 &= \Phi_2(X_1, X_2, X_3) = X_2 \cap X_3 \\ Z_3 &= \Phi_3(X_1, X_2, X_3) = X_1 \cap \neg X_2 \cap X_3 \\ Z_4 &= \Phi_4(X_1, X_2, X_3) = \neg X_1 \cap X_2 \cap X_3\,,\end{aligned}$$

as stipulated in Th. 11.57. Consequently the histograms Hist_{Z_r}, $r \in \{1, 2, 3, 4\}$, can be represented by $(m + 1)$-dimensional arrays H_r^+ and H_r^- defined in accordance with (11.82) and (11.83). Recalling that the cardinality coefficients ℓ_r, u_r are defined by $\ell_r = \ell_r(j) = |Z_r|_{\overline{\gamma}_j}^{\min}$ and $u_r = u_r(j) = |Z_r|_{\overline{\gamma}_j}^{\max}$ for $j \in \{0, \ldots, m-1\}$, we can now utilize the incremental rule presented in Th. 11.48 in order to compute these coefficients efficiently. The particular algorithms so obtained, that implement cardinal comparatives based on floating-point arithmetics in the three prototype models, are shown in Table 11.14 ($\mathcal{F}_{\mathrm{owa}}$), Table 11.15 ($\mathcal{M}$) and Table 11.16 (for $\mathcal{M}_{CX}$), respectively.

These algorithms provide a generic solution for arbitrary cardinal comparatives, because they implement the full variety of these quantifiers without any special requirements on monotonicity properties. Some tags have been included into the pseudo-code, which indicate how the algorithms can be taylored to the prominent monotonicity type that was considered in Th. 11.59. In this typical case of a natural language comparative, a more efficient implementation is possible due to the simplified description of $q^{\min}$ and $q^{\max}$

Table 11.14. Algorithm for evaluating cardinal comparatives in $\mathcal{F}_{\mathrm{owa}}$ (based on floating-point arithmetics)

Algorithm for computing $\mathcal{F}_{\mathrm{owa}}(Q)(X_1, X_2, X_3)$
(floating-point arithmetics)

```
INPUT: X1, X2, X3
Compute G, m, H1+, H2+, H3+, H4+, H1-, H2-, H3-, H4-; // see text
// initialize ℓ, u:
for( r := 1; r ≤ 4; r := r+1 )
    { ℓr := Σ_{j=1}^{m} Hr+[j]; ur := ℓr + Hr+[0]; }
sum := G[1] * (q^min(ℓ1,ℓ2,ℓ3,ℓ4,u1,u2,u3,u4) + q^max(ℓ1,ℓ2,ℓ3,ℓ4,u1,u2,u3,u4));
for( j := 1; j<m; j := j + 1 ) {
    // update clauses for ℓ1,ℓ2,ℓ3,ℓ4,u1,u2,u3,u4
    for( r := 1; r ≤ 4; r := r+1 )
        { ℓr := ℓr - Hr+[j]; ur := ur + Hr-[j]; }
    sum := sum + (G[j+1] - G[j])
                * (q^min(ℓ1,ℓ2,ℓ3,u4,u1,u2,u3,u4) + q^max(ℓ1,ℓ2,ℓ3,ℓ4,u1,u2,u3,u4));
    }
return sum / 2;
END
```

can be treated along the same lines), and result in a practical implementation of cardinal comparatives which is ready for use in the targeted applications of fuzzy quantifiers.

Having presented a solution for floating-point arithmetics, we will now add variants of the above algorithms for integer arithmetics. Hence let Q be a cardinal comparative (not required to be monotonic) and let $X_1, X_2, X_3 \in \widetilde{\mathcal{P}}(E)$ be given fuzzy arguments. Further suppose that $m' \in \mathbb{N} \setminus \{0\}$ is given, which specifies the available range of integers. The given choice of m' again restricts admissible fuzzy arguments to a finite set of equidistant membership grades in the unit interval, i.e. all legal choices of arguments must satisfy

$$\mu_{X_i}(e) \in U_{m'}$$

for all $e \in E$ and $i \in \{1, 2, 3\}$, where $U_{m'}$ is defined as before, see (11.85). By pursuing the same strategy that already proved useful for deriving the integer-based implementation of unary quantifiers (see pp. 315–317) and also of proportional quantifiers, we have also developed the corresponding solutions which implement cardinal comparatives based on integer arithmetics. The resulting variants of the above algorithms are shown in Table 11.17 ($\mathcal{F}_{\text{owa}}$), Table 11.18 ($\mathcal{M}$), and Table 11.19 (for $\mathcal{M}_{CX}$). The listings are again annotated by tags `[*]`, `[1]`, `[2]`, in order to indicate the possible optimizations for monotonic quantifiers. In this case, the update clauses annotated by `[*]` can again be deleted, and lines marked with `[1]` should be replaced by the simplified variants `[2]`.

Table 11.17. Algorithm for evaluating cardinal comparatives in $\mathcal{F}_{\text{owa}}$ (based on integer arithmetics)

Algorithm for computing $\mathcal{F}_{\text{owa}}(Q)(X_1, X_2, X_3)$
(integer arithmetics)

```
INPUT: X1, X2, X3
// initialise Hr, ℓ, u
H1 := Hist_{X1∩X3};
H2 := Hist_{X2∩X3};
H3 := Hist_{X1∩¬X2∩X3};
H4 := Hist_{¬X1∩X2∩X3};
for( r := 1; r ≤ 4; r := r+1 )
     { ℓr := Σ_{j=1}^{m} Hr[m+j]; ur := ℓr + Hr[m]; }
cq := q^min(ℓ1,ℓ2,ℓ3,ℓ4,u1,u2,u3,u4) + q^max(ℓ1,ℓ2,ℓ3,ℓ4,u1,u2,u3,u4);
sum := cq;
for( j := 1; j<m; j := j + 1 ) {
     ch := false; // "change"
     // update clauses for ℓ and u:
     for( r := 1; r ≤ 4; r := r+1 ) {
          if( Hr[m+j] ≠ 0 )
               { ℓr := ℓr - Hr[m+j]; ch := true;}
          if( Hr[m-j] ≠ 0 )
               { ur := ur + Hr[m-j]; ch := true; }
          }
     if( ch )
          // one of the ℓr or ur has changed
          { cq := q^min(ℓ1,ℓ2,ℓ3,ℓ4,u1,u2,u3,u4) + q^max(ℓ1,ℓ2,ℓ3,ℓ4,u1,u2,u3,u4); }
     sum := sum + cq;
     }
return sum / m'; // where m' = 2*m
END
```

Table 11.18. Algorithm for evaluating cardinal comparatives in $\mathcal{M}$ (based on integer arithmetics)

Algorithm for computing $\mathcal{M}(Q)(X_1, X_2, X_3)$
(integer arithmetics)

```
INPUT: X_1, X_2, X_3
// initialise H_r, ℓ, u
H_1 := Hist_{X_1 ∩ X_3};
H_2 := Hist_{X_2 ∩ X_3};
H_3 := Hist_{X_1 ∩ ¬X_2 ∩ X_3};
H_4 := Hist_{¬X_1 ∩ X_2 ∩ X_3};
for( r := 1; r ≤ 4; r := r+1 )
    { ℓ_r := Σ_{j=1}^{m} H_r[m+j]; u_r := ℓ_r + H_r[m]; }
⊤ := q^max(ℓ_1, ℓ_2, ℓ_3, ℓ_4, u_1, u_2, u_3, u_4);
⊥ := q^min(ℓ_1, ℓ_2, ℓ_3, ℓ_4, u_1, u_2, u_3, u_4);
if( ⊥ > 1/2 ) {
    sum := ⊥;
    for( j := 1; j<m; j := j + 1 ) {
        nc:= true; // "no change"
        // update clauses for ℓ, u:
        if( H_1[m+j] ≠ 0 )
            { ℓ_1 := ℓ_1 - H_1[m+j]; nc:= false; }
        if( H_2[m+j] ≠ 0 ) // [*]
            { ℓ_2 := ℓ_2 - H_2[m+j]; nc:= false; } // [*]
        if( H_3[m+j] ≠ 0 )
            { ℓ_3 := ℓ_3 - H_3[m+j]; nc:= false; }
        if( H_4[m+j] ≠ 0 ) // [*]
            { ℓ_4 := ℓ_4 - H_4[m+j]; nc:= false; } // [*]
        if( H_1[m-j] ≠ 0 )
            { u_1 := u_1 + H_1[m-j]; nc:= false; }
        if( H_2[m-j] ≠ 0 )
            { u_2 := u_2 + H_2[m-j]; nc:= false; }
        if( H_3[m-j] ≠ 0 ) // [*]
            { u_3 := u_3 + H_3[m-j]; nc:= false; } // [*]
        if( H_4[m-j] ≠ 0 )
            { u_4 := u_4 + H_4[m-j]; nc:= false; }
        if( nc)
            { sum := sum + ⊥; continue; }
        // one of the ℓ_r, u_r has changed
        ⊥ := q^min(ℓ_1, ℓ_2, ℓ_3, ℓ_4, u_1, u_2, u_3, u_4); // [1]
        // ⊥ := q^min(ℓ_1, ℓ_3, u_1, u_2, u_4); // [2]
        if( ⊥ ≤ 1/2 ) break;
        sum := sum + ⊥;
        }
    }
else if( ⊤ < 1/2 ) {
    sum := ⊤;
    for( j := 1; j<m; j := j + 1 ) {
        nc:= true;
        // update clauses for ℓ and u:
        if( H_1[m+j] ≠ 0 )
            { ℓ_1 := ℓ_1 - H_1[m+j]; nc:= false; }
        if( H_2[m+j] ≠ 0 )
            { ℓ_2 := ℓ_2 - H_2[m+j]; nc:= false; }
        if( H_3[m+j] ≠ 0 ) // [*]
            { ℓ_3 := ℓ_3 - H_3[m+j]; nc:= false; } // [*]
        if( H_4[m+j] ≠ 0 )
            { ℓ_4 := ℓ_4 - H_4[m+j]; nc:= false; }
        if( H_1[m-j] ≠ 0 )
            { u_1 := u_1 + H_1[m-j]; nc:= false; }
        if( H_2[m-j] ≠ 0 ) // [*]
            { u_2 := u_2 + H_2[m-j]; nc:= false; } // [*]
        if( H_3[m-j] ≠ 0 )
            { u_3 := u_3 + H_3[m-j]; nc:= false; }
        if( H_4[m-j] ≠ 0 ) // [*]
            { u_4 := u_4 + H_4[m-j]; nc:= false; } // [*]
        if( nc)
            { sum := sum + ⊤; continue; }
        // one of the ℓ_r, u_r has changed
        ⊤ := q^max(ℓ_1, ℓ_2, ℓ_3, ℓ_4, u_1, u_2, u_3, u_4); // [1]
        // ⊤ := q^max(ℓ_1, ℓ_2, ℓ_4, u_1, u_3); // [2]
        if( ⊤ ≥ 1/2 ) break;
        sum := sum + ⊤;
        }
    }
else
    { return 1/2; }
return (sum + 1/2*(m-j)) / m;
END
```

Table 11.19. Algorithm for evaluating cardinal comparatives in $\mathcal{M}_{CX}$ (based on integer arithmetics)

Algorithm for computing $\mathcal{M}_{CX}(Q)(X_1, X_2, X_3)$
(integer arithmetics)

```
INPUT: X1, X2, X3
// initialise Hr, ℓ, u
H1 := Hist_{X1∩X3};
H2 := Hist_{X2∩X3};
H3 := Hist_{X1∩¬X2∩X3};
H4 := Hist_{¬X1∩X2∩X3};
for( r := 1; r ≤ 4; r := r+1 )
      { ℓr := Σ_{j=1}^{m} Hr[m+j]; ur := ℓr + Hr[m]; }
⊤ := q^max(ℓ1, ℓ2, ℓ3, ℓ4, u1, u2, u3, u4);
⊥ := q^min(ℓ1, ℓ2, ℓ3, ℓ4, u1, u2, u3, u4);
if( ⊥ > 1/2 ) {
      for( j := 1; j<m; j := j + 1 ) {
            ch := false; // "change"
            // update clauses for ℓ and u:
            if( H1[m+j] ≠ 0 )
                  { ℓ1 := ℓ1 - H1[m+j]; ch := true; }
            if( H2[m+j] ≠ 0 ) // [*]
                  { ℓ2 := ℓ2 - H2[m+j]; ch := true; } // [*]
            if( H3[m+j] ≠ 0 )
                  { ℓ3 := ℓ3 - H3[m+j]; ch := true; }
            if( H4[m+j] ≠ 0 ) // [*]
                  { ℓ4 := ℓ4 - H4[m+j]; ch := true; } // [*]
            if( H1[m-j] ≠ 0 )
                  { u1 := u1 + H1[m-j]; ch := true; }
            if( H2[m-j] ≠ 0 )
                  { u2 := u2 + H2[m-j]; ch := true; }
            if( H3[m-j] ≠ 0 ) // [*]
                  { u3 := u3 + H3[m-j]; ch := true; } // [*]
            if( H4[m-j] ≠ 0 )
                  { u4 := u4 + H4[m-j]; ch := true; }
            if( ch )
                  // one of the ℓr, ur has changed
                  { ⊥ := q^min(ℓ1, ℓ2, ℓ3, ℓ4, u1, u2, u3, u4); } // [1]
                  // { ⊥ := q^min(ℓ1, ℓ3, u1, u2, u4); } // [2]
            if( ⊥ ≤ m+j )
                  { return (m+j)/m'; }
      }
      return 1;
      }
else if( ⊤ < 1/2 ) {
      for( j := 1; j<m; j := j + 1 ) {
            ch := false;
            // update clauses for ℓ and u:
            if( H1[m+j] ≠ 0 )
                  { ℓ1 := ℓ1 - H1[m+j]; ch := true; }
            if( H2[m+j] ≠ 0 )
                  { ℓ2 := ℓ2 - H2[m+j]; ch := true; }
            if( H3[m+j] ≠ 0 ) // [*]
                  { ℓ3 := ℓ3 - H3[m+j]; ch := true; } // [*]
            if( H4[m+j] ≠ 0 )
                  { ℓ4 := ℓ4 - H4[m+j]; ch := true; }
            if( H1[m-j] ≠ 0 )
                  { u1 := u1 + H1[m-j]; ch := true; }
            if( H2[m-j] ≠ 0 ) // [*]
                  { u2 := u2 + H2[m-j]; ch := true; } // [*]
            if( H3[m-j] ≠ 0 )
                  { u3 := u3 + H3[m-j]; ch := true; }
            if( H4[m-j] ≠ 0 ) // [*]
                  { u4 := u4 + H4[m-j]; ch := true; } // [*]
            if( ch )
                  // one of the ℓr, ur has changed
                  { ⊤ := q^max(ℓ1, ℓ2, ℓ3, ℓ4, u1, u2, u3, u4); } // [1]
                  // { ⊤ := q^max(ℓ1, ℓ2, ℓ4, u1, u3); } // [2]
            if( ⊤ ≥ m-j )
                  { return (m-j)/m'; }
      }
      return 0;
      }
return 1/2;
END
```

Finally let us consider a special case of comparative quantifiers which permits additional simplifications.[6] In many cases, a cardinal comparative $Q(Y_1, Y_2, Y_3)$ can be expressed in terms of the difference $c_1 - c_2 = |Y_1 \cap Y_3| - |Y_2 \cap Y_3|$. For example, this is the case with "more than", cf. (11.97) above. The simplicity of these quantifiers translates into rules of computation that are equally simple. In particular, we need not know the complete relation

$$R_\gamma^{\Phi_1,\Phi_2}(X_1, X_2, X_3) = \{(|Y_1 \cap Y_3|, |Y_2 \cap Y_3|) : (Y_1, Y_2, Y_3) \in \mathcal{T}_\gamma(X_1, X_2, X_3)\}$$

in order to compute $\top_{Q,X_1,X_2,X_3}$ and $\bot_{Q,X_1,X_2,X_3}$. By contrast, it is sufficient to know the set of differences

$$D = \{|Y_1 \cap Y_3| - |Y_2 \cap Y_3| : (Y_1, Y_2, Y_3) \in \mathcal{T}_\gamma(X_1, X_2, X_3)\} .$$

Hence let us describe the latter set directly, which might pay off from the efficiency point of view because all choices of arguments

$$(Y_1, Y_2, Y_3), (Y_1', Y_2', Y_3') \in \mathcal{T}_\gamma(X_1, X_2, X_3)$$

with

$$|Y_1 \cap Y_3| - |Y_2 \cap Y_3| = |Y_1' \cap Y_3'| - |Y_2' \cap Y_3'|$$

now become identified, i.e. it is sufficient to consider one representative only. The new format therefore lets us eliminate some computational redundancy. The next theorem shows how to compute the set of differences:

Theorem 11.60.

Let $E \neq \varnothing$ be a finite base set, $X_1, X_2, X_3 \in \widetilde{\mathcal{P}}(E)$ and $\gamma \in \mathbf{I}$. Then

$$\begin{aligned} &\{c_1 - c_2 : (Y_1, Y_2, Y_3) \in \mathcal{T}_\gamma(X_1, X_2, X_3), c_1 = |Y_1 \cap Y_3|, c_2 = |Y_2 \cap Y_3|\} \\ &= \{d : \ell_3 - u_4 \leq d \leq u_3 - \ell_4\} , \end{aligned}$$

referring to the cardinality coefficients ℓ_3, ℓ_4, u_3 and u_4 introduced in Th. 11.57.

The latter theorem has some obvious applications. If a cardinal comparative which depends on D only is convex in its second argument, then $q : \mathbb{Z} \longrightarrow \mathbf{I}$ with $Q(Y_1, Y_2, Y_3) = q(|Y_1 \cap Y_3| - |Y_2 \cap Y_3|)$ is also convex. Hence there exists a maximum $d_{\mathrm{pk}} \in \mathbb{Z}$ such that q is nondecreasing for $d \leq d_{\mathrm{pk}}$ and nonincreasing for $d \geq d_{\mathrm{pk}}$. We then have

$$\top_j = q^{\max}(\ell_3, \ell_4, u_3, u_4) = \begin{cases} q(d_{\mathrm{pk}}) & : \ \ell_3 - u_4 \leq d_{\mathrm{pk}} \leq u_3 - \ell_4 \\ q(u_3 - \ell_4) & : \ d_{\mathrm{pk}} > u_3 - \ell_4 \\ q(\ell_3 - u_4) & : \ d_{\mathrm{pk}} < \ell_3 - u_4 \end{cases}$$

$$\bot_j = q^{\min}(\ell_3, \ell_4, u_3, u_4) = \min(q(\ell_3 - u_4), q(u_3 - \ell_4)) .$$

[6] These coincide with the narrower concept of absolute comparative quantifiers proposed in [29].

Table 11.20. Complexity of the algorithms for $\mathcal{F}_{\text{owa}}$, $\mathcal{M}$ and $\mathcal{M}_{CX}$: Main loop

Type (Monot.)	Example	Complexity
unary quant.		
(general case)	There is an even number of X's	$\mathcal{O}(m\,N)$
(convex)	There are about 50 X's	$\mathcal{O}(m)$
(monotonic)	There are at least 20 X's	$\mathcal{O}(m)$
absolute	About 50 X_1's are X_2's	like *unary quantitative*
exception	All except ten 10 X_1's are X_2's	like *unary quantitative*
proportional		
(general case)	$\lvert X_1 \cap X_2\rvert/\lvert X_1\rvert$ is either 0.1 or 0.2	$\mathcal{O}(m\,N^2)$
(convex)	About 50% of the X_1's are X_2's	$\mathcal{O}(m\,N)$
(monotonic)	Most X_1's are X_2's	$\mathcal{O}(m)$
abs. comparative		
(general case)	$\lvert X_1 \cap X_3\rvert - \lvert X_2 \cap X_3\rvert$ is prime	$\mathcal{O}(m\,N^2)$ resp. $\mathcal{O}(m\,N)^\dagger$
(convex)	Twice the X_1 than X_2's are X_3's	$\mathcal{O}(m\,N)$ resp. $\mathcal{O}(m)^\dagger$
(monotonic)	More X_1's than X_2's are X_3's	$\mathcal{O}(m)$

to the search criterion, and to present them sorted by their relevance to the user query. In this process, fuzzy quantifiers can be useful as aggregation operators which support importance qualification. We then have a number of subcriteria which comprise the domain of quantification, E. Each criterion $e \in E$ has an associated importance $\mu_W(e) \in \mathbf{I}$ and an associated grade of applicability $\mu_{A,d}(e)$ with respect to the considered document d. These importance grades can be organized into fuzzy subsets $W, A_d \in \widetilde{\mathcal{P}}(E)$. Based on a semi-fuzzy quantifier $Q : \mathcal{P}(E)^2 \longrightarrow \mathbf{I}$, e.g. $Q = \textbf{almost all}$, we can then express the aggregation score ϱ_d which corresponds to the NL statement

"Q important criteria apply to the document d".

The ranking score then becomes

$$\varrho_d = \widetilde{Q}(W, A_d)\,,$$

where $\widetilde{Q} = \mathcal{F}(Q)$ for the chosen DFS $\mathcal{F}$. The documents $d \in D$ of the document base D will then be sorted according to their relevance grade ϱ_d and presented in decreasing order of relevance (most relevant first). For example, we can determine the documents to which most relevant criteria $e \in E$ apply by using the ranking criterion

$$\varrho_d = \widetilde{\textbf{most}}(W, A_d)\,.$$

In this case, the fuzzy quantifier $\widetilde{\textbf{most}} = \mathcal{F}(\textbf{most})$ is independent of the assumed standard model, see (2.2) and Th. 5.22.

This basic approach has been utilized for improving retrieval performance in a meta-search engine for bibliographic databases [65]. In this system, the possible combinations of search terms are no longer limited to Boolean operators; apart from the Boolean connectives, the users can also choose quantifiers like "almost all", "many" and "a few". In addition, all terms can be weighted by their importance with respect to the aggregation.

Similar techniques have been employed by Bordogna and Pasi [19] in order to support more powerful queries to information retrieval systems which also include fuzzy quantification. Further examples which demonstrate the benefits of fuzzy quantification to the querying of information systems are [23, 84, 86, 135]. It should be emphasized, though, that these applications resort to the Σ-count or OWA approaches, and therefore do not profit from the improvements of fuzzy quantification achieved in this work.

The proposed method for weighted aggregation with fuzzy quantifiers is not restricted to textual documents and classical information retrieval. The following example starts from a meaningful query to an image database, which involves the use of fuzzy quantification. The document base assumed in the example again consists of digitized satellite images, from which corresponding images of cloudiness situations have been extracted. The goal is to rank these cloudiness situations according to a criterion like "Q of Southern Germany is cloudy", where Q is a quantifier like "more than 20 percent" or "as much as possible". We thus have a uniform criterion, i.e. the fuzzy predicate "cloudy", which is applied repeatedly and evaluated for all pixels in the image of cloudiness grades. The global criterion therefore expresses some kind of aggregation over the spatial region of Southern Germany (see Fig. 1.2). The criterion results in a single scalar evaluation for each image. Consequently, the aggregation totally eliminates the spatial dimension. It results in a single numerical score ϱ_d which determines the relevance of each image under the query.

Let us now consider the case that $Q =$ **as much as possible**, which corresponds to the ranking criterion

"As much as possible of Southern Germany is cloudy".

Because we are working with discrete, digitized images, we will first translate the original criterion into a corresponding condition on discrete pixels in the digitized image, viz

"As many pixels as possible that belong to Southern Germany
are classified as cloudy".

The ranking is then computed by determining the document score ϱ_d of each image d in the database, and by presenting the images in decreasing order of their associated scores. The processing of the quantifier is delegated to the model $\mathcal{M}$ presented in Chap. 7. Thus the score ϱ_d of the considered image d is given by

$$\varrho_d = \mathcal{M}(\textbf{as many as possible})(\textbf{Southern Germany}, \textbf{cloudy}_d)\,,$$

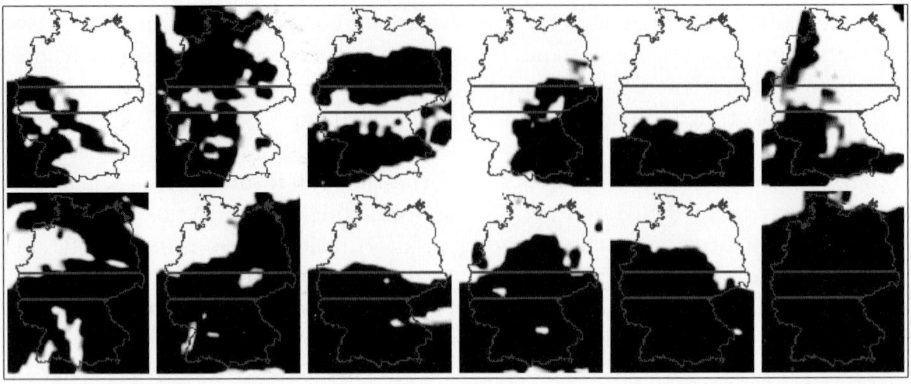

Fig. 11.1. Ranking computed for query "As much as possible of Southern Germany is cloudy". Most relevant results in top left corner, quality of match decreases from left to right and from top to bottom

where **as many as possible** is the quantifier defined by (2.9), which denotes the relative share, and $\mathbf{cloudy}_d$ is the fuzzy region of cloudiness grades depicted in image d. The example query was then run on twelve images, a random sample of the images in the document base. The resulting ranking of cloudiness situations under the criterion "As much as possible of Southern Germany is cloudy" is presented in Fig. 11.1. As opposed to the results of existing approaches in the image ranking task shown in the introduction, the computed ordering appears to be in good conformance with our expectations, i.e. the resulting image sequence shows a steady decrease of cloudiness in Southern Germany. This positive result is not suprising, though, because the important expectations on quantification results have been encoded in the axiom system.

The example also provides experimental data on processing times. Due to the large size of the base set (all pixel positions in the images) and the modest demands on accuracy, we applied the algorithm based on integer arithmetics with 8 bit accuracy (see Table 11.12). In order to compute the ranking, a quantification result had to be calculated for each image, based on the domain of 219063 (411x533) pixel coordinates. This took approx. 0.03 seconds processing time per image on an AMD Athlon 600 MHz PC running Linux, assuming that the image is already loaded into the working memory.

Fuzzy quantifiers are not only useful for weighted aggregation and importance qualification; they are also important tools for *linguistic data summarization*. Following Yager [172], the linguistic summary is usually constructed from a linguistic quantifier, which represents the *quantity in agreement*, and from a natural language predicate, which functions as the *summarizer*. For example, consider the following linguistic summary cited from Kacprzyk and Strykowski [83, p. 31], who detail a system for generating summaries of sales data at a computer retailer.

"Much sales of components is with a high commission".

The summary can be split into the domain 'sales of components', the quantity in agreement 'much' and finally the summarizer, '[sales] with a high commission'. We can therefore express the summary in terms of the quantifying expression

$$\widetilde{\textbf{much}}(\textbf{sales with a high commission}),$$

where $\widetilde{\textbf{much}}$ must be a suitable fuzzy quantifier defined on the universe of component sales.
Of course, it is also possible to use two-place quantifiers for constructing summaries. For example, imagine a summary

"Much recent sales of components is with a high commission".

In this case, we have an additional summarizer 'recent [sales of components]' which expresses a soft constraint on those sales of components covered by the summary. Such relative summaries can be modelled by a quantifying expression built from a two-place quantifier, viz

$$\widetilde{\textbf{much}}(\textbf{recent sales of components}, \textbf{sales with a high commission}).$$

Experimental systems for linguistic data summarization which fit into this general framework are described in [82, 83, 135, 202]. These systems rely on structured, relational representations, and are therefore taylored to classical database applications. However, data summarization with fuzzy quantifiers needs not be based on relational data. The same techniques can also be used to summarize image sequences, for example.

In order to illustrate this, let us consider a large collection of intensity images (each also annotated by a 'time stamp'), which represent cloudiness situations. We will assume that users of a weather information system are interested in the weather in certain 'natural' geographic areas and the distribution and flow of weather in certain 'natural' time intervals. Here 'natural' means that these regions of interest have an associated cognitive representation, and can therefore be *expressed linguistically*.

Hence consider a user interested in characteristics of the recent weather situation. Let us further assume that the user expresses the time of interest in natural language by specifying the temporal summarizer "in the last days". The chosen term "in the last days" then imposes a soft constraint on those images in the database which must be considered in the summary. For purposes of demonstration, we shall limit ourselves to a small sequence of cloudiness situations, which is depicted in the upper row of Fig. 11.2. The relevance of these images with respect to the criterion "in the last days" is also shown in the figure. The individual pieces of information must now be condensed by generating image summaries. Fuzzy two-place quantifiers are suitable operators to accomplish this summarization because they can incorporate the importances expressed by "in the last days". The criteria for generating the summary images thus take the form

"Q-times cloudy in the last days",

where Q is the chosen quantifier. In order to describe the data from several perspectives, the system is supposed to apply several quantifiers Q, each of which highlights a certain characteristic of the images.

The summary images will be generated as follows. We start from the base set E of images representing cloudiness situations. It is understood that these images share the same set of pixel coordinates, i.e. all images have the same dimensions. The images are also annotated with a time stamp. The time stamp can be used to define a fuzzy set $X_1 \in \widetilde{\mathcal{P}}(E)$, which judges membership of the images in the fuzzy time interval "in the last days". The summary criterion must be specified in terms of a semi-fuzzy quantifier $Q : \mathcal{P}(E)^2 \longrightarrow \mathbf{I}$. The target image is defined on the same set of pixel coordinates as the original images. It is generated by pixel-wise application of the following procedure. We have already fixed the quantifier Q and the restriction $X_1 \in \widetilde{\mathcal{P}}(E)$ which encodes the importances. In dependence on the given pixel p, we define a second fuzzy subset $X_{2,p} \in \widetilde{\mathcal{P}}(E)$, where $\mu_{X_{2,p}}(e)$ is the degree of cloudiness observed in cloudiness situation e at the location of p, i.e. the degree to which p is classified as 'cloudy'. By applying the quantifier to the restriction X_1 and scope $X_{2,p}$, we obtain the intensity of the considered pixel in the result image, which is the score of the criterion

"Q-times cloudy in pixel p in the last days".

In other words, the resulting image R has pixel intensities

$$R_p = \mathcal{F}(Q)(X_1, X_{2,p}),$$

for all pixels p, where $\mathcal{F}$ is the chosen model of fuzzy quantification. Here we shall use $\mathcal{F} = \mathcal{M}$, which was the only known DFS at the time the experiments were carried out, but the other models $\mathcal{F}_{\mathrm{owa}}$ and $\mathcal{M}_{CX}$ are also conceivable choices. Noticing that every quantifier will highlight a certain regularity observed in the images, it makes sense to use more than one quantifier to achieve a richer description of the data. Here we will use the universal quantifier for modelling "always"; the existential quantifier for modelling "at least once"; and some intermediate quantifiers for modelling "sometimes" (i.e., at least a few times), "often" (assumed to increase linearly with frequency) and "almost always". The latter quantifiers were all derived from a trapezoidal proportional quantifier $\mathbf{trp}_{a,b,c} : \mathcal{P}(E)^2 \longrightarrow \mathbf{I}$ defined by

$$\mathbf{trp}_{a,b,c}(Y_1, Y_2) = \begin{cases} t_{a,b}(|Y_1 \cap Y_2|/|Y_1|) & : \quad Y_1 \neq \varnothing \\ c & : \quad Y_1 = \varnothing \end{cases}$$

$$t_{a,b}(z) = \begin{cases} 0 & : \quad z < a \\ \frac{z-a}{b-a} & : \quad a \leq z \leq b \\ 1 & : \quad z > b \end{cases}$$

Data:

0.2	0.4	0.6	0.7	0.8	0.9	1.0	1.0

Results:

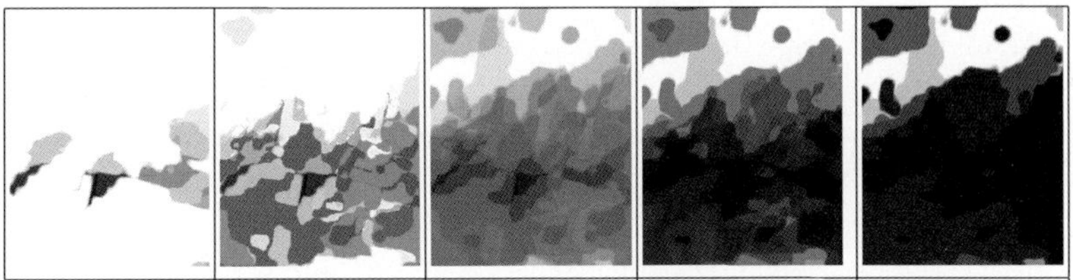

at least once	sometimes	often	almost always	always
some	$\mathbf{trp}_{0,0.4,0}$	$\mathbf{trp}_{0,1,0.5}$	$\mathbf{trp}_{0.6,1,1}$	all

Fig. 11.2. Image sequence and summarization results for various choices of the criterion "Q-times cloudy in the last days". Regions that meet the criterion are depicted white

where $Y_1, Y_2 \in \mathcal{P}(E)$, $a, b, c, z \in \mathbf{I}$, $a < b$. The precise definition of the derived quantifiers in terms of the parameters $a, b, c \in \mathbf{I}$ as well as the corresponding summaries are shown in Fig. 11.2. In the generated summary images, one immediately recognizes those areas that were sunny all of the time (those depicted black in the leftmost summary), those areas that were overcast with clouds all of the time (those shown white in the rightmost image) and the exact graduation of the areas in between (center images). In the crisp case, it would be possible to display all of this information in one single image, which depicts frequency. However, this cannot be done in the fuzzy case, because the importances of the summarized images (specified by "in the last days") and also the cloudiness grades are both a matter of tendency. This is why several images are needed to summarize the data.

Finally let us consider the processing times measured in the second experiment. In the example shown in Fig. 11.2, a total number of 219063 quantification results involving a domain of 8 images had to be computed for each result image (i.e. one quantifying expression for each of the 411x533 pixels). We again used the integer-based implementation of proportional quantifiers in $\mathcal{M}$ with eight bits of accuracy. The summarisation process took a total of about 5 seconds per result image on the same computer platform as above. These numbers suggest that the proposed algorithms are indeed very efficient, and not likely to create a performance bottleneck in applications.

12

Multiple Variable Binding and Branching Quantification

12.1 Motivation and Chapter Overview

Observing that Zadeh's traditional framework is too limited in scope, the framework for fuzzy quantification proposed in Chap. 2 was explicitly designed to capture a substantial class of linguistic quantifiers. The main part of the book was then concerned with the development of a complete and practical theory of fuzzy quantification within the proposed framework by defining the intended models, by devising constructions for such models, and finally by presenting algorithms for implementing quantifiers in these models which make them useful for applications.

All of these efforts were directed at the modelling of individual quantifiers. However, it is not only the diversity of possible quantifiers in NL which poses difficulties to a systematic and comprehensive modelling of linguistic quantification. Further difficulties arise when the quantifiers are no longer treated in isolation. Even in simple cases like "most", the ways in which these quantifiers interact when combined in meaningful propositions can be complex and sometimes even puzzling. Consider the quantifying proposition *"Most men and most women admire each other"*, for example, in which we find a reciprocal predicate, "admire each other". Barwise [6] argues that so-called branching quantification is needed to capture the meaning of propositions involving reciprocal predicates. Without branching quantifiers, the above example must be linearly phrased as either a. or b.,

a. $[\mathrm{most}\, x : \mathrm{men}(x)][\mathrm{most}\, y : \mathrm{women}(y)]\, \mathrm{adm}(x, y)$

b. $[\mathrm{most}\, y : \mathrm{women}(y)][\mathrm{most}\, x : \mathrm{men}(x)]\, \mathrm{adm}(x, y)$.

(Obviously, we need a logic which supports generalized quantifiers in order for these expressions to make sense. For a suitable system of first-order logic with generalized quantifiers see Barwise and Feferman [8, Ch. 2]). Neither interpretation captures the expected symmetry with respect to the men and

I. Glöckner: *Fuzzy Quantifiers*, StudFuzz **193**, 355–373 (2006)
www.springerlink.com
© Springer-Verlag Berlin Heidelberg 2006

and Q are in general not identical: it is true that the semi-fuzzy L-quantifier Q' is a also an ordinary n-ary semi-fuzzy quantifier, but it is defined on a different base set, i.e. E^1 rather than E. Here, E^1 is the set of all one-tuples $E^1 = \{(e) : e \in E\}$, i.e. the set of all mappings $f : \{1\} \longrightarrow E$, which is usually different from the set $E = \{e : e \in E\}$. A similar situation occurs when we turn attention to fuzzy quantifiers vs. fuzzy L-quantifiers. In this case, a fuzzy quantifier $\widetilde{Q} : \widetilde{\mathcal{P}}(E)^n \longrightarrow \mathbf{I}$ is also different from its possible counterparts $\widetilde{Q}' : \widetilde{\mathcal{P}}(E^1)^n \longrightarrow \mathbf{I}$, essentially for the same reason. We shall now consider the obvious one-to-one correspondence between the elements of these sets. Let $\beta : E \longrightarrow E^1$ denote the mapping which maps the elements $e \in E$ to one-tuples

$$\beta(e) = (e) \tag{12.2}$$

The mapping which maps the one-tuples $(e) \in E$ to their first component $(e)_1 = e$ will be denoted $\vartheta : E^1 \longrightarrow E$, i.e.

$$\vartheta((e)) = e \tag{12.3}$$

for all $(e) \in E^1$. Obviously, β and ϑ are isomorphisms, and ϑ is the inverse mapping of β. Based on β and ϑ, we are now able to express the correspondence between ordinary (semi-)fuzzy quantifiers and their generalized counterparts in formal terms. Hence let $Q : \mathcal{P}(E)^n \longrightarrow \mathbf{I}$ be a semi-fuzzy quantifier. The semi-fuzzy L-quantifier Q' of type $\underbrace{\langle 1, \dots, 1\rangle}_{n \text{ times}}$ on E which corresponds to Q can be expressed as

$$Q' = Q \circ \mathop{\times}_{i=1}^{n} \widehat{\vartheta}, \tag{12.4}$$

i.e.

$$Q'(Y_1, \dots, Y_n) = Q(\widehat{\vartheta}(Y_1), \dots, \widehat{\vartheta}(Y_n)) \tag{12.5}$$

for all $Y_1, \dots, Y_n \in \mathcal{P}(E)$. The inverse of ϑ, i.e. β, can be used in a similar way to translate back from semi-fuzzy L-quantifiers of arity n and type $\langle 1, \dots, 1\rangle$ to corresponding ordinary semi-fuzzy quantifiers, i.e. assuming crisp arguments. In the fuzzy case, we can also proceed analogously. In order to translate a fuzzy quantifier $\widetilde{Q} : \widetilde{\mathcal{P}}(E)^n \longrightarrow \mathbf{I}$ into an n-ary fuzzy L-quantifier of type $\langle 1, \dots, 1\rangle$ on E, we proceed as in (12.4); but now, the arguments range over fuzzy sets, so we will replace the crisp powerset mapping $\widehat{\vartheta}$ with the fuzzy image mapping $\hat{\hat{\vartheta}}$ obtained from the standard extension principle. Again, we can also relate a fuzzy L-quantifier $\widetilde{Q}$ of arity n and type $\langle 1, \dots, 1\rangle$ on E to a corresponding ordinary fuzzy quantifier $\widetilde{Q}' : \widetilde{\mathcal{P}}(E)^n \longrightarrow \mathbf{I}$, which then becomes

$$\widetilde{Q}' = \widetilde{Q} \circ \mathop{\times}_{i=1}^{n} \hat{\hat{\beta}}$$

i.e.

$$\widetilde{Q}'(X_1, \ldots, X_n) = \widetilde{Q}(\hat{\hat{\beta}}(X_1), \ldots, \hat{\hat{\beta}}(X_n))$$

for all $X_1, \ldots, X_n \in \widetilde{\mathcal{P}}(E)$. In the following definition, we combine these constructions in order to accomplish an interpretation of ordinary quantifiers in L-QFMs.

Definition 12.8.
Let $\mathcal{F}$ be an L-QFM. The corresponding ordinary QFM $\mathcal{F}_R$ is defined by

$$\mathcal{F}_R(Q) = \mathcal{F}(Q \circ \mathop{\times}_{i=1}^{n} \widehat{\vartheta}) \circ \mathop{\times}_{i=1}^{n} \hat{\hat{\beta}}\,,$$

for all semi-fuzzy quantifiers $Q : \mathcal{P}(E)^n \longrightarrow \mathbf{I}$.

The construction of a corresponding 'ordinary' QFM $\mathcal{F}_R$ for a given L-QFM $\mathcal{F}$ permits us to define the induced truth functions and the induced extension principle of an L-DFS $\mathcal{F}$ as the fuzzy truth functions and the extension principle, respectively, which are induced by the ordinary DFS $\mathcal{F}_R$:

Definition 12.9.
Let $\mathcal{F}$ be an L-QFM and $f : \mathbf{2}^n \longrightarrow \mathbf{I}$ be a given mapping (i.e. a 'semi-fuzzy truth function') of arity $n \in \mathbb{N}$. The induced fuzzy truth function $\widetilde{\mathcal{F}}(f) : \mathbf{I}^n \longrightarrow \mathbf{I}$ is defined by

$$\widetilde{\mathcal{F}}(f) = \widetilde{\mathcal{F}_R}(f),$$

where $\mathcal{F}_R$ is the ordinary QFM defined by Def. 12.8.

Notes 12.10.

- whenever $\mathcal{F}$ is clear from the context, we will again use abreviations $\widetilde{\neg} = \widetilde{\mathcal{F}}(\neg)$, $\widetilde{\vee} = \widetilde{\mathcal{F}}(\vee)$ etc.
- The definition of induced truth functions is extended to the induced fuzzy set operations of complementation $\widetilde{\neg}$, intersection $\widetilde{\cap}$ and $\widetilde{\cup}$ in the obvious ways, i.e. again by elementwise application of the induced negation, conjunction and disjunction on membership grades.

Based on the induced fuzzy truth functions and fuzzy set operations of an L-QFM, we can now develop constructions on semi-fuzzy and fuzzy L-quantifiers which parallel my definitions of antonyms, negation and duals for ordinary quantifiers. Here, we will confine ourselves to exemplifying the definition of dualization, which will be needed for generalizing (Z-3) to L-QFMs; the other concepts can be developed in a similar way but will not be significant in the following.

Definition 12.11.
Given a semi-fuzzy L-quantifier Q of type $t = \langle t_1, \ldots, t_n \rangle$, $n > 0$ on some base set $E \neq \varnothing$, the semi-fuzzy L-quantifier $Q\widetilde{\Box}$ of type t on E is defined by

$$Q\widetilde{\Box}(Y_1, \ldots, Y_n) = \widetilde{\neg} Q(Y_1, \ldots, Y_{n-1}, \neg Y_n)$$

for all $Y_1 \in \mathcal{P}(E^{t_1}), \ldots, Y_n \in \mathcal{P}(E^{t_n})$. $Q\widetilde{\Box}$ is called the dual of Q. The dual of a fuzzy L-quantifier $\widetilde{Q}$ of type t on E is defined analogously.

In order to express a condition constraining the behaviour of $\mathcal{F}$ for unions of argument sets which parallels (Z-4), we must also generalize the construction of quantifiers from such unions. The definition of semi-fuzzy and fuzzy L-quantifiers resulting from unions should be apparent:

Definition 12.12.
Given a semi-fuzzy L-quantifier Q of type $t = \langle t_1, \ldots, t_n \rangle$, $n > 0$ on some base set $E \neq \varnothing$, the semi-fuzzy L-quantifier $Q\cup$ of arity $n + 1$ and type $\langle t_1, \ldots, t_n, t_n \rangle$ on E is defined by

$$Q\cup(Y_1, \ldots, Y_{n+1}) = Q(Y_1, \ldots, Y_{n-1}, Y_n \cup Y_{n+1})$$

for all $Y_1 \in \mathcal{P}(E^{t_1}), \ldots, Y_n \in \mathcal{P}(E^{t_n})$ and $Y_{n+1} \in \mathcal{P}(E^{t_n})$. For fuzzy L-quantifiers $\widetilde{Q}$ of type t on E, the $(n + 1)$-ary fuzzy L-quantifier $\widetilde{Q}\widetilde{\cup}$ of type $\langle t_1, \ldots, t_n, t_n \rangle$ on E is defined analogously.

Monotonicity of semi-fuzzy and fuzzy L-quantifiers is defined as in the ordinary case. We only need to make sure that all arguments involved are chosen from their appropriate ranges $Y_i \in \mathcal{P}(E^{t_i})$ for semi-fuzzy L-quantifiers, and $X_i \in \widetilde{\mathcal{P}}(E^{t_i})$ in the fuzzy case:

Definition 12.13.
A semi-fuzzy L-quantifier Q of type $t = \langle t_1, \ldots, t_n \rangle$, $n > 0$ on some base set E is nondecreasing in its i-th argument, *$i \in \{1, \ldots, n\}$, if*

$$Q(Y_1, \ldots, Y_n) \leq Q(Y_1, \ldots, Y_{i-1}, Y_i', Y_{i+1}, \ldots, Y_n)$$

for all $Y_1 \in \mathcal{P}(E^{t_1}), \ldots, Y_n \in \mathcal{P}(E^{t_n})$ and $Y_i' \in \mathcal{P}(E^{t_i})$ with $Y_i \subseteq Y_i'$. Q is said to be nonincreasing in its i-th argument *if under the same conditions, it always holds that*

$$Q(Y_1, \ldots, Y_n) \geq Q(Y_1, \ldots, Y_{i-1}, Y_i', Y_{i+1}, \ldots, Y_n) .$$

The corresponding definitions for fuzzy quantifiers $\widetilde{Q}$ of type t on E are analogous. In this case, the arguments range over $\widetilde{\mathcal{P}}(E^{t_i})$, and '$\subseteq$' is the fuzzy inclusion relation.

Based on semi-fuzzy L-quantifiers which are nonincreasing in their last argument, we will be able to express a condition on L-QFMs which parallels (Z-5).

Finally, let us extend the condition that $\mathcal{F}$ be compatible with functional composition (Z-6) towards the new type of fuzzification mechanisms. We will resort to the ordinary QFM $\mathcal{F}_R$ associated with $\mathcal{F}$ in order to define the induced extension principle of an L-QFM.

Definition 12.14.
The induced extension principle $\widehat{\mathcal{F}}$ of an L-QFM $\mathcal{F}$ is defined by $\widehat{\mathcal{F}} = \widehat{\mathcal{F}_R}$, i.e. $\widehat{\mathcal{F}}$ is the induced extension principle of the QFM $\mathcal{F}_R$ defined by Def. 12.8.

12.5 The Class of Plausible L-Models

Based on the above constructions, we can now define my axiom system for a class of plausible L-QFMs, called 'L-DFSes' or 'L-models' of fuzzy quantification. It should be obvious from the earlier characterization of DFSes and the generalization of the relevant constructions how the criteria for L-DFSes which support Lindström-like quantifiers can be expressed:

Definition 12.15.
An L-QFM $\mathcal{F}$ is called an L-DFS *if the following conditions are satisfied for all semi-fuzzy L-quantifiers $Q : \times_{i=1}^{n} \mathcal{P}(E^{t_i}) \longrightarrow \mathbf{I}$ of arbitrary types $t = \langle t_1, \ldots, t_n \rangle$ and on arbitrary base sets $E \neq \varnothing$:*

Correct generalisation	$\mathcal{U}(\mathcal{F}(Q)) = Q$ *if* $t \in \{\langle\rangle, \langle 1 \rangle\}$	(L-1)
Projection quantifiers	$\mathcal{F}(Q) = \widetilde{\pi}_{(e)}$ *if* $Q = \pi_{(e)}$ *for some* $e \in E$	(L-2)
Dualisation	$\mathcal{F}(Q\widetilde{\square}) = \mathcal{F}(Q)\widetilde{\square}$ $\quad n > 0$	(L-3)
Internal joins	$\mathcal{F}(Q\cup) = \mathcal{F}(Q)\widetilde{\cup}$ $\quad n > 0$	(L-4)
Preservation of monotonicity	*If Q is nonincreasing in the n-th arg, then* $\mathcal{F}(Q)$ *is nonincreasing in the n-th arg,* $n > 0$	(L-5)
Functional application	$\mathcal{F}(Q \circ \times_{i=1}^{n} \widehat{f_i}) = \mathcal{F}(Q) \circ \times_{i=1}^{n} \widehat{\mathcal{F}}(f_i)$ *where* $f_i : E'^{t'_i} \longrightarrow E^{t_i}$, $t' \in \mathbb{N}^n$ *(same n),* $E' \neq \varnothing$.	(L-6)

The axiom system is constructed in total analogy to my basic system for ordinary QFMs. Thus it is not surprising that the generalized L-models are also suitable for carrying out 'ordinary' quantification, i.e. the model $\mathcal{F}_R$ associated with $\mathcal{F}$ is indeed a DFS.

Theorem 12.16.
Let $\mathcal{F}$ be an L-QFM and $\mathcal{F}_R$ the corresponding QFM defined by Def. 12.8. If $\mathcal{F}$ is an L-DFS, then $\mathcal{F}_R$ is a DFS.

There will be no further comments on the axioms here because they are so closely modelled after the conditions (Z-1)–(Z-6) imposed on plausible choices of QFMs. Ample motivation for these was given in Chap. 3, and there is no point repeating these arguments here. However, another issue needs clarification. As witnessed by the previous chapters, the comprehensive discussion of ordinary DFSes, the development of constructive principles for useful models, and finally the implementation of such models, required considerable effort. Obviously, there would be a lot to gain if we could reuse these results on ordinary models. Ideally, it would be possible to express all L-DFSes in terms of suitable ordinary DFSes. In this case, my results on properties of the models, constructive classes of plausible models and finally the techniques developed for implementation would directly translate into corresponding results on the new L-models.

In order to effect the desired reduction, we need a generic method which lets us construct a 'suitable' L-QFM $\mathcal{F}_L$ from a given ordinary QFM $\mathcal{F}$. Here 'suitable' means that: (a) $\mathcal{F}_L$ properly extends $\mathcal{F}$, i.e. $\mathcal{F} = \mathcal{F}_{LR}$; and (b) $\mathcal{F}_L$ is an L-DFS whenever $\mathcal{F}$ is a DFS, i.e. the construction 'works'. In addition, the new construction should be compatible with the earlier construction of $\mathcal{F}_R$ from a given L-QFM. Hence we should also have (c) $\mathcal{F} = \mathcal{F}_{RL}$ for all L-DFSes $\mathcal{F}$, i.e. $\mathcal{F}$ can be recovered from $\mathcal{F}_R$. A generic construction of L-models which answers these requirements, will eliminate the problem of establishing classes of L-DFSes altogether, because every L-DFS $\mathcal{F}'$ can then be expressed as $\mathcal{F}' = \mathcal{F}_L$ for an ordinary DFS $\mathcal{F}$. we propose the following construction to accomplish this.

Let $\mathcal{F}$ be an ordinary QFM and let Q be a semi-fuzzy L-quantifier of type $t = \langle t_1, \dots, t_n \rangle$ on some base set E. we will abbreviate

$$m = \max\{t_1, \dots, t_n\}, \tag{12.6}$$

in particular $t_i \leq m$ for all $i \in \{1, \dots, n\}$. We can therefore introduce injections $\zeta_i : E^{t_i} \longrightarrow E^m$ which embed E^{t_i} into a common extended domain E^m. These are defined by

$$\zeta_i(e_1, \dots, e_{t_i}) = (e_1, \dots, e_{t_i}, \underbrace{e_{t_i}, \dots, e_{t_i}}_{(m - t_i) \text{ times}}) \tag{12.7}$$

for all $(e_1, \dots, e_{t_i}) \in E^{t_i}$, $i \in \{1, \dots, n\}$. Hence ζ_i fills in the $(m - t_i)$ missing components by repeating e_{t_i}. It should be apparent that ζ_i is indeed an injection. We also need mappings in the reverse direction. These will be written $\kappa_i : E^m \longrightarrow E^{t_i}$, $i \in \{1, \dots, n\}$. The definition is as follows,

$$\kappa_i(e_1, \dots, e_m) = (e_1, \dots, e_{t_i}) \tag{12.8}$$

for all $(e_1, \ldots, e_m) \in E^m$, $i \in \{1, \ldots, n\}$. Hence κ_i is the projection on the first t_i components of $(e_1, \ldots, e_m)$, i.e. it simply drops the final $(m - t_i)$ components. In particular, the κ_i's are onto (surjective). It should be apparent from (12.7) and (12.8) that subsequent application of ζ_i (which adds $m - t_i$ components) and κ_i (which drops these components) will take us back to the original t_i-tuple. Hence for all $i \in \{1, \ldots, n\}$,

$$\kappa_i \circ \zeta_i = \mathrm{id}_{E^{t_i}} . \tag{12.9}$$

In dependence on the given semi-fuzzy L-quantifier Q, we now define a mapping $Q' : \mathcal{P}(E^m)^n \longrightarrow \mathbf{I}$ by

$$Q'(Y_1, \ldots, Y_n) = Q(\widehat{\kappa}_1(Y_1 \cap \operatorname{Im} \zeta_1), \ldots, \widehat{\kappa}_n(Y_n \cap \operatorname{Im} \zeta_n)) , \tag{12.10}$$

for all $Y_1, \ldots, Y_n \in \mathcal{P}(E^m)$. It is instructive to observe that Q' is, at the same time, an n-ary semi-fuzzy L-quantifier of type $\langle m, \ldots, m \rangle$ on E, and an ordinary semi-fuzzy quantifier $Q : \mathcal{P}(E^m)^n \longrightarrow \mathbf{I}$ on a different base set, E^m. In fact, this is the basic idea underlying the following construction of L-QFMs from ordinary QFMs.

Definition 12.17.
The L-QFM $\mathcal{F}_L$ associated with a given QFM $\mathcal{F}$ is defined as follows. For all semi-fuzzy L-quantifiers Q of some type $t = \langle t_1, \ldots, t_n \rangle$, on a base set $E \neq \varnothing$,

$$\mathcal{F}_L(Q) = \mathcal{F}(Q') \circ \mathop{\times}_{i=1}^{n} \hat{\hat{\zeta}}_i$$

where Q' is the semi-fuzzy quantifier defined by (12.10), *and the ζ_i's given by* (12.7) *are extended to mappings $\hat{\hat{\zeta}}_i : \widetilde{\mathcal{P}}(E^{t_i}) \longrightarrow \widetilde{\mathcal{P}}(E^m)$ by applying the standard extension principle.*

Now let us consider the above criteria (a)–(c) for a successful reduction of L-DFSes to ordinary DFSes in turn.

Theorem 12.18.
If $\mathcal{F}$ is a DFS, then $\mathcal{F}_{LR} = \mathcal{F}$.

($\mathcal{F}_L$ properly generalizes the original model $\mathcal{F}$.)

Theorem 12.19.
If $\mathcal{F}$ is a DFS, then the corresponding L-QFM $\mathcal{F}_L$ is an L-DFS.

(The proposed construction results in meaningful interpretations.)

Theorem 12.20.
If $\mathcal{F}$ is an L-DFS, then $\mathcal{F}_{RL} = \mathcal{F}$.

The latter theorem is of particular relevance because it eliminates the problem of establishing classes of L-DFSes altogether. To sum up, my criteria (a)–(c) are all valid and by the above reasoning, we need not be concerned with developing classes of L-DFSes any longer. The ordinary models $\mathcal{F}$ that have already been studied in some depth are sufficient to represent every L-DFS $\mathcal{F}'$ as $\mathcal{F}' = \mathcal{F}_L$, given a suitable choice of $\mathcal{F}$. This is of invaluable help because the investigation of constructive principles for plausible models, like those presented in Chap. 3, turned out to be a rather complex matter. The canonical construction of $\mathcal{F}_L$, then, permits the reuse of the computational models $\mathcal{M}$, $\mathcal{M}_{CX}$ and $\mathcal{F}_{\text{owa}}$ to handle the new cases of fuzzy L-quantification. In the following, we will identify these models with their extensions for simplicity, thus writing $\mathcal{M}_{CX}$ rather than $(\mathcal{M}_{CX})_L$ etc.

12.6 Nesting of Quantifiers

The axioms for L-DFSes are directly modelled after the corresponding axioms for ordinary DFSes; they do not refer to details of multiple variable binding (on the syntactic level), or handling t_i-ary (fuzzy) relations which occur as arguments of L-quantifiers (i.e. on the level of modelling devices). It is an interesting question if additional axioms specifically concerned with L-quantifiers might prove useful for identifying those L-DFSes best suited for modelling branching quantification.

A construction that comes to mind is *quantifier nesting*. Consider the formula $Q'xQ''y\varphi(x,y)$, for example. It should be apparent how to develop a semantical interpretation of such formulas in the proposed framework. Based on semi-fuzzy L-quantifiers, we can either analyze this in terms of two quantifiers of type $t_1 = t_2 = \langle 1 \rangle$ applied in succession. Alternatively, we can look at the whole block $Q_{xy} = Q'xQ''y$, which corresponds to a single semi-fuzzy L-quantifier of type $t = \langle 2 \rangle$. Now let $R \in \widetilde{\mathcal{P}}(E^2)$ be the interpretation of $\varphi(x,y)$ as a fuzzy relation. The analysis in terms of two separate quantifiers will result in an interpretation $\mathcal{F}(Q')(Z)$, where $\mu_Z(e) = \mathcal{F}(Q'')(eR)$ and $\mu_{eR}(e') = \mu_R(e,e')$ for all $e, e' \in E$. This analysis thus corresponds to a fuzzy L-quantifier $\mathcal{F}(Q') \mathbin{@} \mathcal{F}(Q'')$ of type $\langle 2 \rangle$ defined by

$$\mathcal{F}(Q') \mathbin{@} \mathcal{F}(Q'')(R) = \mathcal{F}(Q')(Z)$$

for all fuzzy relations $R \in \widetilde{\mathcal{P}}(E^2)$. The quantifier block Q_{xy}, on the other hand, corresponds to a semi-fuzzy L-quantifier $Q' \mathbin{\widetilde{@}} Q''$ of type $\langle 2 \rangle$ defined by

$$Q' \mathbin{\widetilde{@}} Q''(S) = \mathcal{F}(Q')(Z)\,,$$

$\mu_Z(e) = Q''(eS)$, $eS = \{e' : (e,e') \in S\}$ for all *crisp* relations $S \in \mathcal{P}(E^2)$, and it results in a second interpretation, $\mathcal{F}(Q' \mathbin{\widetilde{@}} Q'')(R)$. (The 'tilde'-notation $\widetilde{@}$ is used here to signify that $Q' \mathbin{\widetilde{@}} Q''$ depends on the chosen L-QFM). It is natural to require that the two interpretations coincide, i.e.

$$\mathcal{F}(Q' \mathbin{\widetilde{@}} Q'') = \mathcal{F}(Q') \mathbin{@} \mathcal{F}(Q'') .$$

More generally, we can define a nesting operation for semi-fuzzy L-quantifiers and fuzzy L-quantifiers of arbitrary types (the quantifier Q' must offer an argument slot to nest in, though, and thus needs positive arity $n > 0$).

Definition 12.21.
Let $E \neq \varnothing$ be a given base set and Q' a semi-fuzzy L-quantifier on E of type $t = \langle t_1, \ldots, t_n \rangle$, $n > 0$. Further let Q'' be a semi-fuzzy L-quantifier on E of arbitrary type $t' = \langle t'_1, \ldots, t'_{n'} \rangle$, $n' \in \mathbb{N}$. The semi-fuzzy L-quantifier $Q' \mathbin{\widetilde{@}} Q''$ on E of type $t^* = \langle t_1, \ldots, t_{n-1}, t_n + t'_1, \ldots, t_n + t'_{n'} \rangle$ which results from the nesting of Q'' into the last argument of Q' is defined by

$$Q' \mathbin{\widetilde{@}} Q''(Y_1, \ldots, Y_{n-1}, S_1, \ldots, S_{n'}) = \mathcal{F}(Q')(Y_1, \ldots, Y_{n-1}, Z)$$

for all $Y_i \in E^{t_i}$, $i \in \{1, \ldots, n-1\}$ and $S_j \in E^{t_n + t'_j}$, $j \in \{1, \ldots, n'\}$, where $Z \in \widetilde{\mathcal{P}}(E^{t_n})$ is defined by

$$\mu_Z(e_1, \ldots, e_{t_n}) = \mathcal{F}(Q'')((e_1, \ldots, e_{t_n})S_1, \ldots, (e_1, \ldots, e_{t_n})S_{n'}) \qquad (12.11)$$

for all $e_1, \ldots, e_{t_n} \in E$, and where the t'_j-ary relation $(e_1, \ldots, e_{t_n})S_j \in \mathcal{P}(E^{t'_j})$ is given by

$$(e_1, \ldots, e_{t_n})S_j = \{(e'_1, \ldots, e'_{t'_j}) : (e_1, \ldots, e_{t_n}, e'_1, \ldots, e'_{t'_j}) \in S_j\} \qquad (12.12)$$

for $j \in \{1, \ldots, n'\}$.

A similar construction is possible for fuzzy L-quantifiers:

Definition 12.22.
Let $E \neq \varnothing$ be a given base set and $\widetilde{Q}'$ a fuzzy L-quantifier on E of type $t = \langle t_1, \ldots, t_n \rangle$, $n > 0$. Further let $\widetilde{Q}''$ be a fuzzy L-quantifier on E of arbitrary type $t' = \langle t'_1, \ldots, t'_{n'} \rangle$, $n' \in \mathbb{N}$. The fuzzy L-quantifier $\widetilde{Q}' \mathbin{\widetilde{@}} \widetilde{Q}''$ of type $t^* = \langle t_1, \ldots, t_{n-1}, t_n + t'_1, \ldots, t_n + t'_{n'} \rangle$ on E is defined by

$$\widetilde{Q}' \mathbin{@} \widetilde{Q}''(X_1, \ldots, X_{n-1}, R_1, \ldots, R_{n'}) = \widetilde{Q}'(X_1, \ldots, X_{n-1}, Z)$$

for all $X_i \in \widetilde{\mathcal{P}}(E^{t_i})$, $i \in \{1, \ldots, n-1\}$ and $R_j \in \widetilde{\mathcal{P}}(E^{t_n + t'_j})$, $j \in \{1, \ldots, n'\}$, where $Z \in \widetilde{\mathcal{P}}(E^{t_n})$ is defined by

$$\mu_Z(e_1, \ldots, e_{t_n}) = \widetilde{Q}''((e_1, \ldots, e_{t_n})R_1, \ldots, (e_1, \ldots, e_{t_n})R_{n'}) \qquad (12.13)$$

for all $e_1, \ldots, e_{t_n} \in E$, and where $(e_1, \ldots, e_{t_n})R_j \in \widetilde{\mathcal{P}}(E^{t'_j})$, $j \in \{1, \ldots, n'\}$, is the fuzzy relation defined by

$$\mu_{(e_1, \ldots, e_{t_n})R_j}(e'_1, \ldots, e'_{t'_j}) = \mu_{R_j}(e_1, \ldots, e_{t_n}, e'_1, \ldots, e'_{t'_j}) \qquad (12.14)$$

for all $e'_1, \ldots, e'_{t'_j} \in E$.

In this general case, too, we would expect that plausible choices of $\mathcal{F}$ comply with the nesting operation.

Definition 12.23.
An L-QFM $\mathcal{F}$ is said to be compatible with quantifier nesting *(in the last argument) if the equality*

$$\mathcal{F}(Q' \mathbin{\widetilde{@}} Q'') = \mathcal{F}(Q') \mathbin{@} \mathcal{F}(Q'') \tag{QN}$$

is valid for all semi-fuzzy L-quantifiers Q' of type $t = \langle t_1, \ldots, t_n\rangle$, $n > 0$ and Q'' of arbitrary type $t' = \langle t'_1, \ldots, t'_{n'}\rangle$, $n' \in \mathbb{N}$ defined on the same base set $E \neq \varnothing$.

Judging from some first tests, it appears that quantifier nesting expresses a very restrictive condition and excludes many useful models. In fact, we have the following result on general quantifier nestings:

Theorem 12.24.
Suppose that an L-DFS $\mathcal{F}$ is compatible with quantifier nesting. Then $\mathcal{F}_R$ is compatible with fuzzy argument insertion.

In other words, the only standard model which potentially admits the nesting of quantifiers is $\mathcal{M}_{CX}$, as shown by Th. 7.80 and Th. 7.81. It has not been established yet whether $\mathcal{M}_{CX}$ validates the criterion but I would rather guess that this is not the case. It is perfectly possible – as in the case of conservativity Def. 6.20 or convexity Def. 6.8 – that the unrestricted or 'strong' requirement of compatibility with quantifier nestings conflicts with other reasonable criteria and must thus be weakened to a compatible 'core' requirement. Obviously, further research should be directed at these issues in order to determine the status of $\mathcal{M}_{CX}$ and develop suitable weakenings of the nesting criterion if necessary.

Let us briefly discuss the limitation of (QN) to nestings in the last argument of the quantifier Q'. Nestings in another argument position k can be reduced to this case by flipping argument positions k and n, performing the nesting in the last argument, and then reordering the arguments again into their intended target positions. Every DFS is known to be compatible with such permutations of argument positions, a property which apparently generalizes to the L-models. Thus the compatibility of a DFS with nestings in the last argument position is sufficient to ensure that it also comply with nestings in other argument positions. By repeating this process, (QN) ensures the compatibility of conforming L-DFSes with multiple nestings of arbitrary depths. Thus, the condition is sufficient to cover general nestings involving quantifiers of arbitrary types with an arbitrary number of embedded quantifiers.

It would be useful however, to relate the proposed general nestings in the last argument to the simple nesting of unary quantifiers considered earlier, or to other simplified nesting criteria. This might facilitate the test that a model of interest (not necessary a standard model like $\mathcal{M}_{CX}$) comply with the nesting condition, and also eliminate some redundancy from the axiom system. The general nestings described by (QN) can probably be reduced to a simpler criterion, but further research is necessary to clarify this matter. Should a substantial weakening of the original requirement be necessary, the investigation of simplified criteria might also guide us to a useful weakening which still covers the nestings of linguistic relevance.

12.7 Application to the Modelling of Branching Quantification

Let us now elucidate how the motivating example *"Many young and most old people respect each other"* can be interpreted in the proposed framework. In this case, we have semi-fuzzy quantifiers $Q_1 = \mathbf{many}$, defined by $\mathbf{many}(Y_1, Y_2) = |Y_1 \cap Y_2|/|Y_1|$, say, and $Q_2 = \mathbf{most}$, defined as above. Both quantifiers are nondecreasing in their second argument, i.e. we can adopt the analysis proposed by Barwise. The modification of equation (12.1) towards gradual truth values will be accomplished in the usual way, i.e. by replacing existential quantifiers with sup and conjunctions with min (non-standard connectives are not possible here). The semi-fuzzy L-quantifier Q of type $\langle 1, 1, 2\rangle$ constructed from Q_1, Q_2 then becomes

$$Q(A, B, R) = \sup\{\min(Q_1(A, U), Q_2(B, V)) : U \times V \subseteq R\}$$

for all $A, B \in \mathcal{P}(E)$ and $R \in \mathcal{P}(E^2)$. Hence *"Many men and many women are relatives of each other"*, for example, which rests on crisp arguments, is now taken to denote the maximum degree to which many men and many women belong to a group $U \times V \subseteq R$ of mutual relatives. This analysis appears to be conclusive – but we still need to extend it to fuzzy arguments. To this end, it is sufficient to apply the chosen L-DFS $\mathcal{F}$. We then obtain the fuzzy L-quantifier $\mathcal{F}(Q)$ of type $\langle 1, 1, 2\rangle$ suited to handle this case. Returning to the original example involving young and old people, we have fuzzy subsets $\mathbf{young}, \mathbf{old} \in \widetilde{\mathcal{P}}(E)$ of young and old persons, respectively, and a fuzzy relation $\mathbf{rsp} \in \widetilde{\mathcal{P}}(E^2)$ of persons who respect each other. Thus, a meaningful interpretation of *"Many young and most old people respect each other"* is now given by $\mathcal{F}(Q)(\mathbf{young}, \mathbf{old}, \mathbf{rsp})$.

In the event that either Q_1 or Q_2 fail to be increasing in the second argument, we must adopt Westerståhl's generic method for interpreting branching quantifiers. Let us now discuss how the method can be applied in the fuzzy case. Suppose that Q_1, Q_2 are arbitrary semi-fuzzy quantifiers of arity $n = 2$ on some base set E. Following Westerståhl, we introduce nondecreasing and nonincreasing approximations of the Q_i's, defined by $Q_i^+(Y_1, Y_2) = \sup\{Q_i(Y_1, L) : L \subseteq Y_2\}$ and $Q_i^-(Y_1, Y_2) = \sup\{Q_i(Y_1, U) : U \supseteq Y_2\}$, respectively. With the usual replacement of existential quantification with sup and conjunction with min, Westerståhls interpretation formula [167, p. 281, Def. 3.1] becomes:

$$Q(A,B,R) = \sup\{\min\{Q_1^+(A,U_1), Q_2^+(B,V_1), Q_1^-(A,U_2), Q_2^-(B,V_2)\} : (U_1 \cap A) \times (V_1 \cap B) \subseteq R \cap (A \times B) \subseteq (U_2 \cap A) \times (V_2 \cap B)\}$$

for all $A, B \in \mathcal{P}(E)$ and $R \in \mathcal{P}(E^2)$. Application of an L-DFS then determines the corresponding fuzzy L-quantifier $\mathcal{F}(Q)$ of type $\langle 1, 1, 2 \rangle$ suitable for interpretation.

As shown by Westerståhl [167, p.284], his method results in meaningful interpretations provided that (a) the Q_i's are 'logical', i.e. $Q_1(Y_1, Y_2)$ and $Q_2(Y_1, Y_2)$ can be expressed as a function of $|Y_1|$ and $|Y_1 \cap Y_2|$, see van Benthem [9, p. 446] and [10, p. 458]; and (b), the Q_i's are *convex* in their second argument, or 'CONT' in Westerståhl's terminology, i.e. $Q_i(Y_1, Y_2) \geq \min(Q_i(Y_1, L), Q_i(Y_1, U)$ for all $L \subseteq Y_2 \subseteq U$ according to Def. 6.8. The latter condition ensures that Q_1 and Q_2 can be recovered from their nondecreasing approximations Q_i^+ and their nonincreasing approximations Q_i^-, i.e. $Q_i = \min(Q_i^+, Q_i^-)$. This is generally the case when Q_1 and Q_2 are either nondecreasing in their second argument ("many"), nonincreasing ("few"), or of unimodal shape ("about ten", "about one third"). An example of branching quantification with unimodal quantifiers, which demand the generic method, is *"About fifty young and about sixty old persons respect each other"*.

12.8 Chapter Summary

Recognizing the potential of branching quantifiers to linguistic modelling, this chapter was concerned with a generalization of the quantification framework which incorporates these cases. The extension comprises fuzzy L-quantifiers (generalizations of Lindström quantifiers to approximate quantifiers and fuzzy arguments), semi-fuzzy L-quantifiers (uniform specifications of such quantifiers) and plausible models of fuzzy quantification involving these quantifiers, called L-DFSes. The criteria for plausible L-models of fuzzy quantification parallel the requirements on QFMs, and the rationale for these conditions is the same as in the case of ordinary QFMs. It has further been shown that the new L-DFSes are in one-to-one correspondence to the ordinary models. Thus

(a) every plausible model of 'ordinary' fuzzy quantification (DFS) can be extended to a unique L-DFS for quantifiers with multiple variable binding;
(b) no other L-DFSes exist beyond those obtained from (a).

Westerståhl's analysis of branching NL quantification in terms of Lindström quantifiers is easily generalized to semi-fuzzy L-quantifiers and thus, branching Type III quantifications. By applying the chosen model of fuzzy quantification, one then obtains a meaningful interpretation for branching Type IV quantifications, i.e. both approximate quantifiers and fuzzy arguments can be handled. At this point, it should be remarked that Westerståhl also extends his method to other types of branching quantification in NL, which in general can span more than two quantifiers. However, the formulas he proposes to cope with these cases can be adapted to our analysis in terms of (semi-)fuzzy L-quantifiers in total analogy to the example chosen for demonstration. Thus, the methods must be extended to semi-fuzzy L-quantifiers in the apparent ways; applying the L-QFM $\mathcal{F}$ then lets us fetch the final interpretation.

The identification of those L-DFSes specifically suited for modelling branching quantification in NL is an advanced topic that should be tackled by future research. A possible strengthening of the system based on quantifier nesting has already been sketched. The nesting construction is linked to branching quantification in the following way: In the motivating example of this chapter, the introduction of the complex quantifier Q of type $\langle 1, 1, 2 \rangle$ rests on the grouping of several (independent) quantifiers into a block of quantifiers, which is then treated as an integral, singular quantifier. Applying the same idea to a linear sequence of quantifiers takes us to the construction of nested quantifiers discussed in Sect. 12.6. Naturally, a model suited for branching quantification, which rests on the grouping of quantifiers into blocks, should also be compatible with such nestings of linear quantifiers. However, the resulting criterion appears to be extremely restrictive and it cannot be adopted without sacrificing useful models like $\mathcal{M}$ and $\mathcal{F}_{\mathrm{owa}}$. It is not even clear if the full compatibility with quantifier nestings is consistent with the basic axioms at all, and it might therefore be necessary to weaken the criterion to typical classes of linguistic quantifiers (e.g. quantitative). Obviously a more thorough discussion of quantifier nestings is necessary to clarify these issues.

The proposed analysis of reciprocal constructions in terms of fuzzy branching quantifiers is of particular relevance to linguistic data summarization [83, 182]. Many summarizers of interest express mutual (or symmetric) relationships and can therefore be verbalized by a reciprocal construction. (Asymmetrical relations R' can also be used in reciprocal constructions after symmetrization, i.e. they must be replaced with $R = R' \cap R'^{\mathrm{op}}$). An ordinary summary like *"Q_1 X_1's are strongly correlated with Q_2 X_2's"* does not capture the symmetrical nature of correlations, and it neglects the resulting groups of mutually correlated objects. The proposed modelling in terms of branching quantifiers, by contrast, supports a novel type of linguistic summaries specifically suited for describing groups of interrelated objects. Branching quantification, in this sense, is a natural language technique for detecting such groups in the data. A possible summary involving a reciprocal predicate is *"The intake of most vegetables and many health-related indicators are strongly associated with each other"*.

terms of cardinalities), and they are often not definable in first-order predicate logic. The generalized quantifiers of TGQ are sufficiently powerful to express such quantifiers. However, they are essentially limited to precise quantifications based on crisp arguments.
Research in fuzzy set theory, on the other hand, has achieved a modelling of Type IV quantifications [108] where both the quantifier and its arguments can be fuzzy. Zadeh's framework for analyzing fuzzy quantification [197, 199] is restricted to absolute and proportional quantifiers, though. Moreover there is negative evidence against the traditional approaches with regard to linguistic adequacy (see also discussion in Sects. 1.9.1 to 1.10, and the evaluation of existing approaches detailed in Appendix A).

The new framework for fuzzy quantification described in this monograph was expected to establish a unifying perspective on both traditions of analyzing NL quantification which accounts for the linguistic considerations but also incorporates a fuzzy-sets based modelling of vagueness. In order to achieve the intended coverage of linguistic phenomena, we started from the linguistic analysis. The concept of a two-valued generalized quantifier known from TGQ was considered a suitable point of departure for developing the framework.

The generalization to *fuzzy quantifiers* had to support gradual interpretations and fuzzy arguments in order to cope with arbitrary Type IV quantifications. We have adopted Zadeh's view of fuzzy quantifiers as fuzzy second-order predicates to achieve this. However, the expressiveness of fuzzy quantifiers made it difficult to establish a suitable choice of fuzzy quantifier for a linguistic quantifier of interest. This is because fuzzy quantifiers must deal with fuzzy arguments, which forbids a simple definition of their input-output behaviour in terms of crisp cardinalities.

In order to cope with this *modelling problem* of explaining the relationship between linguistic quantifiers and the available modelling devices, the specification of a quantifier was separated from its operational form used for interpretation. We therefore introduced simplified descriptions which only capture the essential aspects of a quantifier. This specification format makes it easy to define the quantifier of interest. It is the responsibility of fuzzy quantification theory to establish the matching operational quantifier for a given specification.

Following this general strategy, we introduced the concept of a *semi-fuzzy quantifier* which offers a compact description of the linguistic quantifiers of interest. It thus plays a similar role as membership functions μ_Q in the earlier approaches. Semi-fuzzy quantifiers isolate the modelling of approximate quantifiers like "almost all" from the problem of dealing with fuzzy arguments, i.e. they support gradual interpretations but only accept crisp arguments. This simplification corresponds to the Type III quantifications of Liu and Kerre [108]. Still, semi-fuzzy quantifiers are sufficiently powerful to embed two-valued crisp generalized quantifiers known to the linguistic theory. Moreover, the phenomena of linguistic interest that were previously studied in TGQ can all be discussed on the level of semi-fuzzy quantifiers as well. Due to the restriction to crisp arguments, it is usually easy to grasp the meaning of a given semi-fuzzy quantifier. In particular, semi-fuzzy quantifiers can be conveniently defined in terms of the usual cardinality because they only accept crisp arguments. For the same reason, however, semi-fuzzy quantifiers are essentially limited to their function as specifications because fuzzy arguments cannot be handled.

By connecting the specification of a linguistic quantifier and its associated operational interpretation, the notion of a *quantifier fuzzification mechanism* (QFM) completes the new framework for fuzzy quantification.

The structuring into specifications and interpretations, which is fundamental to the proposed framework, splits the modelling problem into application-dependent and independent parts. The domain-specific task consists in selecting the semi-fuzzy quantifier which best fits the meaning of the quantifier of interest in a given application context. The central problem of fuzzy quantification, which is application-independent, then consists in explaining how this description in terms of a semi-fuzzy quantifier should be extended to fuzzy arguments. The proposed QFMs account for this problem. By isolating the core competence of handling fuzzy arguments, part of the complexity of the modelling problem is effectively delegated to the QFM. This division of responsibilities makes it possible to analyse the essential aspects of fuzzy quantification without getting stuck in discussions which particular interpretation should be assigned to a given linguistic quantifier.

The particular choice of a semi-fuzzy quantifier suited for modelling a given linguistic quantifier can apparently depend on the context. To be sure, this is not really surprising because the precise meaning of quantifiers like "many" or "almost all" is obviously context-dependent as well and determined by contextual factors like an assumed standard of comparison. Thus, the *problem of context dependence* it is not an artefact of the proposed modelling devices, but simply reflects a characteristic of language itself. In particular, the problem will not go away if we switch to the familiar μ_Q-based representations. The notion of semi-fuzzy quantifiers has been designed in such a way that all contextual factors fall into the realm of specification. Hence when describing the quantifiers of interest, the users must resolve these factors and commit to a single unambiguous interpretation. In this way, the context-dependent factors are effectively isolated from the core problems of fuzzy quantification, i.e. from those factors that can be subjected to a general solution. This organization of fuzzy quantification made it possible to develop the theory of fuzzy quantification independently of the intricacies of context dependence. From a practical perspective, though, it makes sense to assist the user or application designer who must decide on the concrete interpretation of these context-dependent quantifiers in terms of a membership function. This problem of determining a suitable choice of quantifiers can be considered as a special case of *knowledge acquisition*. There are several techniques for determining membership grades of fuzzy sets which might also be used to construct membership functions of fuzzy quantifiers. For a survey of these techniques, see Klir and Yuan [99,

Chap. 10], who also give pointers into the specialized literature. In the future, it might be possible to integrate some experimental findings on the effects of external factors like size or shape on the denotation of quantifiers which depend on the context [121, 122]. However, these ideas have not yet reached practical significance.

The proposed QFMs merely serve as a *placeholder* which lets us talk about candidate models of fuzzy quantification on a formal level. They do not account for any considerations regarding the linguistic quality of interpretations. It is the responsibility of a *theory of fuzzy quantification* to identify those models of fuzzy quantification which best answer the linguistic expectations. To achieve this, it was necessary to develop formal criteria which ensure that the QFM preserve the main characteristics of quantifiers as well as relationships between quantifiers connected by linguistic constructions like dual or antonym. This can be likened to the familiar mathematical concept of a homomorphism, i.e. of a structure-preserving mapping compatible with a number of relevant constructions. Some of the concepts needed to express these regularities were already known from TGQ, like the above constructions of antonym and dual which only had to be generalized to the fuzzy case. However, some new concepts were also needed which account for cross-domain and cross-arity coherence. These requirements then comprise a catalogue of formal criteria against which every model of interest can be checked. In order to avoid unnecessary effort in proofs, such a catalogue should describe the target class of models in terms of a minimal (i.e. irreducible) axiom system. This will ensure that the axioms capture independent aspects of linguistic adequacy. In Chap. 3, we presented a complete system of such criteria which also exhibits the desired minimality.

The semantical postulates (Z-1) to (Z-6) account for a number of very elementary requirements on plausible and coherent interpretations. The first criterion, (Z-1), requires the models to correctly generalize the original specification for a special case of quantifiers. The second criterion (Z-2) asserts that the models be compatible with membership assessments. Based on a canonical construction of induced fuzzy connectives, (Z-3) requires the compatibility of the interpretations with dualisation. The criterion (Z-4), in turn, makes sure that the models comply with unions of arguments. The requirement (Z-5) accounts for the desired monotonicity of interpretations, which is enforced for a special case. Finally (Z-6) links the interpretations obtained for different universes of quantification by requiring the compatibility of the model with the canonical choice of extension principle. Taken together the axiom system establishes a type of plausible model for which we coined the term *determiner fuzzification scheme*, or *DFS* for short.

The proposed class of models was then evaluated against a catalog of additional semantic desiderata that can legitimately be expected of interpretations of fuzzy quantifiers. First of all, we were concerned with the induced choice of fuzzy connectives, which was shown to be plausible in every DFS. We then considered various aspects of argument structure. It was shown that every DFS is compatible with permutations of arguments, cylindrical extensions (augmentation with vacuous arguments), complementation (formation of antonyms), but also external negation and the construction of duals. In addition, it admits intersections of arguments, their symmetrical difference, the insertion of crisp arguments and hence, crisp adjectival restriction. The models further preserve monotonicity properties of quantifiers, including 'local' monotonicity for restricted ranges of arguments. The induced extension principle as well as the induced fuzzy inverse images are also reasonable. The models will interpret both quantitative and non-quantitative cases of quantifiers in a plausible way, and they preserve the property of 'having extension' which is shared by most linguistic quantifiers. In addition, the models are 'contextual', i.e. insensitive to the behaviour of a quantifier outside the context given by the fuzzy arguments. Finally the standard quantifiers are interpreted in a plausible way, judging from Thiele's characterization of fuzzy universal and existential quantifiers in terms of T- and S-quantifiers [158]. This explicit evaluation shows that the proposed models are justified from a linguistic perspective and internally consistent. It should be pointed out that the catalog of semantic desiderata can also be applied to other kinds of models. For example, the criteria made it possible to evaluate the existing approaches to fuzzy quantification – in this case with negative results (see Appendix A). This demonstrates that the formalization of desiderata for plausible interpretations is an important research topic in its own right.

It is also instructive to elucidate the *structure of natural subclasses*. We therefore considered collections of models grouped by some common property. The relative homogeneity of members in such a subclass made it possible to define interesting constructions. First of all the models were grouped by their induced negation. It was then shown that without loss of generality, one can focus on those models which induce the standard negation. This result was established by means of a *model translation scheme* which permits a bidirectional transformation between models.
Now grouping these $\neg$-DFSes by their induced disjunction, we obtained the class of $\widetilde{\vee}$-DFSes. These classes are sufficiently homogeneous to permit the *aggregation* of models. In particular, the classes are closed under symmetric sums, convex combinations and similar averaging operators. In order to gather more knowledge on the structure of the classes, we investigated the natural *specificity order* on these models. Due to the symmetry of a DFS with respect to negation, it is not possible to use the linear order $\leq$, i.e. there is nothing like a 'least' or 'greatest' $\widetilde{\vee}$-DFS. However, the models are partially ordered by specificity, i.e. under the usual fuzziness order introduced by Mukaidono [118]. It was shown that for non-empty collections of $\widetilde{\vee}$-models, there is always a greatest lower bound in terms of specificity which can be expressed in terms of the generalized fuzzy median. We further discussed the criterion of *specificity consistency*. It was shown that a non-empty collection of $\widetilde{\vee}$-DFSes has a least upper bound in terms of specificity exactly if it is specificity consistent. The resulting model can again be given an explicit description.

$\mathcal{F}_\xi$-type. However, these $\mathcal{F}_\Omega$-models coincide with the $\mathcal{F}_\psi$-DFSes, i.e. with the natural class of models defined in terms of the standard extension principle. In other words, the most general construction of models that was described in this book is mainly of theoretical interest.

It is an issue of future research to relate the $\mathcal{F}_\Omega/\mathcal{F}_\psi$-models to the full class of standard DFSes, and to develop even more general constructions if necessary to catch hold of arbitrary standard models. Moreover, it would be important to develop constructions which give us a grip on non-standard models as well. This might provide a positive answer to the problem if non-standard models exist at all. In fact, Díaz-Hermida et al [31] have described a probabilistic model of fuzzy quantification which satisfies the axiomatic requirements; the model is interesting in this context because it induces the algebraic product rather than the standard fuzzy conjunction. The method is only defined for finite base sets, though, and does not qualify as a non-standard DFS for that reason. It might be useful to introduce a corresponding notion of QFMs and DFSes restricted to finite universes, in particular to investigate the possible junctures of fuzzy quantification and probability theory.

For practical purposes it is also important to identify those examples of the known models which are likely most useful in applications. The models $\mathcal{M}_{CX}$ and $\mathcal{F}_{\mathrm{owa}}$ as well as a third example $\mathcal{M}$, which is some kind of compromise between $\mathcal{M}_{CX}$ and $\mathcal{F}_{\mathrm{owa}}$, were chosen as prototypes which serve as the candidates for practical implementation.

The level of abstraction achieved by the theoretical skeleton and the clear division of responsibilities, i.e. specification versus interpretation and model versus implementation, fostered the rapid development of fuzzy quantification theory. The strict separation of theoretical analysis and computational aspects made it possible to resolve the semantical issues in a more elegant solution compared to previous approaches. Moreover, it became possible to identify broad classes of prototypical models without getting involved into the practical concern how these models can eventually be implemented. But of course, the new 'nice' models will not be useful in practice unless we can apply them to compute real quantification results. In order to provide the theory with sufficient computational backing, we developed efficient algorithms for implementing the main types of quantifiers in the new models. To this end, it was first shown that for finite E, the computation of quantification results in $\mathcal{F}_\xi$-models can always be based on quantities $\top_j$, $\bot_j \in \mathbf{I}$ obtained from a finite sample of cut levels γ_j, i.e. there is no need to consider all choices of γ in the continuous range $[0, 1]$. We then presented an analysis of 'quantitative' quantifiers in terms of cardinality information sampled from the arguments, and went on explaining how the computation of $\top_j$ and $\bot_j$ can be optimized based on this analysis. Finally, it was shown how the required cardinality information can be computed efficiently from histograms of the involved fuzzy arguments and their Boolean combinations.

We then presented concrete algorithms for implementing the main types of quantifiers in the prototypical models $\mathcal{M}$, $\mathcal{M}_{CX}$ and $\mathcal{F}_{\mathrm{owa}}$. These algorithms are assembled from the above building blocks (like histogram-based computation of the cardinality coefficients) and thus demonstrate the practical utility of this analysis for implementing quantifiers in $\mathcal{F}_\xi$-models. The issue which of these computational models should be preferred in a given application should now be decided by practical experiments. In particular, it will be instructive to compare $\mathcal{M}_{CX}$, which generalizes the Sugeno integral, and $\mathcal{F}_{\mathrm{owa}}$, which abstracts from the Choquet integral, and gather some evidence which model is perceived more natural by the users.

13.2 Comparison to Earlier Work on Fuzzy Quantification

Let us now compare the novel theory of fuzzy quantication to existing work in order to highlight the main contributions of this monograph. One of the main reasons for introducing the new framework was the limited coverage of linguistic phenomena inherent to Zadeh's analysis. The comprehensive classification of natural language quantifiers presented by Keenan and Stavi [92, pp. 253-256] distinguishes sixteen main classes of quantifiers (including quantifiers of exception like "all except one", definites like "the ten", bounding quantifiers like "only" etc). All of these heterogeneous quantifiers are covered by TGQ (assuming two-valued quantifications), while the traditional framework is essentially limited to the absolute and proportional kinds. The proposed theory of fuzzy quantification embeds all of these cases and demonstrates how the basic approach of TGQ can be generalized to arbitrary Type IV quantifications.

It must be pointed out, though, that the absolute and proportional quantifiers are indeed cases of special relevance. The list of main types further includes quantifiers of exception and cardinal comparatives, which are also significant from the linguistic perspective and of obvious utility to applications. These are the quantifiers covered by the algorithms presented in Chap. 11.

Further types of quantifiers, like proportional comparatives, are described in the literature on TGQ, cf. Keenan and Moss [91], Keenan and Stavi [92], or Keenan and Westerståhl [93]. An example is *"a greater proportion of French students than workers"* [91, p. 78]. The *classification of semi-fuzzy quantifiers* into categories of linguistic and practical value can be developed in total analogy to the existing classification of TGQ for precise quantifications; see Díaz-Hermida et al. [30] for a description of several interesting classes. An implementation of these new kinds of quantifiers in the prototypical models is straightforward from the general analysis presented in Chap. 11.

As opposed to the novel framework, existing approaches are closely intertwined with the specifics of absolute and proportional quantifiers. Zadeh's framework does not introduce uniform specifications suited for all quantifiers, but rather stipulates different kinds of membership functions $\mu_Q : \mathbb{R}^+ \longrightarrow \mathbf{I}$ vs. $\mu_Q : \mathbf{I} \longrightarrow \mathbf{I}$ for respresenting the absolute and proportional types. These first-order representations express the shape of the quantifier, but they do not specify the precise way in which the arguments of the quantifier enter into the quantification results. The responsibility to account for argument structure is therefore delegated to the model of quantification: For each type of quantifier, the model must be equipped with a corresponding interpretation method which knows how to combine the arguments in order to compute the quantification result. Consequently, there will be one rule for two-place proportional quantifiers, another rule for unary absolute quantifiers etc. This impairs extensibility because one needs to introduce one specification format and corresponding interpretation rule per quantifier type. The splitting of interpretation rules also makes it difficult to ensure internal coherence. To sum up, the lack of a uniform representation obstructs the formalization of universal models which are capable of interpreting arbitrary quantifiers.

As witnessed by the proposed representation in terms of semi-fuzzy quantifiers, these deficiencies of the traditional framework can easily be avoided if (a) we use a *uniform specification format* for describing quantifiers; and (b) if irregular details like the specifics of argument structure are encoded in these specifications. Based on the uniform representation, it then becomes possible to formulate *universal models* which assign interpretations to such specifications. Without any need to introduce new specifications or extend the interpretation rules, we can then support unanticipated cases of quantifiers even if these do not match the known patterns of argument structure. Obviously the uniform representation also facilitates the development of models which show the desired coherence across quantifiers types.

The limited coverage of the traditional analysis becomes visible, in particular, in its poor support for multi-place quantification. Apart from unary and binary quantifiers, which correspond to unrestricted vs. restricted quantification, one also finds ternary quantifiers in natural languages. An example is "more than" in the quantifying proposition *"More men than women are smokers"*. Such quantifications cannot be interpreted in terms of the absolute and proportional quantifiers supported by the traditional models, simply because "more than" and similar cases are irreducible to two-place quantifiers, as shown by Hamm [69, pp. 23+]. We already mentioned proportional comparatives which can depend on four arguments. And compound quantifiers like *"Most* Y_1*'s or* Y_2*'s are* Y_3*'s and* Y_4*'s"* can involve a potentially unbounded number of arguments. Even if one does not view such constructions as quantifiers proper, a framework for fuzzy quantification artificially restricted to quantifiers of arities $n \leq 4$ would likely be not any simpler than a solution for arbitrary $n \in \mathbb{N}$. Consequently, the novel framework was designed for general multi-place quantifiers.

A model of quantification in this general sense cannot be taylored to special quantifiers like the absolute or proportional kinds. This means that we had to give up Zadeh's idea of using some measure of absolute or relative cardinality to evaluate quantifiers, and we could not use specialized concepts like the 'degree of orness' either [165, 179, 183]. By contrast, such models may *only* be concerned with the effect of fuzzy arguments on the interpretation of quantifying expressions. In this way, it becomes possible to support quantifiers with an arbitrary number of arguments. Because multi-place quantification includes restricted proportional quantification as a special case, the notorious problem of proportional quantification with importance qualification will has been solved as a by-product of treating all n-place quantifiers. To existing approaches, however, the complexity of restricted quantification was prohibitive (see counterexamples in Chap. 1). An extension of these approaches to ternary quantifiers was suggested by Zadeh, who mentions the "third kind" of quantifiers in [197, p. 149] and [199, p. 757]. However, it appears that nobody pursued this direction further, probably due to the difficulties of existing approaches with importance qualification which involves only two (rather than three or more) arguments.

Existing approaches to fuzzy quantification usually assume that the base set be finite.[1] It is true that the universe of quantification will typically be finite in practical applications. However, *infinite base sets* will be essential to the modelling of fuzzy mass quantification. The examples of quantifiers in this book usually involved count nouns ("men", "patients", "Swedes",...) and were thus concerned with concrete objects. Mass nouns like "water" or "wine", however, call for a different kind of quantification which ranges over continuous, possibly atomless masses like "some water" or "much wine". An adequate modelling of quantification over such masses will necessarily involve infinite sets (of which masses are a special case). The proposed notion of QFMs and the plausible models (DFSes) developed from it support infinite universes of quantification and thus provide a good starting point for approaching fuzzy mass quantification. However, it is likely necessary to add some more structure in order to deal with mass quantification. For example, it might be convenient to assume that E be a measurable space with a measure P defined on it. We can then define a proportional quantifier Q, intended to model "more than 30 percent" on E, by

$$Q(Y_1, Y_2) = \begin{cases} 1 & : \quad P(Y_1 \cap Y_2) \geq 0.3\, P(Y_1) \\ 0 & : \quad \text{else} \end{cases}$$

assuming that $Y_1, Y_2 \subseteq E$ be measurable subsets. In the case that Y_1 or Y_2 is not measurable, it is most natural to declare $Q(Y_1, Y_2)$ undefined. This suggests that quantification over masses will involve an extension to partial (rather than total) semi-fuzzy quantifiers, and a suitable notion of QFM for mapping partial semi-fuzzy quantifiers to fuzzy quantifiers. Hence the support for quantifiers on infinite base sets, which is already available in the current proposal, is an important prerequisite of treating mass quantification but it

[1] The Σ-Count approach can be generalised to measurable fuzzy subsets of sets of infinite cardinality by replacing summation with integration [193, p. 167]. However, as we have seen in Chap. 1, the Σ-count approach does not offer a suitable departure for developing plausible models of fuzzy NL quantification.

needs to be enhanced with a model of partial quantifiers and basic concepts of measure theory.

The novel theory also achieves an adequate treatment of quantifiers which lack automorphism invariance (or 'quantitativity'). By Th. 4.43, every model of the DFS axioms maps quantitative semi-fuzzy quantifiers like "almost all" to quantitative fuzzy quantifiers. But the theorem also states that $\mathcal{F}$ maps non-quantitative semi-fuzzy quantifiers to corresponding *non-quantitative* fuzzy quantifiers. Some examples are proper names like **john** $= \pi_{\text{John}}$, and non-quantitative compound quantifiers like *"All X's except the men are Y's"* or *"All male X's are Y's"*. Existing approaches to fuzzy quantification, by contrast, are defined in terms of absolute or relative cardinality measures for fuzzy sets, which must be invariant under automorphisms to capture the notion of 'cardinality'. Consequently, the traditional models *always* result in quantitative quantifiers, i.e. they cannot model any non-quantitative cases. This failure of existing approaches gains some weight as soon as infinite base sets are admitted (e.g. in mass quantification). In this case most of the important quantifiers (including "most", for example) become non-quantitative in the sense of automorphism invariance, see van Benthem [9, p.473+].

Taken together, these observations demonstrate that from the perspective of coverage, only the new proposal accounts for the linguistic data because it embeds the established Theory of Generalized Quantifiers. The traditional approaches to fuzzy quantification, however, are too narrow from a linguistic perspective and for structural reasons, they bear the risk of producing incoherent results. The coverage of the new theory with respect to quantificational phenomena compared to that of TGQ and that of existing approaches to fuzzy quantification is summarized in Fig. 13.1.

With the notable exception of H. Thiele's analysis of specialized classes of quantifiers [158, 159], existing approaches to fuzzy quantification are generally based on interpretation rules the linguistic plausibility of which was 'proven' by presenting a few positive examples. The novel theory of fuzzy quantification, by contrast, rests on an axiomatic foundation. Considerable effort has been spent on the discovery and formalization of the intuitive requirements on plausible interpretations of quantifiers, and the resulting criteria will ensure meaningful quantification results in all conforming models. The coherence of interpretations can also be guaranteed across quantifier types. This is possible due to the use of uniform specifications and generic models which treat all quantifiers in basically the same way. Concerning specific constructions of linguistic relevance, the proposed models are fully compatible with Aristotelian squares, i.e. they admit the formation of external negation, antonyms, and duals. This property is offered by *none* of the earlier approaches. Furthermore, every DFS is compatible with the Piaget group of transformations, the significance of which stems from empirical findings in developmental psychology (see Dubois and Prade [41, p. 158+]). The models further preserve extension and thus reproduce the insensitivity of NL against the precise choice of domain (see Th. 4.45). It should also be pointed out that by committing to a

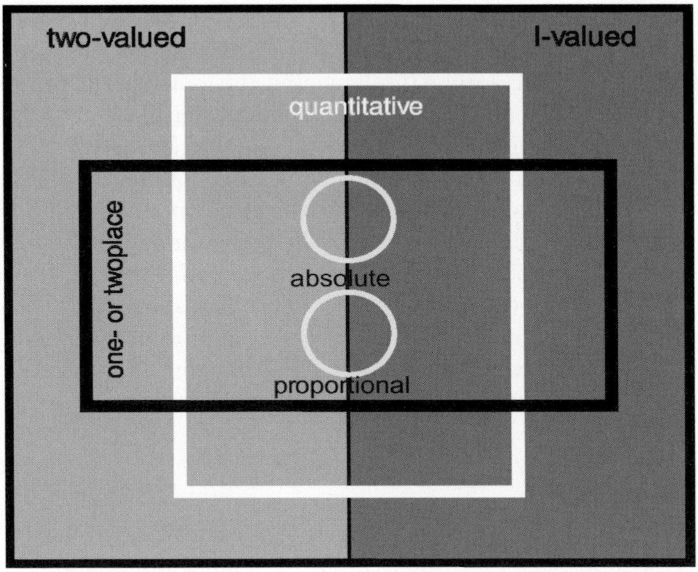

Fig. 13.1. Coverage of linguistic quantifiers: DFS (outer rectangle) vs. TGQ (left half) and existing approaches (circles)

model of fuzzy quantification, a unique definition of all fuzzy connectives will be induced. For example, every DFS is compatible with exactly one choice of conjunction, disjunction, or implication – there is no degree of freedom left in the definition of these connectives. Every DFS further induces a unique choice of extension principle and of fuzzy inverse images. Thus, the proposed models are not only capable of interpreting arbitrary quantifiers in the sense of TGQ and their generalizations to Type IV quantifications. By fixing a coherent choice of fuzzy connectives and other basic constructs of fuzzy set theory, they also establish a consistent view of fuzzy set theory. The approaches described in the literature, by contrast, are essentially limited compared to TGQ and struggling with internal difficulties that stem from their lack of an axiomatic foundation.

13.3 Future Perspectives

The proposed theoretical framework is much more complete and linguistically faithful compared to earlier work. However, there are also a few cases not yet treated into which future research efforts should be directed. We already mentioned the problem of quantification over masses (e.g. "much wine"), which suggests an extension of the basic framework by measure-theoretic notions.

More attention should also be payed to *fuzzy branching quantification*. Some first steps towards the modelling of branching quantification have been made in Chap. 12 which extends the fuzzy sets framework by Lindström quantifiers capable of binding several variables. However it is not yet clear how branching quantifications can be implemented efficiently in the proposed models. In addition a better understanding of these powerful quantifications must be gained and additional semantical criteria should be developed to control them. The postulate of compatibility with nestings of quantifiers is a natural choice, but it appears to be extremely restrictive (possibly even conflicting with the core axioms). This means that the criterion should be restricted to cases of special linguistic relevance if necessary, for example to quantifiers which are quantitative, conservative or monotonic. In order to achieve unrestricted compatibility with the nesting of quantifiers, it would also be possible to explore alternatives to the core axioms. Specifically, the following criterion of *compatibility with inverse images* might be useful in this context, viz

$$\mathcal{F}(Q \circ \mathop{\times}_{i=1}^{n} f_i^{-1}) = \mathcal{F}(Q) \circ \mathop{\times}_{i=1}^{n} \widehat{\mathcal{F}}^{-1}(f_i)$$

for all semi-fuzzy quantifiers $Q : \mathcal{P}(E)^n \longrightarrow \mathbf{I}$ and all mappings $f_i : E \longrightarrow E'$, $i \in \{1, \ldots, n\}$, see Def. 4.63. Like 'functional application' (Z-6), this criterion will ensure the systematicity of $\mathcal{F}$ across domains. Thus, it would be instructive to investigate an alternative axiom system where (Z-6) is replaced by the novel postulate (and additional criteria if necessary).

Another topic worth investigating is concerned with the epistemic difficulty of ascertaining the precise values of attributes like height, age, temperature etc. Rather than assuming exact knowledge of these attributes and thus infinite precision, a more realistic approach would admit *imperfect knowledge of attribute values.* The research into fuzzy quantification under these weaker assumptions has been initiated by H. Prade [126, 127], who approaches the problem in the setting of possibility theory. It would likely be useful to incorporate similar ideas into the new framework and consider its possible extensions to interval-valued quantifications.

To sum up, this book has presented the fundamentals of a theory of fuzzy quantification. Although a few special cases like intensional quantifiers had to be ignored to keep things manageable, the proposed analysis achieves a greater coverage than earlier work. In addition, it demonstrates a clear and rigorous treatment of its subject. The axiomatic procedure, i.e. the formalization of semantical postulates which constrain the admissible models, made it possible to investigate fuzzy quantification with unparalleled formal rigor. The resulting axioms (Z-1) to (Z-6) give us a grip on several dimensions of fuzzy quantification which are relevant with regard to linguistic plausibility. Future research into such criteria will elucidate more and more dimensions of fuzzy quantification in natural language. If the basic method proposed in this work will be pursued further, we might eventually be able to *fully* reproduce the semantics of natural language quantifiers. And, we will also gain some insights into linguistic regularities which clarify the role of vagueness in human language and eventually, cognition.

The basic analysis proposed in this book has already been developed into a computational theory of fuzzy quantification which is ready for practical application. The new algorithms avail us with a computational model of absolute and proportional quantifiers which better conforms to the linguistic expectations. They are easily substituted for the existing methods which proved to be unreliable. Furthermore, it now becomes possible to support new types of quantifiers like cardinal comparatives, which are also of significance to applications. Due to their axiomatic foundation, the new models of fuzzy quantification avoid the problems of earlier approches. The proposed catalogue of semantical postulates accounts for the essential dimensions of linguistic plausibility and coherence of interpretation. For the first time, it now becomes possible to build reliable applications based on fuzzy quantifiers. The applications that directly profit from the improved models include fuzzy querying interfaces for databases, fuzzy information retrieval and linguistic data summarization.

A

An Evaluation Framework for Approaches to Fuzzy Quantification

A.1 Motivation and Chapter Overview

Fuzzy quantifiers will not realize their full potential for applications unless the models used for interpretation are linguistically adequate and catch the intended meaning of these quantifiers. For example, if an operator is labelled 'most', it is essential for avoiding misunderstanding that it behave like the NL quantifier "most". However, an evaluation of the existing approaches to fuzzy quantification from this linguistic viewpoint is somewhat difficult due to the different representations chosen by fuzzy set theory and linguistics, i.e. membership functions μ_Q vs. generalized quantifiers in the sense of TGQ. Accordingly, Zadeh [197] mentions TGQ in his seminal paper on fuzzy quantification, but he does not attempt any cross-comparison. In the remaining literature on fuzzy quantifiers, there are sporadic reports concerning linguistic plausibility [3, 42, 132, 134, 184]. However, there is no systematic evaluation of these approaches from the perspective of TGQ. Based on the framework for fuzzy quantification presented in the main part of this work, we will now conduct such an evaluation of the traditional approaches with respect to their linguistic plausibility. For that purpose, these approaches must first be fitted into the QFM framework. This will be accomplished by a canonical construction which associates with each 'traditional' approach a corresponding partial QFM. This translation will make it possible to apply the formal adequacy criteria developed for evaluating QFMs.

The resulting evaluation framework will then be used to discuss the main approaches to fuzzy quantification. The central results of this evaluation have already been anticipated in Chap. 1. Here we will present these examples in full detail. In particular, the images shown in the introduction will now be analyzed in some more depth. Usually we will also describe paper-and-pencil examples which reproduce the critical behaviour for a handful of participating objects. All examples will be analyzed carefully in order to spot the causes of failure, which can usually be traced to a violation of the proposed adequacy criteria for QFMs. Because of its explanatory value, it is hoped that this

I. Glöckner: *Fuzzy Quantifiers*, StudFuzz **193**, 391–415 (2006)
www.springerlink.com
© Springer-Verlag Berlin Heidelberg 2006

analysis will also be of interest to those who take sides with the traditional view of fuzzy quantification.

Note A.1. For the sake of readability, there will be an occasional repetition of concepts already mentioned in the introduction. In particular, the counter-examples will be developed from scratch and presented without any reference to their summaries in the first chapter.

A.2 The Evaluation Framework

Let us now introduce the formal framework for evaluating the traditional approaches to fuzzy quantification. The framework is taylored to those models which rest on Zadeh's representation of fuzzy linguistic quantifiers as fuzzy subsets of the non-negative reals or of the unit interval. In order to make a uniform framework possible, it is convenient to fit these approaches in the following general structure.

The model of interest, which will be symbolized by $\mathcal{Z}$, must be defined both for absolute quantifiers $\mu_Q : \mathbb{R}^+ \longrightarrow \mathbf{I}$ (Zadeh's quantifiers of the first kind) and relative/proportional quantifiers $\mu_Q : \mathbf{I} \longrightarrow \mathbf{I}$ (Zadeh's quantifiers of the second kind).[1] We will consider both the unrestricted use of the quantifier, where the quantification involves only one explicit and possibly fuzzy argument (*"Most things are tall"*), and the two-place use, where the quantification can be fuzzily restricted, as in *"Most young are poor"*. Labelling the absolute and proportional cases by *abs* and *prp*, respectively, and also marking the unrestriced and restricted uses by superscripts (1) and (2), a 'full' model $\mathcal{Z}$ of fuzzy quantification in Zadeh's setting can then be specified by defining the following interpretation mechanisms for all choices of finite domains E:

- $\mathcal{Z}^{(1)}_{\mathrm{abs}}$, which maps $\mu_Q : \mathbb{R}^+ \longrightarrow \mathbf{I}$ to $\mathcal{Z}^{(1)}_{\mathrm{abs}}(\mu_Q) : \widetilde{\mathcal{P}}(E) \longrightarrow \mathbf{I}$, for modelling unrestricted absolute quantification;
- $\mathcal{Z}^{(2)}_{\mathrm{abs}}$, which maps $\mu_Q : \mathbb{R}^+ \longrightarrow \mathbf{I}$ to $\mathcal{Z}^{(2)}_{\mathrm{abs}}(\mu_Q) : \widetilde{\mathcal{P}}(E)^2 \longrightarrow \mathbf{I}$, for modelling restricted absolute quantification;
- $\mathcal{Z}^{(1)}_{\mathrm{prp}}$, which maps $\mu_Q : \mathbf{I} \longrightarrow \mathbf{I}$ to $\mathcal{Z}^{(1)}_{\mathrm{prp}}(\mu_Q) : \widetilde{\mathcal{P}}(E) \longrightarrow \mathbf{I}$, for modelling unrestricted proportional quantification;
- $\mathcal{Z}^{(2)}_{\mathrm{prp}}$, which maps $\mu_Q : \mathbf{I} \longrightarrow \mathbf{I}$ to $\mathcal{Z}^{(2)}_{\mathrm{prp}}(\mu_Q) : \widetilde{\mathcal{P}}(E)^2 \longrightarrow \mathbf{I}$, for modelling restricted proportional quantification.

The approaches described in the literature do not always instantiate the complete scheme, and we will therefore allow that certain cases be omitted. In addition, the definition of the absolute restricted case,

$$\mathcal{Z}^{(2)}_{\mathrm{abs}}(\mu_Q)(X_1, X_2) = \mathcal{Z}^{(1)}_{\mathrm{abs}}(\mu_Q)(X_1 \cap X_2)\,, \tag{A.1}$$

[1] See Zadeh [197, p. 149] for the distinction of quantifiers of the first kind (based on absolute counts) and those of the second kind (based on relative counts).

is usually assumed without mention, and the two-place use of absolute quantifiers only shows up in examples which demonstrate the application of these approaches. Among other things, the above equation will reduce *"About fifteen children cry"* to *"There are about fifteen crying children"*, because

$$\mathcal{Z}^{(2)}_{\mathrm{abs}}(\mu_{\mathbf{about\ 15}})(\mathbf{children}, \mathbf{cry}) = \mathcal{Z}^{(1)}_{\mathrm{abs}}(\mu_{\mathbf{about\ 15}})(\mathbf{children} \cap \mathbf{cry})\,.$$

Finally, $\mathcal{Z}^{(1)}_{\mathrm{prp}}$ can be expressed in terms of $\mathcal{Z}^{(2)}_{\mathrm{prp}}$ whenever the latter mechanism is available, and then becomes

$$\mathcal{Z}^{(1)}_{\mathrm{prp}}(\mu_Q)(X) = \mathcal{Z}^{(2)}_{\mathrm{prp}}(\mu_Q)(E, X)\,.$$

For example, "Most things are expensive" can be interpreted by

$$\mathcal{Z}^{(1)}_{\mathrm{prp}}(\mu_{\mathbf{most}})(\mathbf{expensive}) = \mathcal{Z}^{(2)}_{\mathrm{prp}}(\mu_{\mathbf{most}})(E, \mathbf{expensive})\,,$$

where $\mu_{\mathbf{most}}(x) = 1$ for $x > 0.5$ and 0 otherwise, and $X \in \mathbf{expensive} \in \widetilde{\mathcal{P}}(E)$.

Now that we have a uniform notation for the approaches of interest, we can set up the evaluation framework. The problem is that the concepts developed in the main part of the book cannot directly be applied to existing approaches. This is because existing approaches are not formulated as quantifier fuzzification mechanisms, i.e. they apply to μ_Q's and not to semi-fuzzy quantifiers. We have to bridge the gap between semi-fuzzy quantifiers and fuzzy linguistic quantifiers in a systematic way. Hence let us recall the notion of an *underlying semi-fuzzy quantifier* $\mathcal{U}(\widetilde{Q}) : \mathcal{P}(E)^n \longrightarrow \mathbf{I}$, which is nothing but the restriction of the given fuzzy quantifier $\widetilde{Q} : \widetilde{\mathcal{P}}(E)^n \longrightarrow \mathbf{I}$ to crisp arguments (see Def. 2.11). Now let us consider one of the approaches $\mathcal{Z}$ based on fuzzy linguistic quantifiers. We cannot expect that $\mathcal{Z}$ give rise to a 'full' (totally defined) quantifier fuzzification mechanism. Compared to the scope of a QFM, which aims at NL quantification in general, the quantificational phenomena addressed by existing approaches are simply too limited, and cover only a fragment of the possible cases. In other words, $\mathcal{Z}$ is not sufficient to determine a total QFM defined for arbitrary quantifiers. However, it is often possible to reconstruct a *partially defined* quantifier fuzzification mechanism $\mathcal{F}$ based on $\mathcal{Z}$ as follows. Given the membership function μ_Q of a fuzzy linguistic quantifier, we first obtain the corresponding semi-fuzzy quantifier relative to $\mathcal{Z}$ as $Q = \mathcal{U}(\mathcal{Z}(\mu_Q))$, and use this to define $\mathcal{F}(Q) = \mathcal{Z}(\mu_Q)$. Obviously, the construction of $\mathcal{F}$ succeeds only if $\mathcal{U}(\mathcal{Z}(\mu_Q)) \mapsto \mathcal{Z}(\mu_Q)$ is functional, but this is a plausible requirement anyway. It will be called the *evaluation framework assumption* (EFA).

- The EFA is very closely related to the QFA, which underlies the quantification framework described in the main part of the book (see Sect. 2.6). To be specific, the QFA states that every base quantifier $\widetilde{Q}$ of interest is uniquely determined by its behaviour on crisp arguments. By contrast, the EFA requires that those quantifiers $\widetilde{Q} = \mathcal{Z}(\mu_Q)$ which result from the considered interpretation mechanism $\mathcal{Z}$ are fully specified by their quantification results for crisp arguments. If we assume that the mechanism $\mathcal{Z}$ is semantically plausible, i.e. all $\widetilde{Q} = \mathcal{Z}(\mu_Q)$ are potential denotations of NL quantifiers, then the EFA demands that the QFA be valid for all quantifiers in the responsibility of $\mathcal{Z}$, and hence comes out as the QFA in disguise.

- In case the EFA holds unconditionally for $\mathcal{Z}$, we can use the constructed partial fuzzification mechanism $\mathcal{F}$ to check the preservation and homomorphism properties of interest. We only need to take care of the fact that we now have a partial QFM, rather than a totally defined QFM. In order to reject a criterion of interest, it is sufficient to spot a violation of the criterion by quantifiers for which $\mathcal{F}$ is defined. In order to establish the validity of a plausibility criterion, we must first extend its definition to partial QFMs. For example, a partial QFM $\mathcal{F}$ is said to preserve negation if $\mathcal{F}(Q)$ is defined exactly if $\mathcal{F}(\neg Q)$ is defined; and in case both are defined, we have $\mathcal{F}(\neg Q) = \neg\mathcal{F}(Q)$. The required modifications to the other criteria are also apparent.
  When evaluating the existing approaches, we will not insist on the use of the induced connectives, because the literature on these approaches generally assumes the standard set of connectives. The use of the standard connectives further avoids the problem that the construction of induced connectives might no longer work when the EFA is violated. For this reason, we will generally assume the standard truth functions (negation $1-x$, conjunction min, disjunction max etc.) in this chapter, and it is understood that all constructions on (semi-) fuzzy quantifiers like dualisation etc. be defined in terms of these connectives.
- If the EFA is violated by $\mathcal{Z}$, then the concepts developed for QFMs can still be useful to assess certain properties of $\mathcal{Z}$. In this case, the EFA can always be enforced by restricting the set of considered membership functions μ_Q, i.e. by eliminating the conflicting cases. To see this, suppose $E \neq \varnothing$ is a base set, $Q : \mathcal{P}(E)^n \longrightarrow \mathbf{I}$ is a semi-fuzzy quantifier, and let us introduce the set of possible choices for the interpretation of Q,
  $$\mathrm{Ch}(Q) = \{\widetilde{Q} : \widetilde{\mathcal{P}}(E)^n \longrightarrow \mathbf{I} \;:\; \mathcal{U}(\widetilde{Q}) = Q \wedge \text{ exists } \mu_Q \text{ s.th. } \widetilde{Q} = \mathcal{Z}(\mu_Q)\}\,.$$
  If the EFA holds, then $\mathrm{Ch}(Q)$ is either empty, in which case $\mathcal{F}(Q)$ is undefined, or it is a singleton set $\mathrm{Ch}(Q) = \{\widetilde{Q}\}$, in which case $\mathcal{F}(Q)$ is defined and $\mathcal{F}(Q) = \widetilde{Q}$. If the EFA does not hold, then $\mathrm{Ch}(Q)$ has more than one element in certain cases. By reducing the set of considered μ_Q, we can always obtain a reduced system in which all $\mathrm{Ch}(Q)$ – now determined from the restricted set of μ_Q – are singleton or empty. We can hence view $\mathrm{Ch}(Q)$ as providing a number of choices for $\mathcal{F}(Q) \in \mathrm{Ch}(Q)$.
  We shall say that $\mathcal{Z}$ can *represent* a semi-fuzzy quantifier $Q : \mathcal{P}(E)^n \longrightarrow \mathbf{I}$ if there exists some μ_Q such that $Q = \mathcal{U}(\mathcal{Z}(\mu_Q))$. In order to reject the validity of a plausibility criterion, it is sufficient to prove that $\mathcal{Z}$ cannot represent Q without violating the property.

Now that the basic framework has been introduced, it will be applied in order to assess the plausibility of the main approaches to fuzzy quantification, viz the Σ-count, FG-count, OWA- and FE-count models. When presenting counter-examples in the evaluation to follow, we will generally prefer precise quantifiers like "all" or $[\mathbf{rate} \geq r]$, because there are stronger intuitions about the intended behaviour of these quantifiers, and the claim that certain results are counter-intuitive should be uncontroversial. Of course, this does not mean that we are only interested in two-valued quantifiers.

A.3 The Sigma-Count Approach

Let us first consider the Σ-count approach already mentioned in the introduction. The model rests on the notion of a *Σ-count*[2] [36], which is is defined as the sum of the membership values of a given fuzzy set $X \in \widetilde{\mathcal{P}}(E)$ (E finite), thus

$$\Sigma\text{-Count}(X) = \sum_{e \in E} \mu_X(e) \,.$$

The Σ-count is claimed to provide a (coarse) summary of the cardinality of the fuzzy set X, expressed as a non-negative real number. A corresponding scalar definition of fuzzy proportion, the *relative Σ-count* or *relative cardinality* [189], is defined by

$$\Sigma\text{-Count}(X_2/X_1) = \Sigma\text{-Count}(X_1 \cap X_2) / \, \Sigma\text{-Count}(X_1) \,.$$

The Σ-count approach to fuzzy quantification, introduced by Zadeh [193, 197], uses Σ-Count(X) and Σ-Count(X_2/X_1) to model the absolute and proportional kinds of fuzzy linguistic quantifiers. The two types of quantifiers are hence treated differently. As explained above, the unrestricted and restricted interpretations of both types of quantifiers must be discerned, i.e. quantification relative to the domain as a whole, or relative to an explicitly given, and possibly fuzzy, restriction. Referring to the notation for the mechanism $\mathcal{Z} = \mathrm{SC}$ ('Sigma Count'), the definition of the Σ-count approach can now be stated as:

$$\begin{aligned}
\mathrm{SC}^{(1)}_{\mathrm{abs}}(\mu_Q)(X) &= \mu_Q\left(\Sigma\text{-Count}(X)\right) \\
\mathrm{SC}^{(2)}_{\mathrm{abs}}(\mu_Q)(X_1, X_2) &= \mathrm{SC}^{(1)}_{\mathrm{abs}}(\mu_Q)(X_1 \cap X_2) \\
\mathrm{SC}^{(1)}_{\mathrm{prp}}(\mu_Q)(X) &= \mathrm{SC}^{(2)}_{\mathrm{prp}}(\mu_Q)(E, X) \\
\mathrm{SC}^{(2)}_{\mathrm{prp}}(\mu_Q)(X_1, X_2) &= \mu_Q\left(\Sigma\text{-Count}(X_2/X_1)\right)
\end{aligned}$$

[2] also known as the 'power' of a fuzzy set

Note A.2. The split definition in terms of four separate formulas simply instantiates the generic scheme for a given mechanism $\mathcal{Z}$ that was presented above. In particular, the 'abs'-versions apply to the absolute kind where $\mu_Q : \mathbb{R}^+ \longrightarrow \mathbf{I}$, and the 'prp'-versions to the proportional kind $\mu_Q : \mathbf{I} \longrightarrow \mathbf{I}$. Furthermore, the superscripts (1) and (2) denote unrestricted and restricted quantification, respectively.

Now that the definition of the Σ-count approach has been presented in the assumed format, we can set out to investigate the linguistic plausibility of the approach. To begin with, the Σ-count approach is somewhat peculiar because (unlike the other approaches), it does not comply with the EFA. This is most apparent with absolute fuzzy linguistic quantifiers $\mu_Q : \mathbb{R}^+ \longrightarrow \mathbf{I}$: The assumption is then violated by any pair of membership functions $\mu_Q, \mu_{Q'} : \mathbb{R}^+ \longrightarrow \mathbf{I}$ with $\mu_Q|_\mathbb{N} = \mu_{Q'}|_\mathbb{N}$ but $\mu_Q \neq \mu_{Q'}$. The trouble is that with absolute quantifiers, the Σ-count approach requires the specification of quantification result $\mu_Q(x)$ in the case that the computed Σ-count x is not a cardinal number. The decision on which quantification results to assign for such x affects the results obtained, but intuitions are scarce in this unfamiliar case.

Next we will consider an example which recasts Yager's observations [184, p.257] on counter-intuitive behaviour of the Σ-count approach. Hence suppose that $E = \{\text{Hans}, \text{Maria}, \text{Anton}\}$ is a set of persons, and using Zadeh's notation, that the fuzzy subset **blond** $\in \widetilde{\mathcal{P}}(E)$ is defined by **blond** $= \frac{1}{3}/\text{Hans} + \frac{1}{3}/\text{Maria} + \frac{1}{3}/\text{Anton}$, i.e. all three are blond to a degree of $\frac{1}{3}$. Now consider the quantifying statement "There is exactly one blond person". Its interpretation in the Σ-count approach is $\mathrm{SC}^{(1)}_{\mathrm{abs}}(\mu_Q)(\mathbf{blond})$, where Q is a quantifier suited to model (the unrestricted use of) "exactly one". In order for the condition of *correct generalisation* to be respected, we are forced to have $\mu_Q(1) = 1$. It follows that the above statement evaluates to 1 (fully true), although there is clearly *not* exactly one blond person in the base set (which one should that be?) but rather *a total amount of blondness* of one, as one might say, i.e. nothing of linguistic significance.[3]

A similar example from the domain of meteorological images already mentioned in the introduction, is presented in Fig. A.1. In this case, the Σ-count approach is used to evaluate the condition that "About ten percent of Southern Germany are cloudy"; and it is assumed that $\mu_Q : \mathbf{I} \longrightarrow \mathbf{I}$ is chosen such that, say, $\mu_Q(x) = 1$ in the range $x \in [0.08, 0.12]$, and that μ_Q decays to zero outside this range (the precise shape of μ_Q in latter case is inessential to the example). As shown in Fig. A.1, both cloudiness situations (a) and (b) are considered cloudy to a degree of one (fully true) if the Σ-count approach is used to evaluate "About 10 percent of Southern Germany are cloudy". While a result of 1 is plausible in case (a), this is not the case in situation (b), in

[3] Zadeh [197] proposes to apply some threshold on X in these cases, in order to prevent accumulative effects of small membership grades, but this is clearly an ad-hoc device because there is no obvious way in which any particular choice of threshold can be justified.

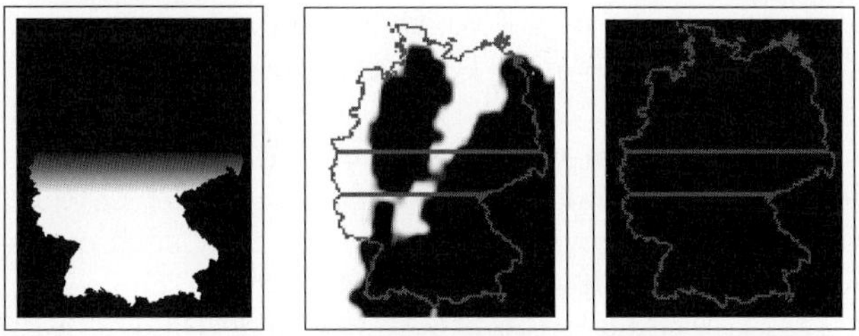

Southern Germany (a) SC: 1 (OK) (b) SC: 1 (implausible)

Fig. A.1. "About 10 percent of Southern Germany are cloudy" (Sigma-Count)

which *all* of Southern Germany is cloudy to a very low degree (viz. one tenth), which certainly does not mean that one tenth of Southern Germany is cloudy.

'Trivial' or 'degenerate' cases often require particular attention. One such case is that of a quantifier supplied with an argument tuple of empty sets. Let us now demonstrate that the Σ-count approach does not treat this case consistently. To see this, let us choose $0 < r_2 < r_1 < 1$ and consider the two-valued proportional quantifiers $Q_1 = [\mathbf{rate} \geq r_1]$, $Q_2 = [\mathbf{rate} > r_2] : \mathcal{P}(E)^2 \longrightarrow \mathbf{2}$ (see Def. 2.5), which have $Q_1(\varnothing, \varnothing) = 1$, but $Q_2(\varnothing, \varnothing) = 0$. The problem is that Zadeh does not specify the denotation of Σ-Count$(\varnothing/\varnothing)$. So let us assume that Σ-Count$(\varnothing/\varnothing) = c \in \mathbf{I}$; further suppose that $\mu_{Q_1}, \mu_{Q_2} : \mathbf{I} \longrightarrow \mathbf{I}$ are used to model Q_1 and Q_2, respectively. By not satisfying the EFA, the Σ-count approach leaves us with some choices which μ_{Q_1}, μ_{Q_2} to use as a model of Q_1 and Q_2. However, every reasonable choice of μ_{Q_1} should have $\mu_{Q_1}(x) = 1$ only if $x \geq r_1$, and $\mu_{Q_2}(x) = 0$ only if $x \leq r_2$.[4] *Correct generalization* then demands that $\mu_{Q_1}(c) = 1$, i.e. $c \geq r_1$, but also that $\mu_{Q_2}(c) = 0$, i.e. $c \leq r_2$, which contradicts the assumption that $r_2 < r_1$.

Let us further observe that the Σ-count approach yields potentially satisfying results only if μ_Q is genuinely fuzzy, because a two-valued quantifier (with corresponding two-valued membership function μ_Q) is mapped to a fuzzy quantifier $\widetilde{Q} : \widetilde{\mathcal{P}}(E)^n \longrightarrow \mathbf{2}$, the results of which are *always crisp*.[5] One might object that, although a two-valued quantifier $Q : \mathcal{P}(E)^2 \longrightarrow \mathbf{2}$ is to be modelled, an adequate choice of μ_Q should be continuous-valued. For example, if $E = \{a, b\}$, can't we model the universal quantifier $\forall$ by

$$\mu_Q(x) = \begin{cases} 0 & : \quad x \leq \frac{1}{2} \\ 2x - 1 & : \quad x > \frac{1}{2} \end{cases}$$

rather than using the two-valued membership function $\mu_\forall : \mathbf{I} \longrightarrow \mathbf{I}$,

[4] it will be shown below that in order for the Σ-count approach to preserve extension, these conditions on $\mu_{Q_1}(x)$ and $\mu_{Q_2}(x)$ must hold for all $x \in \mathbf{I} \cap \mathbb{Q}$. If we also require that monotonicity properties be preserved, this result extends to the full range $x \in \mathbf{I}$.

[5] This problem has been obscured by Zadeh's use of the quantifier **most**, which he views as being genuinely fuzzy.

$$\mu_\forall(x) = \begin{cases} 0 & : & x < 1 \\ 1 & : & x = 1 \end{cases}$$

which results in a crisp operator when applying $\mathrm{SC}^{(1)}_{\mathrm{prp}}$? To see that this objection is invalid, let us assume that the two-valued quantifier Q to be 'fuzzified' is of the proportional type; we will utilize the fact that such quantifiers have extension.[6] Now suppose that $\mu_Q : \mathbf{I} \longrightarrow \mathbf{I}$ is a proper choice for interpreting Q, and $q \in \mathbb{Q} \cap \mathbf{I}$ is some rational number in $\mathbf{I}$. Firstly, we can choose $X_1, X_2 \in \widetilde{\mathcal{P}}(E)$ such that Σ-Count$(X_2/X_1) = q$. Because q is rational and nonnegative, there exist $z, m \in \mathbb{N}$ s.th. $q = z/m$; we may also require that $m \geq |E|$. We extend E by arbitrary elements to some superset $E' \supseteq E$ with $|E'| = m$, and choose an arbitrary *crisp* subset $Z \in \mathcal{P}(E')$ with $|Z| = z$. *Correct generalisation* then yields

$$\mu_{Q_{E'}}(q) = \mu_{Q_{E'}}(\Sigma\text{-Count}(Z/E)) = Q_{E'}(E, Z) \in \mathbf{2}\,.$$

In order to avoid a violation of the semantic requirement of *extensionality* (see Def. 4.44), which would disqualify the Σ-count approach anyway, we are then justified to assume that

$$\mu_{Q_E}(\Sigma\text{-Count}(X_2/X_1)) = \mu_{Q_{E'}}(\Sigma\text{-Count}(X_2/X_1)) = \mu_{Q_{E'}}(q) \in \mathbf{2}\,.$$

The membership functions suited for modelling two-valued proportional quantifiers are therefore restricted to $\{0, 1\}$ on $\mathbf{I} \cap \mathbb{Q}$. The option of selecting intermediate membership grades in the open interval $(0, 1)$ for the remaining 'definition gaps' on $\mathbf{I} \setminus \mathbb{Q}$ has few practical relevance. In particular, the Σ-count approach produces two-valued results, and thus cannot determine useful *gradual* evaluations, in the case of frequently used quantifiers like "all" and $[\mathbf{rate} \geq r]$. This peculiarity of the Σ-count approach is undesirable in at least two ways.

Firstly, there is a well-known relationship between conjunction and universal quantification, which we already mentioned in Sect. 4.16: a conjunction $c_1 \wedge \cdots \wedge c_m$ of m two-valued criteria $c_1, \ldots, c_m \in \mathbf{2}$ corresponds to the quantified statement "All criteria $c_1, \ldots, c_m$ are true", i.e.

$$c_1 \wedge \cdots \wedge c_m = \forall_E(C)$$

where $E = \{1, \ldots, m\}$ and $C = \{j : c_j = 1\}$. We should expect this relationship between conjunction and universal quantification to be preserved in the fuzzy case. (As it is in all models of DFS theory – see theorem Th. 4.55). The point is that $\mu_\forall : \mathbf{I} \longrightarrow \mathbf{I}$, defined by $\mu_\forall(1) = 1$, $\mu_\forall(x) = 0$ otherwise, results in

$$\mathrm{SC}^{(1)}_{\mathrm{prp}}(\mu_\forall)(X) = \begin{cases} 1 & : & X = E \\ 0 & : & \text{else} \end{cases}$$

[6] i.e. f, v_0 in Def. 11.51 can be chosen independently of E. This is apparent from equation (4.2).

In particular, if $E = \{1, 2\}$, we obtain the induced conjunction operator $c : \mathbf{I} \times \mathbf{I} \longrightarrow \mathbf{I}$ defined by $c(x_1, x_2) = \mathrm{SC}^{(1)}_{\mathrm{prp}}(\mu_\forall)(X)$, where $X = x_1/1 + x_2/2$, i.e.

$$c(x_1, x_2) = \begin{cases} 1 & : \quad c_1 = c_2 = 1 \\ 0 & : \quad \text{else} \end{cases}$$

This is certainly not a reasonable fuzzy conjunction operator (in particular, it is not a t-norm).

Apart from this observation that the Σ-count approach assigns an implausible interpretation to the universal quantifier, there is another undesirable consequence of the fact that two-valued μ_Q's are mapped to fuzzy quantifiers $\mathrm{SC}(\mu_Q)$ which always return crisp results: such quantifiers are discontinuous, i.e. very slight changes in the membership grades of their arguments can drastically change the quantification result. More generally, we observe that whenever μ_Q is discontinuous, then the corresponding fuzzy quantifier under the Σ-count approach is also a discontinuous function of the membership grades of its argument sets. In practical applications, there is almost always some amount of noise (e.g. due to the finite precision of floating point operations), which can have drastic effects when using the Σ-count approach for modelling this kind of quantifiers.

An example which demonstrates this sensitivity to noise is presented in Fig. A.2. In this case, image (b) is a slightly modified version of image (a) – all pixels with a cloudiness grade of 0 have been set to a slightly higher grade. Although the difference is small and hardly perceptible, the Σ-count approach 'jumps' from 0 to 1 when we move from the cloudiness situation (a) to the slightly modified situation depicted in (b).

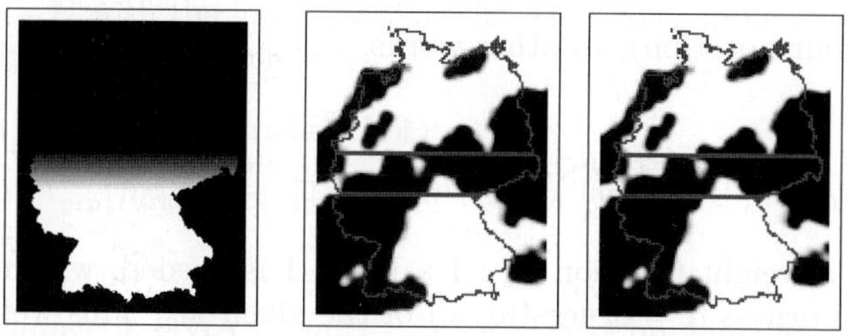

Southern Germany (a) Result: 0 (b) Result: 1

Fig. A.2. "At least 60 percent of Southern Germany are cloudy" (Sigma-Count)

Finally, the Σ-count is not compatible with antonyms, as pointed out by Zadeh [197, p. 167]. Noticing that the Σ-count method admits external negation, we conclude that the approach is also not compatible with dualisation (which can be decomposed into the antonym of the negation).

$$\mathrm{OWA}^{(2)}_{\mathrm{prp}}(\mu_Q)(\varnothing,\varnothing) = 0 \neq (1-\alpha) = \mathrm{OWA}^{(2)}_{\mathrm{prp}}(\mu_Q)(\varnothing, E)\,, \tag{A.3}$$

which shows that, except "all" and "some", no semi-fuzzy quantifier $Q : \mathcal{P}(E)^2 \longrightarrow \mathbf{I}$ which can be represented by OWA is conservative. In particular, *the OWA approach cannot represent any proportional semi-fuzzy quantifiers except for "all" and "some".*[10,11] But this is exactly the type of quantifiers OWA is intended to model.

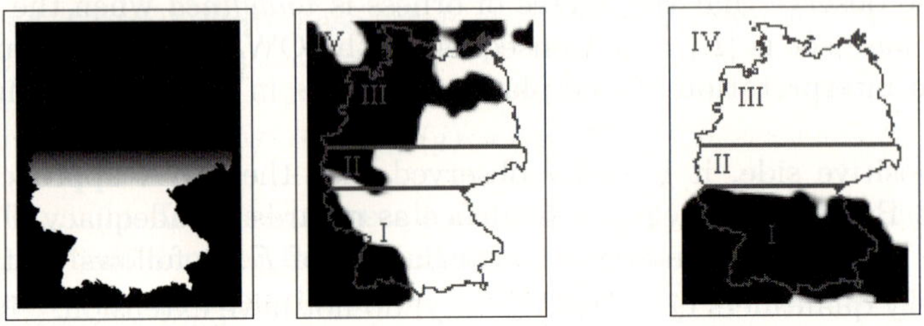

Southern Germany (a) desired: 1, OWA: 0.1 (b) desired: 0, OWA: 0.6

Fig. A.3. At least 60 percent of Southern Germany are cloudy (OWA approach)

If we still try to use OWA for interpreting proportional quantifiers, implausible results must be expected. The canonical construction of $\mathcal{F}$, which has been presented in Sect. A.2 above, is then replaced with direct assertions that a given semi-fuzzy quantifier Q should be mapped to a certain fuzzy quantifier $\mathcal{F}(Q)$ under the OWA approach. The property of correct generalisation, which was enforced by the canonical construction, is not necessarily guaranteed by such ad hoc correspondences. Nevertheless, part of the evaluation framework remains applicable, because we can still use the semantical postulates for plausible approaches in order to judge the adequacy of the resulting model. For example, let us consider the fuzzy linguistic quantifier $\mu_{[\mathrm{rate}\geq 0.6]}$, defined by

$$\mu_{[\mathrm{rate}\geq 0.6]}(x) = \begin{cases} 1 & : \quad x \geq 0.6 \\ 0 & : \quad \text{else} \end{cases}$$

and make the explicit assertion that $[\mathbf{rate} \geq r_1]$ be interpreted by the fuzzy quantifier $\mathcal{F}([\mathbf{rate} \geq r_1]) = \mathrm{OWA}^{(2)}_{\mathrm{prp}}(\mu_{[\mathrm{rate}\geq 0.6]})$. In fact, this is the common

[10] all proportional quantifiers are conservative by Def. 11.51.

[11] It is rather questionable anyway that "all" and "some" are considered proportional in Yager's setting. For example, "all" and "some" are easily defined for infinite domains, which is in clear opposition to the situation with proportional quantifiers. In the author's view, "some" should rather be considered an absolute quantifier, and "all" should be considered a special case of quantifier of exception (i.e., "all except 0", which permits no exceptions at all).

choice for "at least 60 percent" when using the OWA approach. Now let us consider the suitability of the resulting fuzzy quantifier for interpreting "At least 60 percent of Southern Germany are cloudy", given the cloudiness situation displayed in Fig. A.3. In case (a), we expect the result of 1, because so many pixels which fully belong to Southern Germany (I) are classified as fully cloudy that, regardless of whether we view the intermediate cases (II) as belonging to Southern Germany or not, its cloud coverage is always larger than 60 percent. Likewise in (b), we expect a result of 0 because regardless of whether the pixels in (II) are viewed as belonging to Southern Germany, its cloud coverage is always smaller than 60 percent. OWA, however, ranks image (b) much higher than image (a). This counter-intuitive result is explained by OWA's lack of conservativity: the cloudiness grades of pixels in areas (III) and (IV), which do not belong to Southern Germany at all, still have a strong (and undesirable) impact on the computed results.

To see that this failure of OWA is caused by the lack of conservativity (in the weak sense of Def. 6.21) of the quantifier $\mathrm{OWA}^{(2)}_{\mathrm{prp}}(\mu_{[\mathrm{rate}\geq 0.6]})$ let us consider the synthetic examples depicted in Fig. A.4. In images (a) and (b),

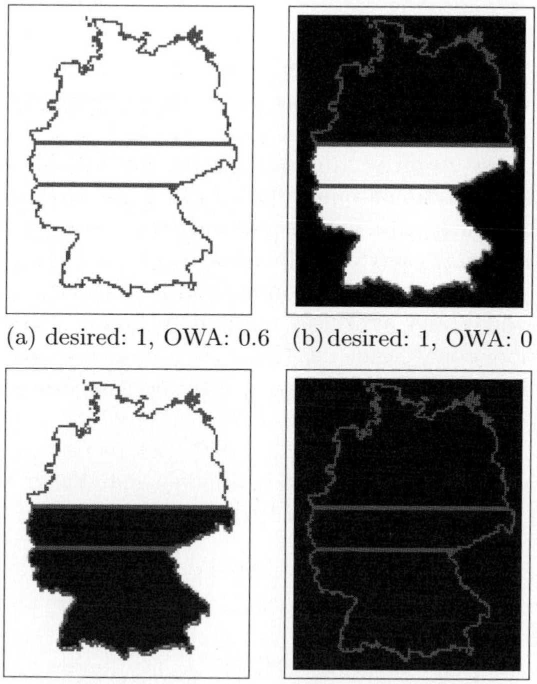

(a) desired: 1, OWA: 0.6 (b) desired: 1, OWA: 0

(c) desired: 0, OWA: 0.6 (d) desired: 0, OWA: 0

Fig. A.4. "More than 60 percent of Southern Germany are cloudy". Synthetic examples: desired results vs. results computed by $\mathrm{OWA}^{(2)}_{\mathrm{prp}}(\mu_{[\mathrm{rate}\geq 0.6]})$

Southern Germany is fully covered by clouds in the sense that each pixel in the support of Southern Germany[12] has a cloudiness grade of 1. We hence expect that in both (a) and (b), the condition "More than 60 percent of Southern Germany are cloudy" is fully satisfied. In particular, image (b) is the intersection of image (a) and the support of Southern Germany. Because "more than 60 percent" is conservative, we should expect that the results for image (a) and (b) coincide.

In images (c) and (d), there is no cloudiness at all in Southern Germany in the sense that each pixel in the support of Southern Germany has a cloudiness grade of 0. We should therefore expect that in both cases, the condition "More than 60 percent of Southern Germany are cloudy" fails completely (i.e. the result should be 0). In particular, the results of (c) and (d) should coincide because (d) is the intersection of (c) with the support of Southern Germany, and "more than 60 percent" is conservative.
As shown in Fig. A.4, the OWA approach neither computes the same results for (a) and (b), nor for (c) and (d). Even worse, its lack of weak conservativity causes OWA not to produce the intended order of results, which is (a) and (b) best, (c) and (d) worst. By contrast, image (c), which shows no cloudiness at all in Southern Germany, and image (a), in which Southern Germany is totally covered by clouds, are both assigned the same (rather high) score of 0.6; while image (b), in which Southern Germany is totally covered by clouds, and image (d), showing now clouds at all, both receive the lowest possible score of 0.

Apart from these negative findings concerning OWA's ability to represent conservative proportional quantifiers in the case of restricted (two-place) quantification, there are some further pecularities of the OWA approach that also deserve attention. It has already been mentioned that the 'core' OWA approach is limited to the regular nondecreasing type of $\mu_Q : \mathbf{I} \longrightarrow \mathbf{I}$. Yager [174] proposes a quantifier synthesis method to handle the remaining quantifiers of the proportional kind, i.e. those choices of μ_Q that are not regular nondecreasing. This construction can be rephrased as follows. For each considered μ_Q, we define $\mu_{\mathrm{ant}\,Q} : \mathbf{I} \longrightarrow \mathbf{I}$ by $\mu_{\mathrm{ant}\,Q}(x) = \mu_Q(1-x)$. $\mu_{\mathrm{ant}\,Q}$ is intended to model the antonym of μ_Q [197]. Let us call $\mu_Q : \mathbf{I} \longrightarrow \mathbf{I}$ *regular nonincreasing* if $\mu_{\mathrm{ant}\,Q}$ is regular nondecreasing. Yager [184] suggests to interpret such quantifiers in terms of their antonyms, i.e. to define

$$\mathrm{OWA}^{(1)}_{\mathrm{prp}}(\mu_Q)(X) = \mathrm{OWA}^{(1)}_{\mathrm{prp}}(\mu_{\mathrm{ant}\,Q})(\neg X)\,,$$

for all $X \in \widetilde{\mathcal{P}}(E)$, provided that μ_Q is regular nonincreasing. Of course, we could also have used the negation, defined by $\mu_{\neg Q}(x) = 1 - \mu_Q(x)$, to state an alternative definition

$$\mathrm{OWA}^{(1)\,\prime}_{\mathrm{prp}}(\mu_Q)(X) = 1 - \mathrm{OWA}^{(1)}_{\mathrm{prp}}(\mu_{\neg Q})(X)\,.$$

[12] i.e. each pixel which belongs to Southern Germany to a degree larger than 0

However, both alternatives coincide because $\mathrm{OWA}^{(1)}_{\mathrm{prp}}$ preserves the duality between the quantifiers which correspond to $\mu_{\mathrm{ant}\,Q}$ and $\mu_{\neg Q}$.

In the general case where $\mu_Q : \mathbf{I} \longrightarrow \mathbf{I}$ is neither regular nondecreasing nor nonincreasing, Yager suggests to decompose μ_Q into a conjunction (or possibly other Boolean combination) of regular nondecreasing and nonincreasing quantifiers. For example, if $\mu_Q = \mu_{Q_1} \wedge \mu_{Q_2}$, where μ_{Q_1} is regular nondecreasing and μ_{Q_2} is regular nonincreasing, the quantifier synthesis method yields

$$\mathrm{OWA}^{(1)}_{\mathrm{prp}}(\mu_Q)(X) = \mathrm{OWA}^{(1)}_{\mathrm{prp}}(\mu_{Q_1})(X) \wedge \mathrm{OWA}^{(1)}_{\mathrm{prp}}(\mu_{\mathrm{ant}\,Q_2})(\neg X)\,,$$

for all $X \in \widetilde{\mathcal{P}}(E)$. Unfortunately, the obtained results depend on the decomposition of μ_Q chosen. For example, consider

$$\mu_Q(x) = \begin{cases} 0 & : \quad x \in [0, 0.25) \\ 0.3 & : \quad x \in [0.25, 0.75] \\ 0 & : \quad x \in (0.75, 1] \end{cases} \tag{A.4}$$

which can be decomposed into a conjunction of

$$\mu_{Q_1}(x) = \begin{cases} 0 & : \quad x \in [0, 0.25) \\ 0.3 & : \quad x \in [0.25, 0.75] \\ 1 & : \quad x \in (0.75, 1] \end{cases} \qquad \mu_{Q_2}(x) = \begin{cases} 1 & : \quad x \in [0, 0.25) \\ 0.3 & : \quad x \in [0.25, 0.75] \\ 0 & : \quad x \in (0.75, 1] \end{cases}$$

but also into a conjunction of

$$\mu_{Q'_1}(x) = \begin{cases} 0 & : \quad x \in [0, 0.25) \\ 0.3 & : \quad x \in [0.25, 0.5) \\ 1 & : \quad x \in [0.5, 1] \end{cases} \qquad \mu_{Q'_2}(x) = \begin{cases} 1 & : \quad x \in [0, 0.5) \\ 0.3 & : \quad x \in [0.5, 0.75] \\ 0 & : \quad x \in (0.75, 1] \end{cases}$$

If we interpret the regular nondecreasing μ_{Q_2} and $\mu_{Q'_2}$ by means of $\mu_{\mathrm{ant}\ Q_2}$ and $\mu_{\mathrm{ant}\ Q'_2}$, as suggested by Yager, and choose $E = \{a, b, c, d\}$, and $X \in \widetilde{\mathcal{P}}(E)$ with $X = 0/a + 0.5/b + 0.5/c + 1/d$, then

$$\begin{aligned} \mathrm{OWA}^{(1)}_{\mathrm{prp}}(\mu_{Q_1})(X) \wedge \mathrm{OWA}^{(1)}_{\mathrm{prp}}(\mu_{\mathrm{ant}\,Q_2})(\neg X) &= 0.3 \\ \mathrm{OWA}^{(1)}_{\mathrm{prp}}(\mu_{Q'_1})(X) \wedge \mathrm{OWA}^{(1)}_{\mathrm{prp}}(\mu_{\mathrm{ant}\,Q'_2})(\neg X) &= 0.65 \end{aligned}$$

i.e. the quantifier synthesis method depends on the decomposition chosen, and is hence ill-defined. These problems carry over to two-place quantification based on $\mathrm{OWA}^{(2)}_{\mathrm{prp}}$. We then have

$$\mathrm{OWA}^{(2)}_{\mathrm{prp}}(\mu_Q)(E, X) = \mathrm{OWA}^{(1)}_{\mathrm{prp}}(\mu_Q)(X)\,.$$

This demonstrates that $\mathrm{OWA}^{(2)}_{\mathrm{prp}}$ embeds all counter-examples for one-place quantification.

In the two-place case, however, there are some additional problems because $\mathrm{OWA}^{(2)}_{\mathrm{prp}}$, unlike $\mathrm{OWA}^{(1)}_{\mathrm{prp}}$, does not preserve duals. Suppose we wish to evaluate the criterion "Less than 60 percent of Southern Germany are cloudy". Noticing that the quantifier is not of the regular nondecreasing type, we must resort to one of the following equivalent statements.

i. "It is not the case that at least 60 percent of Southern Germany are cloudy", i.e. use negation and compute $\neg\mathrm{OWA}^{(2)}_{\mathrm{prp}}(\mu_{[\mathrm{rate}\geq 0.6]})(X_1, X_2)$;
ii. "More than 40 percent of Southern Germany are not cloudy", i.e. use the antonym and compute $\mathrm{OWA}^{(2)}_{\mathrm{prp}}(\mu_{[\mathrm{rate}>0.4]})(X_1, \neg X_2)$.

Unlike in NL, however, these statements are not equivalent under the OWA approach. When applied to the images in Fig. A.3, we obtain different results as shown in Table A.1. In this case, both i. and ii. compute the same (wrong) ranking[13], which demonstrates that neither alternative is correct; in other cases, their rankings can differ. The problem is that OWA, which cannot model "less than 60 percent" directly, forces us to choose one of i., ii.; but due to their expected equivalence, there is no preference for either choice. In an appropriate model, the computed results coincide in both cases and the chosen alternative for computing the result becomes inessential. The OWA approach, however, fails to preserve the duality of "at least 60 percent" vs. "more than 40 percent".

Table A.1. Less than 60 percent of Southern Germany are cloudy

Quantifier	Fig. A.3(a)	Fig. A.3(b)
$\neg\mathrm{OWA}^{(2)}_{\mathrm{prp}}(\mu_{[\mathrm{rate}\geq 0.6]})$	0.9	0.4
$\mathrm{OWA}^{(2)}_{\mathrm{prp}}(\mu_{[\mathrm{rate}>0.4]})\neg$	0.4	0
(desired result)	0	1

A.5 The FG-Count Approach

Apart from the Σ-count approach, Zadeh [197] also introduces a second approach to fuzzy quantification. The FG-count approach rests on the idea that the cardinality of a fuzzy set cannot be fully described by a single scalar number (as in the Σ-count approach). By contrast, the cardinality of a fuzzy set should rather be be viewed as a fuzzy subset of the non-negative integers. If E is a finite base set, and $X \in \widetilde{\mathcal{P}}(E)$ a fuzzy subset of E, then the FG-count of X [197, p.156,p.157], denoted FG-count$(X) \in \widetilde{\mathcal{P}}(\mathbb{N})$, is given by

$$\begin{aligned}\mu_{\text{FG-count}(X)}(j) &= \sup\{\alpha \in \mathbf{I} : |X_{\geq\alpha}| \geq j\} \\ &= \mu_{[j]}(X)\,,\end{aligned}$$

for all $j \in \mathbb{N}$, recalling in particular that $\mu_{[0]}(X) = 1$ and $\mu_{[j]}(X) = 0$ for all $j > |E|$, see Def. 7.91. Intuitively, the FG-count expresses the degree to which $X \in \widetilde{\mathcal{P}}(E)$ has at least j elements. The FG-count can then be used to interpret absolute quantifiers $\mu_Q : \mathbb{R}^+ \longrightarrow \mathbf{I}$ as follows [174]:

[13] i.e. result order with respect to $\geq$

$$\mathrm{FG}^{(1)}_{\mathrm{abs}}(\mu_Q) = \max_{j=0,\dots,m} \mu_Q(j) \wedge \mu_{[j]}(X)$$

Following the canonical strategy (A.1) for defining the restricted use of absolute quantifiers, $\mathrm{FG}^{(2)}_{\mathrm{abs}}$ can be modelled by

$$\mathrm{FG}^{(2)}_{\mathrm{abs}}(\mu_Q)(X_1, X_2) = \mathrm{FG}^{(1)}_{\mathrm{abs}}(\mu_Q)(X_1 \cap X_2)\,.$$

Yager [174] proposes a method for interpreting proportional quantifiers $\mu_Q : \mathbf{I} \longrightarrow \mathbf{I}$, which is obviously related to Zadeh's FG-count approach. $\mathrm{FG}^{(1)}_{\mathrm{prp}}$ then becomes

$$\mathrm{FG}^{(1)}_{\mathrm{prp}}(\mu_Q)(X) = \mathrm{FG}^{(1)}_{\mathrm{abs}}(\mu_{Q,E})(X)\,,$$

(adapted to our notation), where $\mu_{Q,E}(j) = \mu_Q(j/m)$, $m = |E|$, as in the case of the OWA approach. As concerns the restricted use of proportional quantifiers, Zadeh proposes the *relative* FG-count for pairs of fuzzy sets $X_1, X_2 \in \widetilde{\mathcal{P}}(E)$, defined by

$$\text{FG-count}(X_2/X_1) = \sum_{\alpha \in [0,1]} \alpha / \frac{|X_{1 \geq \alpha} \cap X_{2 \geq \alpha}|}{|X_{1 \geq \alpha}|}$$

and emphasises *"that the right-hand member" [of the defining equation] "should be treated as a fuzzy multiset, which implies that terms of the form α_1/u and α_2/u should not be combined into a single term $(\alpha_1 \vee \alpha_2)/u$, as they would be in the case of a fuzzy set"* [197, p.157]. However, if the right-hand member of the equation is a fuzzy multiset, then obviously FG-count(X_2/X_1) is a fuzzy multiset as well. It is therefore *not* possible to formulate $\mathrm{FG}^{(2)}_{\mathrm{prp}}(\mu_Q)$ in analogy to the definition of $\mathrm{FG}^{(1)}_{\mathrm{abs}}(\mu_Q)$, i.e. the following proposal is *not* licensed by Zadeh's definition:

$$\mathrm{FG}^{(2)}_{\mathrm{prp}}(\mu_Q)(X_1, X_2) = \sup\{\mu_Q(r) \wedge \mu_{\text{FG-count}(X_2/X_1)}(r) : r \in \mathbf{I}\}$$

(This definition would be ill-formed because it identifies the cases which Zadeh explicitly wants to be kept separate). To the best of my knowledge, no attempt has been made to provide a definition of $\mathrm{FG}^{(2)}_{\mathrm{prp}}(\mu_Q)$ in terms of the above FG-count(X_2/X_1). In particular, Zadeh's examples on the application of the FG-count approach in [197] only demonstrate its application in the unrestricted case, and can thus be handled by $\mathrm{FG}^{(1)}_{\mathrm{prp}}$.

As in the OWA case, it is easily shown that the FG-count approach complies with the EFA. However, the range of fuzzy quantifiers which can be handled by the FG-count approach is essentially limited, because the two-place use of proportional quantifiers cannot be modelled, which is of obvious relevance to most applications.

Another limitation of the model becomes visible once we investigate the monotonicity properties of the fuzzy quantifiers $\widetilde{Q} = \mathrm{FG}(\mu_Q)$ obtained from

$\mu_{[\text{rate}>\frac{p}{q}]}(x)$ for all $x \in \mathbf{I} \cap \mathbb{Q}$. In the following, we will use $\mu_{[\text{rate}>\frac{p}{q}]}$, but the effects demonstrated will also occur for other possible choices of μ_Q.

Now suppose that E is a base set such that $m = |E| > q$. Then E can be written as $E = \{e_1, \dots, e_m\}$, where the e_i are pairwise distinct elements of E. Let us define $X_1^{(\lambda)} \in \widetilde{\mathcal{P}}(E)$, $X_2 \in \mathcal{P}(E)$ by

$$\begin{aligned} X_1^{(\lambda)} &= 1/e_1 + \dots + 1/e_q + \lambda/e_{q+1} \\ X_2 &= \{e_1, \dots, e_p\} \end{aligned}$$

for all $\lambda \in \mathbf{I}$. We should expect that $\mathrm{FG}^{(2)}_{\mathrm{prp}}(\mu_{[\text{rate}>\frac{p}{q}]})(X_1^{(\lambda)}, X_2) = 0$, because $[\textbf{rate} > \frac{p}{q}]$ is constantly zero in the range (U, V), where

$$\begin{aligned} U &= (\{e_1, \dots, e_q\}, X_2) = (X_1^{(0)}, X_2)\,, \\ V &= (\{e_1, \dots, e_{q+1}\}, X_2) = (X_1^{(1)}, X_2)\,. \end{aligned}$$

However, instead of the desired result stated above, what we really get is the following,

$$\mathrm{FG}^{(2)}_{\mathrm{prp}}(\mu_{[\text{rate}>\frac{p}{q}]})(X_1^{(\lambda)}, X_2) = \begin{cases} 0 & : \lambda = 0 \\ 1 - \lambda & : \lambda > 0 \end{cases} \tag{A.7}$$

In fact, the example reveals that $\mathrm{FG}^{(2)}_{\mathrm{prp}}$ does not preserve local monotonicity properties, because the resulting fuzzy quantifier fails to be locally nondecreasing in (U, V). Moreover, it is not contextual (see Def. 4.47 for the definition of contextuality): its result $1 - \lambda$ in the considered range is nonzero, although the original quantifier, i.e. the specification for crisp sets, is constantly zero in (U, V).

Apart from the failure of $\mathrm{FG}^{(2)}_{\mathrm{prp}}$ to preserve local monotonicity, equation (A.7) also points at another weakness of $\mathrm{FG}^{(2)}_{\mathrm{prp}}$: if μ_Q is discontinuous (in particular if it is two-valued), then $\mathrm{FG}^{(2)}_{\mathrm{prp}}(\mu_Q)(X_1, X_2)$ can be discontinuous in the membership grades of X_1 as well. In the example, a very slight change of $X_1^{(\lambda)}$ (from $\lambda = 0$ to $\lambda' = \varepsilon > 0$) can result in a drastic change of $\mathrm{FG}^{(2)}_{\mathrm{prp}}(\mu_{[\text{rate}>\frac{p}{q}]})(X_1^{(\lambda)}, X_2)$. This is not acceptable in practical applications due to the inevitable presence of noise: a very slight change in X_1, X_2 might cause totally different results of $\mathrm{FG}^{(2)}_{\mathrm{prp}}(\mu_Q)(X_1, X_2)$.

In order to demonstrate these effects in the image ranking task, let us consider the quantifier $[\textbf{rate} \geq 0.05]$, i.e. **at least five percent**. Some results of $\mathrm{FG}^{(2)}_{\mathrm{prp}}$ for the chosen quantifier are shown in Fig. A.5. Let us first consider the case that X_1 is the standard choice for interpreting Southern Germany depicted in A.5(a), and where X_2 is the cloudiness situation depicted in (c). We expect a result of zero in this case because there are no clouds in the support of X_1. The FG-count approach, however, returns a score of 0.55. One might subject that, as in the case of OWA, this implausible result could be

(a) Southern Germany#1 (b) Southern Germany#2 (c) cloudy

Result for Southern Germany#1: 0.55
Result for Southern Germany#2: 0.95
Desired result: 0

Fig. A.5. "At least 5 percent of Southern Germany are cloudy" (FG-Count)

caused by a lack of conservativity – after all, there are some clouds in the upper part of image (c). However, it is easily verified that all fuzzy quantifiers obtained from $\mathrm{FG}^{(2)}_{\mathrm{prp}}$ are conservative. It is the failure of $\mathrm{FG}^{(2)}_{\mathrm{prp}}$ to preserve local monotonicity properties which causes the implausible behaviour.

The sensitivity of $\mathrm{FG}^{(2)}_{\mathrm{prp}}$ to slight changes in X_1 becomes apparent when the standard definition of the fuzzy region 'Southern Germany' (a) is replaced with the slightly modified Southern Germany#2 shown in A.5(b). The image has been generated from (a) by replacing black pixels ($\mu_X(e) = 0$) in the lower two thirds of (a) with a slightly lighter black, $\mu_X(e) = 0.05$. Because the upper third of the image is unchanged, i.e. the clouds in the northern part of (c) are still outside the support of Southern Germany#2, we expect a result of 0 in this case, too. The FG-count approach, however, results in a score of 0.95, again because local monotonicity is not preserved. The large change in the result, although we have modified the representation of Southern Germany in terms of X_1 only very slightly, demonstrates that the operators obtained from the FG-count approach can indeed be very brittle.

We have already seen that $\mathrm{FG}^{(1)}_{\mathrm{prp}}$ is essentially limited to nondecreasing quantifiers. Owing to (A.5), then, the proposed extension to restricted quantification is subject to the same limitations, i.e. $\mathrm{FG}^{(2)}_{\mathrm{prp}}$ will behave reasonably only if μ_Q is nondecreasing. In particular, the failure of the 'quantifier synthesis' method to extend the coverage of $\mathrm{FG}^{(1)}_{\mathrm{abs}}$ and $\mathrm{FG}^{(1)}_{\mathrm{prp}}$ directly carries over to $\mathrm{FG}^{(2)}_{\mathrm{prp}}$ (in this case, E must be used for the first argument, while X becomes the second argument. Let us consider an additional example which shows that in the case of two-place quantification, the decomposition can also fail because $\mathrm{FG}^{(2)}_{\mathrm{prp}}$ is not compatible with dualisation. Hence let us consider

$$\mu_Q(x) = \begin{cases} 1 & : \quad x \geq \frac{1}{6} \\ 0 & : \quad x < \frac{1}{6} \end{cases} \qquad \mu_{Q'}(x) = \begin{cases} 1 & : \quad x > \frac{5}{6} \\ 0 & : \quad x \leq \frac{5}{6} \end{cases}$$

representing [**rate** $\geq \frac{1}{6}$] and its dual [**rate** $> \frac{5}{6}$], respectively, and choose $E = \{e_1, e_2\}$, $X_1 = X_2 = \frac{1}{3}/e_1 + \frac{1}{3}/e_2$. We then have

$$\mathrm{FG}^{(2)}_{\mathrm{prp}}(\mu_Q)(X_1, X_2) = \tfrac{2}{3} \neq \tfrac{1}{3} = 1 - \tfrac{2}{3} = 1 - \mathrm{FG}^{(2)}_{\mathrm{prp}}(\mu_{Q'})(X_1, \neg X_2)\,.$$

As in the case of $\mathrm{OWA}^{(2)}_{\mathrm{prp}}$, the failure of $\mathrm{FG}^{(2)}_{\mathrm{prp}}$ to preserve duals is of particular importance because $\mathrm{FG}^{(2)}_{\mathrm{prp}}$ cannot provide a reasonable interpretation of quantifiers like "less than 30 percent" directly. In order to evaluate this quantifier, we either have to use the negation "at least 30 percent", or the antonym "more than 70 percent", which are duals of each other. However, because duality is not preserved by $\mathrm{FG}^{(2)}_{\mathrm{prp}}$, both choices produce different results as shown in Fig. A.6.

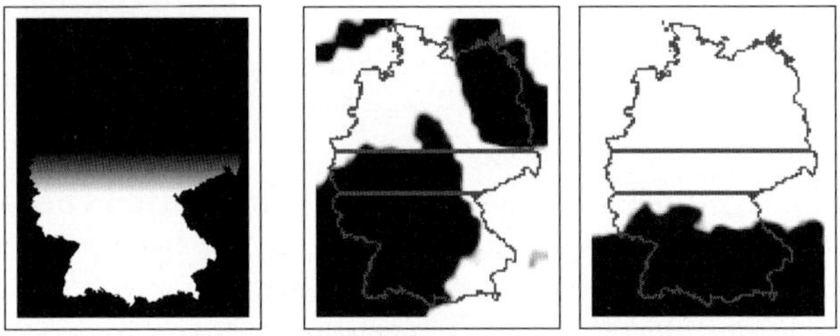

Southern Germany (a) (b)

Quantifier	Image (a)	Image (b)
not at least 30 percent	0.72	0
more than 70 percent not	1	0.29

Fig. A.6. "Less than thirty percent of Southern Germany are cloudy" (FG-count)

A.6 The FE-Count Approach

Apart from introducing the FG-count and suggesting its use for the modelling of fuzzy quantification, Zadeh [197] proposes yet another definition of fuzzy cardinality. The *FE-count* of $X \in \mathcal{P}(E)$ is given by

$$\mu_{\text{FE-count}(X)}(j) = \mu_{[j]}(X) \wedge \neg\mu_{[j+1]}(X)\,.$$

Unlike FG-count(X), which expresses the degree to which $X \in \widetilde{\mathcal{P}}(E)$ has *at least* j elements, the FE-count is intended to express the degree to which X has *exactly* j elements. The adequacy of the FE-count as a measure of fuzzy cardinality has been questioned by Dubois and Prade [42].

Ralescu [132] has proposed a method for interpreting quantifiers which is related to the concept of FE-count. The method is only defined for the unrestricted use of absolute quantifiers $\mu_Q : \mathbb{R}^+ \longrightarrow \mathbf{I}$. In this case

$$\mathrm{FE}^{(1)}_{\mathrm{abs}}(\mu_Q)(X) = \max_{j=0,\ldots,m} \mu_Q(j) \wedge \mu_{[j]}(X) \wedge \neg\mu_{[j+1]}(X)\,,$$

for all $X \in \widetilde{\mathcal{P}}(E)$. Ralescu's model will be called the 'FE-count approach' here because obviously

$$\mathrm{FE}^{(1)}_{\mathrm{abs}}(\mu_Q)(X) = \max_{j=0,\ldots,m} \mu_Q(j) \wedge \mu_{\mathrm{FE\text{-}count}(X)}(j)\,.$$

It should be pointed out, however, that Ralescu develops his proposal from a possibilistic argument and without reference to the FE-count.
The model complies with the EFA, but the phenomenon treated (only unary absolute quantifiers) is of course very limited. Nevertheless, Ralescu's approach explicitly targets at an improvement upon the FG-count model, so we should be curious about its score on the plausibility scale.

Given a domain E with $|E| = m$ and $\mu_Q : \mathbb{R}^+ \longrightarrow \mathbf{I}$, we can conveniently restrict attention to $\mu_{Q,E} : \{0,\ldots,m\} \longrightarrow \mathbf{I}$ (as we have already done in the case of OWA), which for absolute quantifiers becomes $\mu_{Q,E}(j) = \mu_Q(j)$ for all $j = 0,\ldots,m$. We can then view $\mathrm{FE}^{(1)}_{\mathrm{abs}}(\mu_Q)$ as a function of $\mu_{Q,E}$. In the case of a two-element domain, say $E = \{a, b\}$, the existential and universal quantifiers are modelled by

$$\mu_{\exists,E}(j) = \begin{cases} 0 & : \quad j = 0 \\ 1 & : \quad \text{else} \end{cases} \qquad\qquad \mu_{\forall,E}(j) = \begin{cases} 0 & : \quad j < 2 \\ 1 & : \quad j = 2 \end{cases}$$

for $j \in \{0, 1, 2\}$. Letting $X = 1/a + 0/b$ and $Y = 1/a + 0.5/b$, we observe that $\mathrm{FE}^{(1)}_{\mathrm{abs}}(\mu_{\exists,E})(X) = 1$, but $\mathrm{FE}^{(1)}_{\mathrm{abs}}(\mu_{\exists,E})(Y) = 0.5$, although $X \subseteq Y$ (in the sense of inclusion of fuzzy sets). It follows that the FE-count approach does not preserve monotonicity properties of quantifiers. In particular, the example shows that the FE-count does not generate a reasonable interpretation of the existential quantifier, which should be expressable in terms of an s-norm and thus be monotonically nondecreasing.[15] The example further demonstrates that the FE-count approach does not preserve duality, because $\mu_{\forall,E}$ is indeed mapped to the intended

$$\mathrm{FE}^{(1)}_{\mathrm{abs}}(\mu_{\forall,E})(Z) = \min(\mu_Z(a), \mu_Z(b))\,,$$

which is different from the dual of $\mathrm{FE}^{(1)}_{\mathrm{abs}}(\mu_{\exists,E})$. Finally, the FE-count approach also does not preserve negation.

The failure of the FE-count method to preserve monotonicity can also be demonstrated in the example scenario of fuzzy image regions. Hence let us consider the statement that a fuzzy image region X is nonempty.[16] The condition corresponds to existential quantification, and the FE-count formula is hence applied to $\mu_\exists : \mathbb{R}^+ \longrightarrow \mathbf{I}$ defined by $\mu_\exists(0) = 0$, $\mu_\exists(j) = 1$ else. Now consider the results shown in Fig. A.7. The result in case (a) is satisfactory

[15] see Sect. 4.16 on the intended interpretation of fuzzy existential and universal quantification, which reviews Thiele's analysis of T- and S-quantifiers.

[16] Note that we cannot use examples of two-place proportional quantifiers here because the FE-count approach is unable to model these.

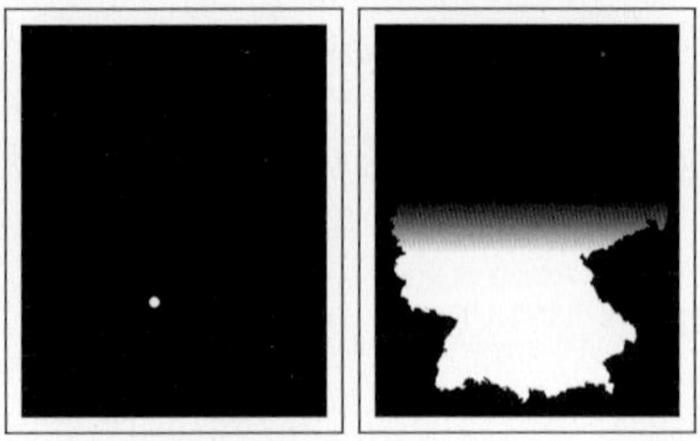

(a) desired: 1, FE: 1 (b) desired: 1, FE: 0.5

Fig. A.7. "The image region X is nonempty" (FE-count)

because X is a crisp nonempty subset of the set of pixel coordinates E. The fuzzy image region X' depicted in (b) is much larger and contains X. Due to the monotonicity of the existential quantifier, we should expect that the fuzzy image region X' depicted in (b) is nonempty, too. However, due to the presence of fuzziness, the FE-count approach results in a score of 0.5, which is clearly implausible.

To sum up, the FE-count is limited in applicability and it even assigns inacceptable interpretations in the simplest case of unrestricted monotonic absolute quantifiers.

A.7 Chapter Summary

The traditional approaches to fuzzy quantification rely on the very simple representation offered by membership functions $\mu_Q : \mathbf{I} \longrightarrow \mathbf{I}$ and $\mu_Q : \mathbb{R}^+ \longrightarrow \mathbf{I}$. Although formally less elaborated than the theory presented here, these approaches are particularly easy to implement, which makes them appealing to applications at first sight. We therefore checked if these approaches perform well in those cases they intend to model, knowing that they are too narrow for general treatment of fuzzy NL quantification (see Sect. 1.10 above).

In order to assess the plausibility of the approaches described in the literature, a canonical construction was introduced which fits these approaches into the framework pursued in this work. In this way, an (incompletely determined) QFM $\mathcal{F}$ is assigned to each of these approaches.

The framework is directly applicable if its underlying assumption is satisfied, the evaluation framework assumption (EFA). Basically, the EFA demands that all fuzzy quantifiers which can be expressed by the considered approach can be discerned on the level of crisp arguments. This requirement is justified by the fundamental assumption of the quantification framework, i.e. the QFA described in Chap. 2, which poses the related requirement that all base quantifiers of interest can also be discerned on the level of crisp arguments. This suggests that beyond its part in the construction of the canonical embedding, the EFA is also a plausibility criterion which is interesting in its own right.

Those approaches that violate the EFA can still be discussed in the analysis framework. Instead of a unique fuzzy quantifier assigned by the construction, we then have a number of available choices. In this case, we can still check if a given adequacy criterion can be fulfilled by making an appropriate selection from the available interpretations. If no verifying choice exists, the considered approach to fuzzy quantification must be considered incompatible with the criterion of interest.

The proposed method permits a rigorous evaluation of existing approaches as to their compatibility with the semantic criteria which were originally introduced for evaluating the plausibility of QFMs. The criteria formalized in the main part of this book can guide a systematic search for critical examples which result in implausible interpretations. However, the framework can also help to better understand the defects of existing approaches because the counter-examples can now be explained as a violation of basic plausibility criteria.

In the chapter, the framework was instantiated for the main approaches to fuzzy quantification, i.e. the Σ-count, FG-count, OWA and the FE-count methods. Based on the known plausibility criteria, these approaches were then checked against the intended semantics of fuzzy NL quantification. This evaluation has raised severe doubts on the utility of the traditional approaches as models of natural language quantification. For each of the approaches in this tradition, we have described several counter-examples in which its results are inacceptable. It should be emphasized that the OWA- and FG-count approaches have a reasonable 'core' for nondecreasing unary quantifiers. However, the proposals to extend these models to non-monotonic quantifiers and importance qualification are not conclusive. Due to the algorithmic simplicity of existing approaches, they are certainly useful as aggregation operators. However, the findings of this chapter discourage the use of these models in applications of fuzzy quantifiers which depend on their natural language meaning. It was this massive evidence against the known models which motivated the author's departure from the traditional view of fuzzy quantification. The novel theory was built on an axiomatic foundation in order to ensure reliable interpretations and avoid the problems of existing approaches on formal grounds.

Th. 4.54: see [156], Th-8.1, p. 42
Th. 4.55: see [55], Th-24, p. 24
Th. 4.59: see [156], Th-8.2, p. 48
Th. 4.61: see [55], Th-25, p. 25
Th. 4.62: see [55], Th-26, p. 25
Th. 4.64: see [53], Th-23, p. 41
Th. 4.70: see [55], Th-33, p. 34
Th. 4.72: see [55], Th-34, p. 36
Th. 4.74: see [55], Th-35, p. 36

Theorems cited in Chap. 5:
Th. 5.3: see [53], Th-27, p. 43
Th. 5.4: see [53], Th-28, p. 44
Th. 5.6: see [53], Th-29, p. 44
Th. 5.11: see [55], Th-30, p. 27
Th. 5.13: see [56], Th-27, p. 22
Th. 5.17: see [53], Th-32, p. 48
Th. 5.19: see [53], Th-33, p. 48
Th. 5.21: see [53], Th-34, p. 49
Th. 5.22: see [55], Th-125, p. 74
Th. 5.25: see [59], Th-47, p. 168

Theorems cited in Chap. 6:
Th. 6.5: see [53], note, p. 50
Th. 6.7: see [53], note, p. 52
Th. 6.10: see [55], Th-76, p. 56
Th. 6.12: see [55], Th-77, p. 56
Th. 6.13: see [55], Th-78, p. 56
Th. 6.14: see [55], Th-79, p. 57
Th. 6.16: see [55], Th-81, p. 57
Th. 6.17: see [55], Th-84, p. 59
Th. 6.19: see [55], Th-85, p. 59
Th. 6.22: see [56], Th-34, p. 27
Th. 6.24: see [56], Th-35, p. 27

Theorems cited in Chap. 7:
Th. 7.3: see [53], Th-40, p. 57
Th. 7.7: see [53], Th-41, p. 57
Th. 7.17: see [59], Th-61, p. 199
Th. 7.20: see [53], Th-42, p. 61
Th. 7.24: see [53], Th-44, p. 63
Th. 7.28: see [55], Th-40, p. 42
Th. 7.29: see [55], Th-41, p. 43
Th. 7.38: see [55], Th-62, p. 49
Th. 7.39: see [55], Th-62, p. 49

Th. 7.40:	see [55], Th-58, p. 49
Th. 7.41:	see [55], Th-65 to Th-70, p. 51/52
Th. 7.43:	see [55], note, p. 47
Th. 7.45:	see [55], Th-64, p. 50
Th. 7.46:	see [55], Th-64, p. 50
Th. 7.47:	see [55], Th-71, p. 52
Th. 7.49:	see [55], note, p. 48
Th. 7.51:	see [55], Th-87, p. 61
Th. 7.54:	see [55], Th-91, p. 62
Th. 7.57:	see [55], Th-93, p. 62
Th. 7.59:	see [55], Th-86, p. 61
Th. 7.60:	see [57], Th-18, p. 23
Th. 7.62:	see [55], Th-88, p. 61
Th. 7.63:	see [55], Th-89, p. 61
Th. 7.64:	see [55], Th-90, p. 62
Th. 7.65:	see [55], Th-92, p. 62
Th. 7.66:	see [55], Th-103, p. 65
Th. 7.67:	see [55], Th-104, p. 65
Th. 7.68:	see [55], Th-105, p. 65
Th. 7.69:	see [55], Th-106, p. 66
Th. 7.70:	see [55], Th-107, p. 66
Th. 7.71:	see [55], Th-108, p. 66
Th. 7.72:	see [55], Th-109, p. 66
Th. 7.73:	see [55], Th-110, p. 67
Th. 7.74:	see [55], Th-111, p. 67
Th. 7.75:	see [55], Th-112, p. 67
Th. 7.76:	see [55], Th-113, p. 67
Th. 7.77:	see [55], Th-44, p. 46
Th. 7.79:	see [55], Th-46, p. 46
Th. 7.80:	see [55], Th-102, p. 65
Th. 7.81:	see [55], Th-124, p. 74
Th. 7.83:	see [59], Th-99, p. 216
Th. 7.85:	see [55], Th-100, p. 65
Th. 7.86:	see [55], Th-101, p. 65
Th. 7.87:	see [55], Th-123, p. 73
Th. 7.90:	see [55], Th-126, p. 74
Th. 7.92:	see [55], Th-132, p. 76
Th. 7.94:	see [55], Th-127, p. 74
Th. 7.95:	see [55], Th-128, p. 75

Theorems cited in Chap. 8:

Th. 8.2:	see [57], Th-20, p. 25
Th. 8.4:	see [57], Th-21, p. 25
Th. 8.7:	see [57], Th-22, p. 26
Th. 8.8:	see [57], Th-23, p. 26

14. M. Black. Vagueness: an exercise in logical analysis. *Philosophy of Science*, 4:427–455, 1937. Reprinted with omissions in [90, p. 69-81].
15. N. Blanchard. Cardinal and ordinal theories about fuzzy sets. In M.M. Gupta and E. Sanchez, editors, *Fuzzy Information and Decision Processes*, pages 149–157. North-Holland, Amsterdam, 1982.
16. I. Bloch. Information combination operators for data fusion: a comparative review with classification. *IEEE Transactions on Systems, Man, and Cybernetics*, 26(1):52–67, 1996.
17. I.M. Bocheński. *A history of formal logic.* Chelsea Publishing Company, New York, 2nd edition, 1970.
18. G. Bordogna, P. Carrara, and G. Pasi. Query term weights as constraints in fuzzy information retrieval. *Information Processing & Management*, 27(1):15–26, 1990.
19. G. Bordogna and G. Pasi. A fuzzy information retrieval system handling users' preferences on document sections. In Dubois et al. [45].
20. P. Bosc, B.B. Buckles, F.E. Petry, and O. Pivert. Fuzzy databases. In D. Dubois, H. Prade, and J. Bezdek, editors, *Fuzzy Sets in Approximate Reasoning and Expert Systems*, pages 403–468. Kluwer, Dordrecht, NL, 1999.
21. P. Bosc and L. Lietard. Monotonic quantified statements and fuzzy integrals. In *Proc. of the NAFIPS/IFI/NASA '94 Joint Conference*, pages 8–12, San Antonio, Texas, 1994.
22. P. Bosc and L. Lietard. On the comparison of the Sugeno and Choquet fuzzy integrals for the evaluation of quantified statements. In *Proceedings of EUFIT '95*, pages 709–716, 1995.
23. P. Bosc and O. Pivert. Extending SQL retrieval features for the handling of flexible queries. In Dubois et al. [45], pages 233–251.
24. N.R. Campbell. The sorites paradox. *Philosophical Studies*, 26:175–191, 1974.
25. J. Cargile. The sorites paradox. *British Journal for the Philosophy of Science*, 1969.
26. S.J. Chen and C.L. Hwang. *Fuzzy Multiple Attribute Decision Making: Methods and Applications.* Springer, Heidelberg, 1992.
27. A. Church. A formulation of the simple theory of types. *Journal of Symbolic Logic*, 5:56–68, 1940.
28. F. Díaz-Hermida, A. Bugarín, P. Cariñena, M. Mucientes, and D.E. Losoda. Fuzzy quantification in two real scenarios: Information retrieval and mobile robotics. *International Journal of Intelligent Systems*, 2005. to appear.
29. F. Díaz-Hermida, P. Cariñena, A. Bugarín, and S. Barro. Probabilistic evaluation of fuzzy quantified sentences: independence profile. *Mathware and Soft Computing*, VIII(3):255–274, 2001.
30. F. Díaz-Hermida, P. Cariñena, A. Bugarín, and S. Barro. Definition and classification of semi-fuzzy quantifiers for the evaluation of fuzzy quantified sentences. *International Journal of Approximate Reasoning*, 34(1):49–88, 2003.
31. F. Díaz-Hermida, P. Cariñena, A. Bugarín, and S. Barro. Random set-based approaches for modelling fuzzy operators. In *Modelling with Words - Learning, Fusion, and Reasoning within a Formal Linguistic Representation Framework*, volume 2873 of *Lecture Notes in Computer Science*, pages 1–25. Springer, 2003.
32. M. Delgado, F. Herrera, E. Herrera-Viedma, and L. Martinez. Combining numerical and linguistic information in group decision making. *Information Sciences*, 107:177–194, 1998.

33. M. Delgado, D. Sánchez, and M.A. Vila. Un enfoque lógico para calcular el grado de cumplimiento de sentencias con cuantificadores linguísticos. In *Actas VII Congreso Español Sobre Tecnologías y Lógica Fuzzy (ESTYLF '97)*, pages 15–20, 1997.
34. M. Delgado, D. Sánchez, and M.A. Vila. A survey of methods for evaluating quantified sentences. In *Proceedings of the First European Society for Fuzzy Logic and Technologies Conference (EUSFLAT '99)*, pages 279–282, 1999.
35. M. Delgado, D. Sánchez, and M.A. Vila. Fuzzy cardinality based evaluation of quantified statements. *International Journal of Approximate Reasoning*, 23(1):23–66, 2000.
36. A. De Luca and S. Termini. A definition of non-probabilistic entropy in the setting of fuzzy sets theory. *Information and Control*, 20(4):301–312, 1972.
37. P. Dienes. On an implication function in many-valued systems of logic. *Journal of Symbolic Logic*, 14:95–97, 1949.
38. J. van der Does. The dynamics of sophisticated laziness. In J. Groenendijk, editor, *Plurals and Anaphora*, DYANA-2 (Dynamic Interpretation of Natural Language) ESPRIT Basic Research Project 6852, Deliverable R.2.2.A, Part I, pages 1–52, 1993.
39. D. Driankov, H. Hellendoorn, and M. Reinfrank. *An Introduction to Fuzzy Control.* Springer, 1996.
40. D. Dubois, L. Godo, R. Lopez de Mantaras, and H. Prade. Qualitative reasoning with imprecise probabilities. *Journal of Intelligent Information Systems*, 2:319–363, 1993.
41. D. Dubois and H. Prade. *Fuzzy Sets and Systems: Theory and Applications.* Academic Press, New York, 1980.
42. D. Dubois and H. Prade. Fuzzy cardinality and the modelling of imprecise quantification. *Fuzzy Sets and Systems*, 16:199–230, 1985.
43. D. Dubois and H. Prade. Fuzzy sets in approximate reasoning I, II. *Fuzzy Sets and Systems*, 40(1):143–202, 202–244, 1991.
44. D. Dubois, H. Prade, and R. Yager. Merging fuzzy information. In D. Dubois, H. Prade, and J. Bezdek, editors, *Fuzzy Sets in Approximate Reasoning and Expert Systems*, pages 335–402. Kluwer, Dordrecht, NL, 1999.
45. D. Dubois, H. Prade, and R.R. Yager, editors. *Fuzzy Information Engineering.* Wiley, New York, 1997.
46. D. Edgington. Vagueness by degrees. In R. Keefe and P. Smith, editors, *Vagueness: A Reader*, pages 294–316. MIT Press/A Bradford Book, Cambridge, MA, 1997.
47. K. Fine. Vagueness, truth and logic. *Synthese*, 1975.
48. G. Frege. Über Sinn und Bedeutung. *Zeitschrift für Philosophie und philosophische Kritik*, 100:25–50, 1892.
49. B.R. Gaines. Foundations of fuzzy reasoning. *Int. Journal of Man-Machine Studies*, 8:623–668, 1978.
50. L.T.F. Gamut. *Intensional Logic and Logical Grammar*, volume II of *Logic, Language, and Meaning.* The University of Chicago Press, Chicago & London, 1984.
51. I. Glöckner. Evaluation of quantified propositions in generalized models of fuzzy quantification. *International Journal of Approximate Reasoning*, 37(2):93–126, 2004.
52. I. Glöckner. *Fuzzy Quantifiers in Natural Language: Semantics and Computational Models.* Der Andere Verlag, Osnabrück, Germany, 2004.

53. I. Glöckner. DFS – an axiomatic approach to fuzzy quantification. TR97-06, Technical Faculty, University Bielefeld, 33501 Bielefeld, Germany, 1997.
54. I. Glöckner. A framework for evaluating approaches to fuzzy quantification. TR99-03, Technical Faculty, University Bielefeld, 33501 Bielefeld, Germany, 1999.
55. I. Glöckner. Advances in DFS theory. TR2000-01, Technical Faculty, University of Bielefeld, 33501 Bielefeld, Germany, 2000.
56. I. Glöckner. An axiomatic theory of quantifiers in natural languages. TR2000-03, Technical Faculty, University of Bielefeld, 33501 Bielefeld, Germany, 2000.
57. I. Glöckner. A broad class of standard DFSes. TR2000-02, Technical Faculty, University of Bielefeld, 33501 Bielefeld, Germany, 2000.
58. I. Glöckner. Standard models of fuzzy quantification. TR2001-01, Technical Faculty, University of Bielefeld, 33501 Bielefeld, Germany, 2001.
59. I. Glöckner. Fundamentals of fuzzy quantification: Plausible models, constructive principles, and efficient implementation. TR2002-07, Technical Faculty, University of Bielefeld, 33501 Bielefeld, Germany, 2003.
60. I. Glöckner and A. Knoll. Fuzzy quantifiers for processing natural language queries in content-based multimedia retrieval systems. TR97-05, Technical Faculty, University of Bielefeld, 33501 Bielefeld, Germany, 1997.
61. I. Glöckner and A. Knoll. Application of fuzzy quantifiers in image processing: A case study. In *Proceedings of KES'99*, Adelaide, Australia, 1999.
62. I. Glöckner and A. Knoll. Natural-language navigation in multimedia archives: An integrated approach. In *Proceedings of the Seventh ACM Multimedia Conference (MM '99)*, pages 313–322, Orlando, Florida, 1999.
63. I. Glöckner and A. Knoll. Query evaluation and information fusion in a retrieval system for multimedia documents. In *Proceedings of Fusion'99*, pages 529–536, Sunnyvale, CA, 1999.
64. I. Glöckner and A. Knoll. Architecture and retrieval methods of a search assistant for scientific libraries. In W. Gaul and R. Decker, editors, *Classification and Information Processing at the Turn of the Millennium*. Springer, Heidelberg, 2000.
65. I. Glöckner and A. Knoll. Architecture and retrieval methods of a search assistant for scientific libraries. In W. Gaul and R. Decker, editors, *Classification and Information Processing at the Turn of the Millennium*, pages 460–468. Springer, Heidelberg, 2000.
66. I. Glöckner and A. Knoll. Fuzzy quantification in granular computing and its role for data summarisation. In W. Pedrycz, editor, *Granular Computing: An Emerging Paradigm*, Studies in Fuzziness and Soft Computing, pages 215–256. Physica-Verlag, Heidelberg, 2001.
67. I. Glöckner, A. Knoll, and A. Wolfram. Data fusion based on fuzzy quantifiers. In *Proceedings of EuroFusion98, International Data Fusion Conference*, pages 39–46, 1998.
68. J.A. Goguen. The logic of inexact concepts. *Synthese*, 19:325–373, 1969.
69. F. Hamm. *Natürlich-sprachliche Quantoren. Modelltheoretische Untersuchungen zu universellen semantischen Beschränkungen.* Number 236 in Linguistische Arbeiten. Niemeyer, Tübingen, 1989.
70. C.J. Harris, C.G. Moore, and M. Brown. *Intelligent Control: Aspects of Fuzzy Logic and Neural Nets.* World Scientific, Singapore, 1993.
71. J. van Heijenoort. *From Frege to Gödel – A Source Book in Mathematical Logic.* Harvard University Press, Cambridge, MA, 1967.

72. I. Heim. *The semantics of definite and indefinite noun phrases.* PhD thesis, University of Massachusetts, Amherst, 1982.
73. H. Helbig. *Die semantische Struktur natürlicher Sprache. Wissensrepräsentation mit MultiNet.* Springer, Berlin, Heidelberg, 2000.
74. L. Henkin. Some remarks on infinitely long formulas. In *Infinitistic methods*, pages 167–183. Pergamon Press, New York, and Polish Scientific Publishers, Warsaw, 1961.
75. F. Herrera and J.L. Verdegay. On group decision making under linguistic preferences and fuzzy linguistic quantifiers. In B. Bouchon, R.R. Yager, and L.A. Zadeh, editors, *Fuzzy Logic and Soft Computing*, pages 173–180. World Scientific, 1995.
76. J. Hintikka. Quantifiers vs. quantification theory. *Dialectica*, 27:329–358, 1973.
77. P. Hájek and L. Kohout. Fuzzy implications and generalized quantifiers. *International Journal of Uncertainty, Fuzziness and Knowledge-Based Systems*, 4(3):225–233, 1996.
78. I. Homann. *Fuzzy-Suchmethoden im Information-Retrieval.* PhD thesis, University of Bielefeld, Bielefeld, Germany, 2004.
79. F. de Jongh and H. Verkuyl. Generalized quantifiers: The properness of their strength. In J. van Benthem and A. ter Meulen, editors, *Generalized Quantifiers in Natural Language.* Foris, Dordrecht, 1984.
80. J. Kacprzyk. *Multistage Decision-Making under Fuzziness.* Verlag TÜV Rheinland, Cologne, 1983.
81. J. Kacprzyk. *Multistage Fuzzy Control: A Model-Based Approach to Control and Decision-Making.* Wiley, Chichester, 1997.
82. J. Kacprzyk. Intelligent data analysis via linguistic data summaries: A fuzzy logic approach. In R. Decker and W. Gaul, editors, *Classification and Information Processing at the Turn of the Millenium*, pages 153–161. Springer, 2000.
83. J. Kacprzyk and P. Strykowski. Linguistic summaries of sales data at a computer retailer: A case study. In *Proc. of 8th International Fuzzy Systems Association World Congress (IFSA '99)*, pages 29–33, 1999.
84. J. Kacprzyk and S. Zadrożny. Fuzzy queries in Microsoft Access V.2. In Dubois et al. [45], pages 223–232.
85. J. Kacprzyk, S. Zadrożny, and M. Fedrizzi. An interactive GDSS for consensus reaching using fuzzy logic with linguistic quantifiers. In Dubois et al. [45], pages 567–574.
86. J. Kacprzyk and A. Ziólkowsi. Database queries with fuzzy linguistic quantifiers. *IEEE Transactions on Systems, Man, and Cybernetics*, 16(3):474–479, 1986.
87. H. Kamp. A theory of truth and semantic representation. In J. Groenendijk, T. Janssen, and M. Stokhof, editors, *Formal Methods in the Study of Language*, Amsterdam: Mathematical Centre, 1981.
88. J. Kamp. Two theories about adjectives. In Edward Keenan, editor, *Formal Semantics of Natural Language.* Cambridge University Press, 1975.
89. A. Kaufmann. *Introduction to the Theory of Fuzzy Subsets*, volume 1: Fundamental theoretical elements. Academic Press, New York, 1975.
90. R. Keefe and P. Smith, editors. *Vagueness: A Reader.* A Bradford Book/MIT Press, Cambridge, MA, 1997.

91. E. Keenan and L. Moss. Generalized quantifiers and expressive power of natural language. In J. van Benthem and A. ter Meulen, editors, *Generalized Quantifiers in Natural Language*, pages 73–124. Foris, Dordrecht, 1984.
92. E. Keenan and J. Stavi. A semantic characterization of natural language determiners. *Linguistics and Philosophy*, 9:253–326, 1986.
93. E. Keenan and D. Westerståhl. Generalized quantifiers in linguistics and logic. In J. van Benthem and A. Ter Meulen, editors, *Handbook of Logic and Language*, chapter 15, pages 837–893. Elsevier, Amsterdam, NL, 1997.
94. H.J. Keisler. Logic with the quantifier 'There Exist Uncountably Many'. *Annals of Mathematical Logic*, 1:1–93, 1969.
95. S.C. Kleene. *Introduction to Metamathematics.* North-Holland, Amsterdam, 1952.
96. E.P. Klement, R. Mesiar, and E. Pap. Quasi- and pseudo-inverses of monotone functions, and the construction of t-norms. *Fuzzy Sets and Systems*, 104:3–13, 1999.
97. G.J. Klir and T.A. Folger. *Fuzzy Sets, Uncertainty, and Information.* Prentice Hall, Englewood Cliffs, NJ, 1988.
98. G.J. Klir, U.H. St. Clair, and B. Yuan. *Fuzzy Set Theory: Foundations and Applications.* Prentice Hall, Upper Saddle River, NJ, 1997.
99. G.J. Klir and B. Yuan. *Fuzzy Sets and Fuzzy Logic: Theory and Applications.* Prentice Hall, Upper Saddle River, NJ, 1995.
100. G.J. Klir and B. Yuan, editors. *Fuzzy Sets, Fuzzy Logic, and Fuzzy Systems: Selected Papers by Lotfi A. Zadeh.* World Scientific, Singapore, 1996.
101. S. A. Kripke. Semantical analysis of modal logic II. In J.W. Addison and A. Tarski, editors, *The theory of models*, pages 206–220. North-Holland, Amsterdam, 1965.
102. S.A. Kripke. A completeness theorem in modal logic. *Journal for Symbolic Logic*, 24:1–14, 1959.
103. R. Kruse, E. Schwecke, and J. Heinsohn. *Uncertainty and Vagueness in Knowledge Based Systems: Numerical Methods.* Springer, Heidelberg, 1991.
104. Y. Lai and C. Hwang. *Fuzzy Mathematical Programming.* Springer, Heidelberg, 1992.
105. G. Lakoff. Hedges: a study in meaning criteria and the logic of fuzzy concepts. *Journal of Philosophical Logic*, 2:458–508, 1973.
106. P. Lindström. First order predicate logic with generalized quantifiers. *Theoria*, 32:186–195, 1966.
107. P. Lindström. On extensions of elementary logic. *Theoria*, 35:1–11, 1969.
108. Y. Liu and E.E. Kerre. An overview of fuzzy quantifiers. (I). Interpretations. *Fuzzy Sets and Systems*, 95:1–21, 1998.
109. Y. Liu and E.E. Kerre. An overview of fuzzy quantifiers. (II). Reasoning and applications. *Fuzzy Sets and Systems*, 96:135–146, 1998.
110. D. Losada, S. Barro, A. Bugarín-Diz, and F. Díaz-Hermida. Experiments on using fuzzy quantified sentences in adhoc retrieval. In *ACM Symposium on Applied Computing*, pages 1059–1066, 2004.
111. K.F. Machina. Truth, belief and vagueness. *Journal of Philosophical Logic*, 5:47–78, 1976. Reprinted in [90, p. 174-203].
112. E.H. Mamdani and S. Assilian. An experiment in linguistic synthesis with a fuzzy logic controller. *Int. Journal of Man-Machine Studies*, 7:1–13, 1975.
113. R. Mesiar and M. Navara. Diagonals of continuous triangular norms. *Fuzzy Sets and Systems*, 104:35–41, 1999.

114. S. Miyamoto. *Fuzzy Sets in Information Retrieval and Cluster Analysis.* Kluwer Academic Publishers, Dordrecht, 1990.
115. R. Montague. The proper treatment of quantification in ordinary English. In J. Hintikka, J. Moravcsik, and P. Suppes, editors, *Approaches to Natural Language*, pages 221–242. Reidel, Dordrecht, 1973.
116. A. Mostowski. On a generalization of quantifiers. *Fundamenta Mathematicae*, 44:12–36, 1957.
117. M. Mucientes, R. Iglesias, C. Regueiro, A. Bugarín, P. Cariñena, and S. Barro. Fuzzy temporal rules for mobile robot guidance in dynamic environments. *IEEE Transactions on Systems, Man, and Cybernetics, Part C*, 66(1), 2002.
118. M. Mukaidono. On some properties of fuzzy logic. *Systems – Computers – Controls*, 6(2):36–43, 1975.
119. H. Nakanishi, I.B. Turksen, and M. Sugeno. A review and comparison of six reasoning methods. *Fuzzy Sets and Systems*, 57(3):257–294, 1993.
120. C.V. Negoita. *Expert Systems and Fuzzy Systems.* Benjamin/Cummings, Menlo Park, CA, 1985.
121. S.E. Newstead and K. Coventry. The role of context and functionality in the interpretation of quantifiers. *European Journal of Cognitive Psychology*, 12:243–259, 2000.
122. S.E. Newstead, P. Pollard, and D. Riezebos. The effect of set size on the interpretation of quantifiers used in rating scales. *Applied Ergonomics*, 18(3):178–182, 1987.
123. H.T. Nguyen and E.A. Walker. *A First Course in Fuzzy Logic.* Chapman & Hall/CRC Press, Boca Raton, FL, 2nd edition, 2000.
124. S.K. Pal and D.K. Majumder. *Fuzzy Mathematical Approach to Pattern Recognition.* John Wiley, New York, 1986.
125. W. Pedrycz. *Fuzzy Control and Fuzzy Systems.* John Wiley, New York, 1989.
126. H. Prade. Lipski's approach to incomplete information data bases restated and generalized in the setting of Zadeh's possibility theory. *Information Sciences*, 9(1):27–42, 1984.
127. H. Prade. A two-layer fuzzy pattern matching procedure for the evaluation of conditions involving vague quantifiers. *Journal of Intelligent and Robotic Systems*, 3:93–103, 1990.
128. H. Prade and C. Testemale. Generalizing database relational algebra for the treatment of incomplete or uncertain information and vague queries. *Information Sciences*, 34(2):115–143, 1984.
129. W.V. Quine. *Word and Object.* Cambridge University Press, Cambridge, Mass., 1960.
130. W.V. Quine. *Ontological Relativity and other Essays.* Columbia University Press, New York, 1969.
131. A. Ralescu, B. Bouchon, and D. Ralescu. Combining fuzzy quantifiers. In W. Chiang and J. Lee, editors, *Fuzzy Logic for the Applications to Complex Systems*, pages 246–252. World Scientific, Singapore, 1995.
132. A.L. Ralescu. A note on rule representation in expert systems. *Information Sciences*, 38:193–203, 1986.
133. Anca Ralescu, Kaoru Hirota, and Dan Ralescu. Evaluation of fuzzy quantified expressions. In *Fuzzy Logic in Artificial Intelligence (IJCAI Workshop)*, volume 1566 of *Lecture Notes in Computer Science*, pages 234–244. Springer, 1999.
134. D. Ralescu. Cardinality, quantifiers, and the aggregation of fuzzy criteria. *Fuzzy Sets and Systems*, 69:355–365, 1995.

axioms for L-DFSes 365

Barwise, J. 355
base quantifier 76
base set 5, 30, 36, 57
base set of infinite cardinality *see* infinite base set
bibliographic databases 348
biimplication *see* equivalence
binary quantifier 9, 31, 58
Black 18
bona fide assumption of linguistic plausibility 53
Boolean algebra 76
borderline case 14, 77
boundary conditions with respect to the crisp case 125
bounded product 113
bounded sum 114
bounding quantifier 383
branching quantification 56, 82, 355, 368, 371, 387
branching quantity in agreement 373
broad notion of quantifiers 6, 30

cardinal comparative 9, 21, 51, 62, 332, 383
cardinal determiners *see* absolute quantifiers
cardinality bounds *see* cardinality coefficients
cardinality coefficients 301, 302, 308
cardinality of fuzzy sets 32, 69
cardinality-based computation of quantitative semi-fuzzy quantifiers 296
case-by-case verification 69, 75
category theory 130
cautiousness 188, 381
cautiousness level 188
characterisation of S-quantifiers 139
characteristic function 18, 88
characterization of $\mathcal{F}_\Omega$-DFSes 241
characterization of $\mathcal{F}_\omega$-DFSes 245
characterization of $\mathcal{F}_\psi$-DFSes 265
characterization of $\mathcal{F}_\varphi$-DFSes 271
characterization of $\mathcal{F}_\xi$-DFSes 224
characterization of $\mathcal{M}_\mathcal{B}$-DFSes 196, 198
characterization of $\mathcal{M}_{CX}$ 207
characterization of 'Gainesian' QFMs 274
characterization of quantitative semi-fuzzy quantifiers 296
characterization of quantitative unary quantifiers 282
characterization of T-quantifiers 137
Choquet integral 222, 226, 381, 383
circularity of a quantifier 63
classification of linguistic quantifiers 63, 383
classification of the models by the induced negation 150
classification of the models by their induced negation and disjunction 151
cloudiness situation 41, 349
co-*t*-norm *see* *s*-norm
coextensional terms 82
coherence VI, 108, 378, 384, 386
comparative determiner *see* cardinal comparative
comparison by specificity 202, 232, 252
compatibility grade 272
compatibility with antonyms VI, 108, 119
compatibility with crisp argument insertion 125
compatibility with cylindrical extensions 118, 379
compatibility with dualisation VI, 96, 108, 378, 379
compatibility with external negation VI, 108, 119, 379
compatibility with functional application 104, 108, 378, 388
compatibility with fuzzy argument insertion 172, 206, 381
compatibility with internal complementation 119, 379
compatibility with intersections of arguments 123
compatibility with inverse images 388
compatibility with partial complementation 122
compatibility with permutations of arguments 118, 379

compatibility with the induced extension principle *see* compatibility with functional application
compatibility with the nesting of quantifiers 370, 388
compatibility with transpositions of arguments 117
compatibility with unions of arguments 97, 108
composite quantifier *see* compound quantifier
compound quantifier 57, 63, 384, 386
computation of $\mathcal{F}_{\text{owa}}$ from finite representation 290
computation of $S_{Q,X_1,\dots,X_n}$ from cardinality coefficients ℓ_r, u_r 299
computation of $\mathcal{M}$ from finite representation 292
computation of $\mathcal{M}_{CX}$ from finite representation 292
computation of cardinality coefficients ℓ, u from histogram information 312
computation of cardinality coefficients ℓ_r, u_r 309
computational analysis of general quantifiers 287
computational models 277, 382
concrete models for fuzzy quantification 381
conjunction of quantifiers 154, 165, 380
conjunctions of convex quantifiers are convex 167
conservativity 170, 297, 380, 403
consistency of the DFS axioms 109, 179
consistent view of fuzzy set theory 387
consistently less specific 151
constant quantifier *see* nullary quantifier
constructed quantifier *see* non-simplex quantifier
construction of L-QFM for some ordinary QFM 367
construction of membership functions 377
construction of partial QFM for existing approaches 393
constructions on models of fuzzy quantification 149, 151, 153, 154
constructive principle 179, 221, 237, 259, 381
context 11, 65, 377
context dependence 11, 15, 64, 73, 377
context range of a fuzzy set 135
contextual equality 135
contextual quantifier fuzzification mechanism 136
contextuality 135, 168, 379
continuity conditions 162
continuos in arguments *see* arg-continuos
continuous in quantifiers *see* Q-continuous
continuous quantifier *see* mass quantification
continuous-valued logic 76
continuous-valued model of vagueness *see* degree theory of vagueness
convex combination 151
convex quantifier 166, 283, 380
core of a fuzzy set 135
correct generalization 87, 108, 111, 144, 378
correlativity 121
correspondence between specifications and operational forms 55, 74, 377
count NP 83
counterexample for FE-count approach 49
covariant functor 130
coverage of linguistic phenomena 376
coverage of quantificational phenomena 51, 52, 73, 80, 386
crisp adjectival restriction 126
crisp argument 71
crisp argument insertion 123, 172
crisp image mapping *see* crisp powerset mapping
crisp powerset mapping 101
crisp range of a three-valued cut 188
crisp set 18
cross-arity coherence 108
cross-domain coherence 105, 108, 378
currying 141
cylindrical extension 77, 118, 128

database interface 28
decomposition of convex quantifiers 167
definite description 3
definite quantifier 3, 56, 62, 79, 383
definition of $\mathcal{F}_{\mathrm{owa}}$ 226
definition of $\mathcal{F}_S$ 227
definition of $\mathcal{F}_A$ 227
definition of $\mathcal{F}_M$ 247
definition of $\mathcal{F}_P$ 247
definition of $\mathcal{F}_K$ 248
definition of $\mathcal{F}_Z$ 248
definition of $\mathcal{M}$ 192
definition of $\mathcal{M}_S$ 200
definition of $\mathcal{M}_U$ 199
definition of $\mathcal{M}_{CX}$ 201
definition of $\mathcal{M}_*$ 201
definition of $\mathcal{M}^*$ 200
degree of appropriateness 27
degree of covering the data 27
degree of imprecision 27
degree of orness 38, 45, 384, 400
degree theory of vagueness 17, 65
determinacy 78
determiner 4
determiner fuzzification scheme 106, 378
DFS *see* determiner fuzzification scheme
DFS axioms 106, 108, 378
dilemma of fuzzy quantification 70, 81
direct interpretation of special quantifiers in $\mathcal{M}_{CX}$ 282
discontinuous behaviour of Σ-count approach 44, 399
discontinuous behaviour of FG-count approach for proportional quantifiers 410
discontinuous behaviour of FG-count approach for restricted proportional quantification 48
discourse representation theory 63
discrete quantifier *see* count NP
discriminative power 221
disjunction of quantifiers 154, 165, 380
disposition 5, 22
distance measure for quantifiers 163
distributivity 161
domain of quantification *see* base set
downward monotonic 99
DQL *see* dynamic quantifier logic
DRT *see* discourse representation theory
dual 94, 108
dual s-norm 114
dual of (semi-)fuzzy L-quantifiers 364
duplication of variables *see* multiple occurrences of variables
dynamic quantifier logic 63

EFA *see* evaluation framework assumption
efficiency considerations 278
epistemic model of vagueness 16, 64
equivalence 114, 165
equivalence of both constructions of induced connectives 115
evaluation framework assumption 393
evaluation framework for existing approaches 391, 392
event 16
exception quantifier *see* quantifier of exception
exclusive or 114, 165
existential quantification 1, 56, 136
existential quantification in a DFS 139
existing approaches to fuzzy quantification 36, 391
explicit quantification 5, 30
expressive power 6, 63, 70
expressiveness *see* expressive power
extension 82
extension principle 102, 259, 272
extensional fuzzy quantifier 134
extensional quantifier 82
extensional quantifier fuzzification mechanism 134
external negation 92

faithful functor 130
FE-count 37, 49, 285, 412
FE-count approach 37, 48, 412
FG-count 36, 37, 47, 285, 406
FG-count approach 36, 37, 47, 213, 381, 406
fields of quantification 22, 70, 80
finite representation of $\top_{Q,X_1,\ldots,X_n}$ and $\bot_{Q,X_1,\ldots,X_n}$ 290

finite scales of membership grades 214
first-order definable 6, 63, 376
first-order predicate 19
first-order proposition 81
first-order representation of fuzzy quantifiers *see* fuzzy linguistic quantifier
fixed context assumption 11, 64, 65
flat representation of generalized quantifiers 57, 59
floating-point arithmetics 313, 325, 337
formal limits on the adequacy of interpretations 161, 370, 380
foundations of fuzzy reasoning 272
framework for fuzzy quantification 29, 55, 376
Frege's principle *see* principle of compositionality
full two-valued L-quantifier 358, 359
functional application 104, 108
functor 130
fuzzification mechanism 74, 75, 108, 274, 378
fuzziness *see* vagueness
fuzziness in quantifiers 21, 376
fuzziness in the arguments 21, 70, 376
fuzzy adjectival restriction 126, 172, 381
fuzzy argument 66, 80
fuzzy argument insertion 172, 206, 370, 381
fuzzy arguments 21
fuzzy boundary 14
fuzzy branching quantification 387
fuzzy cluster analysis 20
fuzzy complement 18
fuzzy conjunction 113
fuzzy control 20
fuzzy data fusion 20, 29
fuzzy database 20, 28
fuzzy decision making 20
fuzzy design and optimization 20
fuzzy determiner *see* fuzzy quantifier
fuzzy disjunction 114
fuzzy equivalence connective 260
fuzzy existential quantifier *see* S-quantifier
fuzzy expert systems 20
fuzzy identity truth function 112
fuzzy image mapping *see* fuzzy powerset mapping
fuzzy inclusion relation 97
fuzzy information retrieval 20, 57, 347
fuzzy intersection 18
fuzzy interval cardinality 284
fuzzy inverse image 141
fuzzy L-quantifier 360
fuzzy linguistic quantifier 34, 377, 383, 392
fuzzy median 152
fuzzy median-based aggregation mechanism 190
fuzzy multi-place quantification VI, 108, 141
fuzzy natural language quantification 86, 388
fuzzy negation 112
fuzzy pattern recognition 20
fuzzy powerset 18, 130
fuzzy powerset mapping 102
fuzzy pre-determiner *see* semi-fuzzy quantifier
fuzzy projection L-quantifier 361
fuzzy projection quantifier 89, 108, 361
fuzzy quantifier 31, 66, 70, 361, 376
fuzzy querying *see* weighted querying
fuzzy relation 18
fuzzy second-order relation 19, 31, 66, 376
fuzzy set *see* fuzzy subset
fuzzy set theory 18
fuzzy subset 18
fuzzy truth function 91
fuzzy union 18
fuzzy universal quantifier *see* T-quantifier

Gainesian fuzzification mechanism 157, 272, 274
general fuzzy quantifier 66
generalized fuzzy median 153, 184, 186, 222, 381
generalized Lindström quantifier 358
generalized quantifier 56, 57, 70, 71, 375, 376
generator of a strong negation 150

granularity of membership values 216
greatest lower specificity bound of $\widetilde{\vee}$-models 153
groups of interrelated objects 373

'have extension' 133
hedge *see* linguistic hedge
hedging *see* linguistic hedge
higher-order logic 81
higher-order proposition 81
higher-order quantifier 81
histogram 310
homogeneity of interpretations 149
homogeneous classes of models 149
homomorphism 86, 378, 394

idempotence 161
identity 121
identity truth function 112
IL *see* intensional logic
image ranking task 41, 349
imperfection of natural language 13
implementation of absolute quantifiers 318
implementation of cardinal comparatives in $\mathcal{F}_{\text{owa}}$ 337, 340
implementation of cardinal comparatives in $\mathcal{M}$ 337, 340
implementation of cardinal comparatives in $\mathcal{M}_{CX}$ 337, 340
implementation of fuzzy quantifiers 277, 382
implementation of proportional quantifiers in $\mathcal{F}_{\text{owa}}$ 325, 328
implementation of proportional quantifiers in $\mathcal{M}$ 325, 328
implementation of proportional quantifiers in $\mathcal{M}_{CX}$ 325, 328
implementation of quantifiers in $\mathcal{M}_{CX}$ based on interval cardinality 285
implementation of quantifiers of exception 318
implementation of unary quantifiers in $\mathcal{F}_{\text{owa}}$ 313, 316
implementation of unary quantifiers in $\mathcal{M}$ 313, 316
implementation of unary quantifiers in $\mathcal{M}_{CX}$ 313, 316
implicit quantification 5, 30
importance qualification 38, 41, 51, 325, 328, 350, 385
inclusion approach to fuzzy quantification 37, 38, 48
incompatibility of Σ-count approach with antonyms 399
incompatibility of OWA approach with dualisation of proportional quantifiers 406
indepencence of the DFS axioms 107
indeterminacy 184
indeterminate truth-value 184
indicator function *see* characteristic function
individual correspondence assertion 75
induced connective *see* induced fuzzy truth function
induced extension principle 103, 107, 129, 379
induced extension principle of an L-QFM 365
induced fuzzy complement 92
induced fuzzy conjunction 92, 114
induced fuzzy connective *see* induced fuzzy truth function
induced fuzzy disjunction 92, 114
induced fuzzy equivalence 155
induced fuzzy exclusive or 155
induced fuzzy existential quantifier 139
induced fuzzy implication 114
induced fuzzy intersection 92
induced fuzzy inverse images 141, 379
induced fuzzy negation 92, 113
induced fuzzy truth function 91, 107, 112
induced fuzzy truth function of an L-QFM 363
induced fuzzy union 92
induced fuzzy universal quantifier 138
induced operation on fuzzy sets 92
induced propositional logic 112
induced truth functions of standard models 156
infinite base set VI, 57, 356, 385, 386
information retrieval 28, 347, 349
injective mapping 130

insertion of crisp arguments *see* crisp argument insertion
insertion of fuzzy argument *see* fuzzy argument insertion
integer arithmetics 315, 316, 328, 340
integral-based approach to fuzzy quantification 38
intension 82
intensional logic 56
intensional quantifier 82, 388
internal coherence 108
internal complementation 93
internal negation *see* internal complementation
interpretation mechanism 29, 35, 36, 51, 392
interpretation of cardinality comparisons 280
interpretation of fuzzy quantifiers *see* modelling problem
interpretation of ordinary quantifiers in an L-QFM 363
interpretation of the standard quantifiers 278
interpretation of two-valued quantifiers in standard models 157
intersection of arguments 97, 122
interval-valued quantifications 388
inverse image mapping 102
IR *see* information retrieval

K-standard sequence logic 156
Kleene's three-valued logic 156, 183
Kleene-Dienes implication 156
Klein group structure 95
knowledge acquisition 377

L-DFS 365
L-DFS axioms 365
L-model *see* L-DFS
L-QFM *see* L-quantifier fuzzification mechanism
L-quantifier *see* two-valued L-quantifier
L-quantifier fuzzification mechanism 360
law of contradiction 161
least upper bound of $\widetilde{\vee}$-models 154
Lindström quantifier 12, 356, 357
linguistic adequacy 53
linguistic data summarization 20, 26, 28, 350, 373
linguistic data summarization based on branching quantification 373
linguistic hedge 78, 81
linguistic plausibility 391
linguistic quantifiers 4, 20
linguistic summary *see* linguistic data summarization, 350, 373
linguistic theory of quantification *see* theory of generalized quantifiers
local monotonicity in arguments 126
local monotonicity of a DFS 129
logic with generalized quantifiers 82, 355
logical quantifier 4, 10, 136, 375
logical symbol 10, 375
lower bound mapping $\perp_{Q,X_1,\ldots,X_n}$ 223

main approaches to fuzzy quantification 38
many-valued logic 24, 313
mass quantification 83, 385–387
mean operator 152
measure theory 386, 387
median-based models 179
membership assessment 88, 89, 108, 361, 378
membership degree 18
membership function 18, 72
meteorological image 41, 349
metric for argument similarity 162
model aggregation scheme 151
model of fuzzy quantification 29, 85
model transformation scheme 150, 214, 379
modelling of vagueness 64
modelling problem 4, 25, 27, 55, 68, 376, 377
models based on the extension principle 259
models defined in terms of the standard extension principle 382
models defined in terms of three-valued cuts 237
models of fuzzy quantification 108
models which induce the standard disjunction 154

models which induce the standard negation 150
monotonic quantifier 283
monotonicity of (semi-)fuzzy L-quantifiers in their arguments 364
monotonicity of a DFS 128
monotonicity of a quantifier in its arguments 98
monotonicity properties of quantifiers 126
Montague, R. 56
Mostowskian quantifier 11
multi-criteria decisionmaking 29, 57
multi-place quantifier VI, 9, 51, 57, 80, 108, 141, 384
multiple occurrences of variables 380
multiple variable binding 355, 357
mutual relationship 373

narrow notion of quantifiers 5
natural language V
natural language interface 28
natural language processing 28
negation 121
negative evidence against the Σ-count approach 395
negative evidence against the FE-count approach 412
negative evidence against the FG-count approach 406
negative evidence against the OWA approach 400
nested representation of generalized quantifiers 58, 59
nesting of quantifiers 368, 369, 388
NL *see* natural language
NL quantifiers *see* linguistic quantifiers
NLI *see* natural language interface
NLP *see* natural language processing
nominal phrase 4
non-logical symbol 10, 375
non-monotonic logic 6
non-monotonic quantifier 50, 285
non-quantitative quantifier VI, 57, 80, 131, 376, 386
non-referring term 64
non-simplex quantifier 77
non-standard model 380
nondecreasing (semi-)fuzzy L-quantifier 364
nondecreasing approximation of a quantifier 371
nondecreasing quantifier 98, 283, 372
nonexistence of α-cut based models 182
nonexistence of models based on the resolution principle 183
nonincreasing (semi-)fuzzy L-quantifier 364
nonincreasing approximation of a quantifier 371
nonincreasing quantifier 98, 283, 372
novel framework for fuzzy quantification 55, 80, 375, 376
NP *see* nominal phrase
nullary quantifier 58, 78, 87

observable behaviour 85
one-place quantifier *see* unary quantifier
opaque context 82
operational form of a linguistic quantifier 29, 55
operations on the argument sets 116
optional semantical properties 161
ordered weighted averaging operator 37, 45
ordered weighted averaging operators 400
OWA approach 37, 38, 45, 222, 226, 381, 400
OWA operator *see* ordered weighted avaraging operator

parametrized quantifiers 72
partial quantifier fuzzification mechanism 393, 394
partially ordered quantifiers 56
permutation 116
persistent 99
Piaget group of transformations 121, 386
plausibility criterion 53, 85, 144, 378
plausibility of induced fuzzy inverse images 141

plausible model of fuzzy quantification 85
possibility theory 388
possible world 83
potential model of fuzzy quantification *see* quantifier fuzzification mechanism
power of a fuzzy set *see* Σ-count
powerset mapping 101
practical $\mathcal{F}_{\Omega}$-models are $\mathcal{F}_{\xi}$-models 255
practical model of fuzzy quantification 179, 277
precise quantifier 21, 57
precisification 17
preservation of Aristotelian squares 96, 119, 156, 386
preservation of Boolean argument structure 97
preservation of contextuality 136
preservation of convexity 168, 380
preservation of extension 134, 379, 386
preservation of inequations between quantifiers *see* monotonicity of a DFS
preservation of local inequalities *see* local monotonicity of a DFS
preservation of local monotonicity 127, 379
preservation of monotonicity in the arguments 100, 108, 126, 378
preservation of monotonicity properties 100
preservation of quantitativity 132, 386
preservation of strong conservativity 171
preservation of symmetry properties 118
preservation of weak conservativity 171
preservation property 86, 394
presupposition 3, 62
principle of compositionality 9, 56, 63, 124
principle of extensionality 82
principle of supervaluation 17, 381
principle of tolerance 16
probability theory 15
processing of fuzzy arguments 377
projection L-quantifier 361
projection quantifier 88, 108, 361
propagation of fuzziness 163, 221, 227, 249, 250, 381
propagation of fuzziness in arguments 164, 205, 229, 250, 251, 380
propagation of fuzziness in quantifiers 164, 205, 228, 249, 250, 380
propagation of unspecificity 229, 250, 251
proper name 4, 62, 89, 108, 131, 386
proper name quantifier 62
properties of $\mathcal{F}_{\mathrm{owa}}$ 381
properties of $\mathcal{F}_{\Omega}$-models 249
properties of $\mathcal{F}_{\xi}$-models 227
properties of $\mathcal{M}_{\mathcal{B}}$-models 201
properties of $\mathcal{M}_{CX}$ 206, 381
properties of induced extension principle 129
properties of induced truth functions 112
proportional comparative 21, 383, 384
proportional determiners *see* proportional quantifier
proportional quantifier VI, 7, 21, 31, 35, 60, 321, 383, 392
propositional function 91
prototypical models 382, 383

Q-continuous 163, 203, 230, 255, 277, 380
QFA *see* quantification framework assumption
QFM *see* quantifier fuzzification mechanism
quantification framework assumption 76, 81, 393
quantifier fuzzification mechanism 74, 85, 377, 393
quantifier interpolation method 38, 41
quantifier nesting *see* nesting of quantifiers
quantifier of comparison 61
quantifier of exception 21, 60, 318, 383
quantifier of the third kind 31, 385
quantifier restricted to E 12
quantifier synthesis method 404, 411
quantifiers for cardinality comparison 280

quantifiers of the first kind *see* absolute quantifiers
quantifiers of the second kind *see* proportional quantifier
quantifiers of the third kind 347
quantitativity 10, 12, 131, 283, 376, 386
quantity in agreement 350
querying of information systems 347, 349

range of quantificational phenomena 29, 55, 73, 80, 383, 386
ranking criterion 348
'raw' approaches to fuzzy quantification *see* quantifier fuzzification mechanism
reasoning with fuzzy quantifiers 26, 27
reciprocal construction 82, 373
reciprocal predicate 355, 373
reciprocal summarizer 373
reciprocity 121
reduction of $\mathcal{F}_\omega$-DFSes to $\mathcal{F}_\psi$-DFSes 266
reduction of $\mathcal{F}_\psi$-DFSes to $\mathcal{F}_\omega$-DFSes 267
reduction of $\mathcal{F}_\psi$-QFMs to $\mathcal{F}_\varphi$-QFMs 270
reduction of $\mathcal{F}_\varphi$-QFMs to $\mathcal{F}_\psi$-QFMs 270
reduction of $S_{Q,X_1,\dots,X_n}$ to cardinality coefficients ℓ_r, u_r 299
reduction of a DFS to a $\neg$-DFS 150
reduction of absolute quantifiers to unary quantifiers 318
reduction of branching quantification to Lindström quantifiers 356
reduction of existential quantification to induced extension principle 131
reduction of fuzzy branching quantification to semi-fuzzy L-quantifiers 371
reduction of fuzzy existential quantification to induced disjunction 140
reduction of fuzzy quantification to cardinality comparisons 35, 37, 52
reduction of fuzzy quantification to first-order representations 32
reduction of fuzzy universal quantification to induced conjunction 140
reduction of induced extension principle to existential quantification 130
reduction of induced fuzzy disjunction to induced extension principle 140
reduction of L-DFSes to ordinary DFSes 367
reduction of multi-place quantification to unary quantification 141
reduction of quantifiers of exception to unary quantifiers 318
reduction of unrestricted proportional quantifiers to unary quantifiers 329
redundant argument *see* cylindrical extension
regular nondecreasing 45, 400
relational structure 357
relationship between quantifiers 63
relative Σ-count 395
relative cardinality 32, 384, 395
relative FG-count 407
relative quantifiers *see* proportional quantifier
reliability 52
Rescher's quantifiers 24, 77, 81
resolution principle 182, 381
restricted proportional quantification 50, 325, 328
restricted quantification *see* restricted use of a quantifier
restricted use of a quantifier 8, 31, 35, 60, 375
restriction of a quantifier 8, 59
robustness against the choice of the full domain 133
robustness of $\mathcal{M}_{CX}$ 211
Russell, B. 3, 56, 62

S-function 33
s-norm 114, 138
S-quantifier 136, 138
scalar cardinality of a fuzzy set 395
scope of a quantifier 8, 36, 58, 59, 332

second-order predicate 19
self-consistency of quantifier interpretations 107
semantic properties of plausible models 111
semantic universal 63
semantical characteristics of quantifiers 63
semantical constraint *see* plausibility criterion
semantical expectation 378
semantical plausibility 55, 386
semantical postulate *see* plausibility criterion
semantical requirement *see* plausibility criterion
semi-fuzzy L-quantifier 359
semi-fuzzy quantifier 71, 361, 376, 393
semi-fuzzy truth function 91, 363
separation into application-dependent and independent parts 377
separation of context-dependent and context-independent factors 377
set of indeterminates 189
set of similarity grades 261
Sigma-count approach *see* Σ-count approach
similarity grade 259
simplex quantifier *see* base quantifier
smoothness *see* continuity conditions
Sorites paradox 14, 17, 65
spatial adverbs 5
spatial quantification 349
specification medium *see* specification of linguistic quantifiers
specification of linguistic quantifiers 29, 55, 71, 376, 384
specificity consistent 153, 202, 233, 252, 253
specificity order 151, 202, 232, 252
SQL 28, 52
square of opposition *see* Aristotelian square
stable symmetric sum 151
standard connectives 107
standard DFS 149, 156
standard dual 155
standard extension principle 103, 105, 107, 140, 382
standard fuzzy complement 113
standard fuzzy conjunction 113
standard fuzzy disjunction 114
standard fuzzy intersection 155
standard fuzzy negation 113
standard fuzzy union 155
standard model of fuzzy quantification *see* standard DFS
standard models of fuzzy quantification 380
standard of comparison 11, 64
standard quantifier 136
strict α-cut 180
strong conservativity 171
strong negation 101, 112
structural characteristics of quantifiers 375
structural weaknesses of the traditional framework 51
structure-preserving mapping *see* homomorphism
subclasses of plausible models 149, 379
substitution approach 39, 208
Sugeno integral 212, 381, 383
summarization *see* linguistic data summarization
summarization of image sequences 352
summarizer 350, 373
supervaluationism 16, 65, 381
support of a fuzzy set 135
syllogism 2, 3, 99
symmetric sum 151, 152
symmetric summarizer 373
symmetrical difference of the arguments 121
symmetrization of asymmetrical relations 373
symmetry 118
symmetry of a quantifier 63, 117
systematicity 51, 384

t-norm 113, 137
T-quantifier 136, 137
target operations *see* operational form of a quantifier
temporal adverb 5
textual documents 349

TGQ *see* theory of generalized quantifiers
theoretical skeleton for fuzzy quantificaton *see* framework for fuzzy quantification
theory of descriptions 3, 56, 62, 79
theory of generalized quantifiers VI, 57, 76, 79, 86, 320, 376
theory of vagueness 13, 64
three-valued arguments 206
three-valued cut 179, 187, 189, 237, 381
three-valued logic 17, 79, 156
three-valued model of vagueness 17, 65
three-valued propositional function 184
three-valued quantifier 206
three-valued subset 184
topological quantifier 11, 12
traditional framework for fuzzy quantification 30, 376, 392
transformations between models of fuzzy quantification 149, 150
transposition 116
trapezoidal proportional quantifier 352
truth status statistics 310
truth value gap 17
truth value glut 17
truthfulness 27
two-place quantifier *see* binary quantifier
two-valued generalized quantifier *see* generalized quantifier
two-valued L-quantifier 358, 359
two-valued quantifier *see* generalized quantifier
two-valued set *see* crisp set
Type I quantification 22
Type II quantification 22, 157
Type III quantification 22, 71, 376
Type IV quantification 22, 24, 32, 37, 70, 376, 383
typed λ-calculus 125

unary quantifier 9, 31, 62, 87, 141
undefined interpretation 58, 62, 79
underlying semi-fuzzy L-quantifier 361
underlying semi-fuzzy quantifier 75, 86, 112, 393
underspecificity 15
uniform representation of quantifiers 63, 71, 383, 384
unimodal quantifier 166, 372, 380
union of arguments 63, 96, 108
unions of arguments for L-QFMs 364
unions of the arguments of (semi-)fuzzy L-quantifiers 364
uniqueness of compatible extension principle 130
universal model 384
universal quantification 1, 56, 68, 74, 136
universal quantification in a DFS 138
universal specification medium 51, 80
universe of discourse 5, 30, 36
universe of quantification *see* base set
unrestricted proportional quantification 329
unrestricted quantification *see* unrestricted use of a quantifier
unrestricted use of a quantifier 8, 31, 35, 60
upper and lower bounds on three-valued cuts 221
upper bound mapping $\top_{Q,X_1,\ldots,X_n}$ 223
upward monotonic 98
user interfaces 28

vacuous argument *see* cylindrical extension
vagueness 13, 78
value judgment det 65
verbal phrase 58, 59, 98, 332

weak conservativity 170
weak preservation of convexity 169, 208, 381
weighted aggregation 349, 350
weighted quantification *see* importance qualification
weighted querying 347

List of Symbols

$\forall$ pp. 3, 60
$\forall^T$ p. 24
$\widetilde{\forall}$ p. 67
$\exists$ pp. 3, 60
$\exists^I$ p. 24
$\widetilde{\exists}$ p. 67
$\widetilde{\exists}^I$ p. 77
$\neg$ crisp negation and complementation
$\neg$ p. 18, standard negation and complementation
$\widetilde{\neg}$ pp. 92, 113, fuzzy negation and complement
$\widetilde{\neg} Q$ p. 93
$\widetilde{\neg} \widetilde{Q}$ p. 93
$\wedge$ crisp conjunction
$\wedge$ p. 113, fuzzy standard conjunction
$\widetilde{\wedge}$ pp. 92, 113, fuzzy conjunction
$\widetilde{\wedge}$ p. 154, $Q \mathbin{\widetilde{\wedge}} Q'$
$\widetilde{\wedge}$ p. 154, $\widetilde{Q} \mathbin{\widetilde{\wedge}} \widetilde{Q}'$
$\widetilde{\wedge}_m$ p. 113
$\widetilde{\wedge}_{\widetilde{Q}}$ p. 137
$\vee$ crisp disjunction
$\vee$ p. 114, fuzzy standard disjunction
$\widetilde{\vee}$ pp. 92, 114, fuzzy disjunction
$\widetilde{\vee}$ p. 154, $Q \mathbin{\widetilde{\vee}} Q'$
$\widetilde{\vee}$ p. 154, $\widetilde{Q} \mathbin{\widetilde{\vee}} \widetilde{Q}'$
$\widetilde{\vee}_{\widetilde{Q}}$ p. 139
$\rightarrow$ implication
$\leftrightarrow$ biimplication/equivalence
$\cap$ crisp intersection
$\cap$ p. 18, standard fuzzy intersection
$\widetilde{\cap}$ p. 92, fuzzy intersection
$\cup$ crisp union
$\cup$ p. 18, standard fuzzy union
$\widetilde{\cup}$ p. 92, fuzzy union
$\setminus$ set difference $A \setminus B = A \cap \neg B$
$\triangle$ p. 122, symmetric difference
$\triangle$ p. 122, $Q \triangle A$
$\widetilde{\triangle}$ p. 122, fuzzy symmetric difference
$\widetilde{\triangle}$ p. 122, $\widetilde{Q} \mathbin{\widetilde{\triangle}} A$
$[\geq k]$ pp. 6, 279
$[\geq k; \leq u]$ pp. 7, 286
$[>k]$ p. 7
$[=k]$ p. 7
$[\leq k]$ p. 7
$\leq_{(U,V)}$ p. 129, $Q \leq_{(U,V)} Q'$

$\leq_{(U,V)}$ p. 129, $\widetilde{Q} \leq_{(U,V)} \widetilde{Q}'$
$\cong$ isomorphic
$\sim$ p. 294, $(Y_1, \dots, Y_n) \sim (Y_1', \dots, Y_n')$
$\sim\forall$ p. 168
$\sim\exists$ p. 169
$\sim_{(X_1,\dots,X_n)}$ p. 135, $Q \sim_{(X_1,\dots,X_n)} Q'$
$\approx$ p. 299, $(Y_1, \dots, Y_n) \approx (Y_1', \dots, Y_n')$
$\subseteq$ subsethood
$\subseteq$ p. 98, $X \subseteq X'$
$\sqsubseteq$ p. 264, $A \sqsubseteq A'$
$\sqsubseteq$ p. 271, $f \sqsubseteq f'$
$\sqsubseteq$ p. 240, $S \sqsubseteq S'$
$\sqsubseteq$ p. 244, $s \sqsubseteq s'$
$\Subset$ p. 249, $S \Subset S'$
$\preceq_c$ p. 151, $\mathcal{F} \preceq_c \mathcal{F}'$
$\preceq_c$ p. 163, $Q \preceq_c Q'$
$\preceq_c$ p. 163, $\widetilde{Q} \preceq_c \widetilde{Q}'$
$\preceq_c$ p. 249, $S \preceq_c S'$
$\preceq_c$ p. 163, $X \preceq_c X'$
$\preceq_c$ p. 151, $x \preceq_c y$
$\preceq_c$ p. 254, $\omega \preceq_c \omega'$
$\trianglelefteq$ p. 245, $s \trianglelefteq s'$
$\triangleleft$ p. 123, $Q \triangleleft A$
$\triangleleft$ p. 123, $\widetilde{Q} \triangleleft A$
$\widetilde{\triangleleft}$ p. 172, $Q \,\widetilde{\triangleleft}\, A$
@ 369, $\widetilde{Q}$ @ $\widetilde{Q}'$
$\widetilde{@}$ p. 369, $Q \,\widetilde{@}\, Q'$
$\varnothing$ empty set, empty tuple
$\{*\}$ p. 115
$\circ$ functional composition
$\times$ cartesian product
$\times_{i=1}^{n} f_i$ p. 104
! pp. 131, 140
$\top_j$ p. 289
$\top_{Q,X_1,\dots,X_n}$ p. 223
$\top_S$ p. 242
$\bot_j$ p. 289
$\bot_{Q,X_1,\dots,X_n}$ p. 223
$\bot_S$ p. 242
$|$ restriction of f to A, $f|_A$
$|\bullet|$ p. 58
$|\bullet|_\gamma^{\min}$ p. 284
$|\bullet|_\gamma^{\max}$ p. 284
$\|\bullet\|_{\mathrm{iv}}$ p. 284
$(\bullet)^\diamond$ p. 204
$(\bullet)_\gamma$ p. 190
$\langle\bullet\rangle$ p. 142, $\langle Q\rangle$
$\langle\bullet\rangle$ p. 142, $\langle\widetilde{Q}\rangle$
$\mathbf{2}$ p. 18
$\boxplus A$ p. 264
$\widehat{\boxplus} A$ p. 264
β p. 362
$\Gamma(X_1, \dots, X_n)$ p. 288
$\overline{\gamma}_j$ p. 289
$\delta_{X,Y}$ p. 260
ζ_i p. 366
ζ_P p. 306
η p. 91
$\widetilde{\eta}$ p. 91
$\varTheta$ p. 273
$\varTheta(X, Y)$ p. 273
ϑ p. 362
$\theta(x, y)$ p. 272
$\imath_i^{n,E}(e)$ p. 142
κ_i p. 366
$\varLambda$ p. 303
$\mu_\exists$ p. 50
$\mu_{\mathrm{ant}\,Q}$ p. 404
$\mu_{[j]}(X)$ pp. 31, 45, 212
μ_Q p. 33
$\mu_{Q,E}$ p. 400
μ_X p. 18

ν_X p. 185
$\Xi_{Y_1,\ldots,Y_n}(X_1,\ldots,X_n)$ p. 260
$\widetilde{\Xi}_{V,W}$ p. 208
ξ p. 224
ξ_A p. 227
ξ_{owa} p. 226
ξ_S p. 227
Π p. 39
π_* p. 115
π_e p. 88
$\widetilde{\pi}_*$ p. 115
$\widetilde{\pi}_e$ p. 89
ϱ_d p. 348
$\sum$ sum
$\sum$ 67
Σ-Count pp. 42, 395
Σ-Count(X_2/X_1) p. 395
σ_{P_*} p. 305
σQ p. 150
σX p. 150
τ p. 42
τ_i p. 116
$\tau_{i,j}$ p. 116
Υ p. 213
$\Phi_{\ell_1,\ldots,\ell_n}$ p. 296
φ p. 269
χ_A pp. 18, 88
$\Psi[(\mathcal{F}_j)_{j\in\mathcal{J}}]$ p. 151
ψ p. 263
Ω p. 240
$\boldsymbol{\Omega}^*$ p. 252
ω p. 243
$\omega \preceq_c \omega'$ p. 254
ω_K p. 248
ω_M p. 247
ω_P p. 247
ω_Z p. 248
$\boldsymbol{\omega}^*$ p. 253
$A \sqsubseteq A'$ p. 264
A^B the set of all mappings $f: B \longrightarrow A$
A_d p. 348
$A_{Q,X_1,\ldots,X_n}$ p. 262
$A^{(n)}_{Q,X_1,\ldots,X_n}$ p. 262
$A(X_1,\ldots,X_n)$ p. 288
A_Γ p. 311
$\mathcal{A}(Q)$ p. 181
$\mathbb{A}$ p. 262
about k p. 320
abs many$_{\rho,\tau}$ p. 72
all p. 59
all except about k p. 320
all except at most k p. 320
all except exactly k p. 320
all except k p. 59
almost all p. 72
almost always p. 353
always p. 353
as many as possible p. 73
at least k p. 59
at least k more than p. 333
at least once p. 353
at least p percent p. 61
at least twice as many p. 333
at most k p. 59
B_j p. 292
$\mathcal{B}$ p. 194
$\mathcal{B}'$ p. 197
$\mathcal{B}'_\int$ p. 199
$\mathcal{B}^{*\prime}$ p. 200
$\mathcal{B}'_*$ p. 201
$\mathcal{B}'_{CX}$ p. 201
$\mathcal{B}'_S$ p. 200
$\mathcal{B}'_U$ p. 199

$\mathbb{B}$ p. 192
$\mathbb{B}^+$ p. 192
$\mathbb{B}^-$ p. 192
$\mathbb{B}^{\frac{1}{2}}$ p. 192
between p and q percent p. 61
between r and s p. 59
C_j p. 289
$\mathrm{C}(Q)$ p. 121
$\mathcal{C}(Q)$ p. 182
$c_\varnothing$ p. 103
c_a p. 193
$c_{\ell_1,\ldots,\ell_n}$ p. 296
$c(P)$ p. 304
$c(p)$ p. 302
$[\mathbf{card} \geq]$ p. 280
$[\mathbf{card} >]$ p. 280
$[\mathbf{card} =]$ p. 280
$(Ch) \int X \, dQ$ p. 226
$\mathrm{Ch}(Q)$ p. 394
$\mathrm{core}(X)$ p. 135
Count p. 35
$\mathrm{cxt}(X)$ p. 135
$D(A)$ p. 262
D_t p. 304
$D(p)$ p. 302
$D_{X_1,\ldots,X_n}$ p. 261
$D^{(n)}_{X_1,\ldots,X_n}$ p. 261
$\mathbb{D}$ p. 261
$d(f,g)$ p. 203
$d(Q,Q')$ p. 163
$d(\widetilde{Q},\widetilde{Q}')$ p. 163
$d((\top,\bot),(\top',\bot'))$ p. 230
$d((X_1,\ldots,X_n),(X'_1,\ldots,X'_n))$ p. 162
$d'(f,g)$ p. 203
$d'((\top,\bot),(\top',\bot'))$ p. 231
E pp. 5, 57
E_n p. 142
$\mathcal{E}$ p. 102
eq p. 280
exactly k p. 59
exactly k more than p. 333
exactly p percent p. 61
exactly twice as many p. 333
F p. 14
$\mathcal{F}$ pp. 74, 360
$\mathcal{F} \preceq_c \mathcal{F}'$ p. 151
$\widehat{\mathcal{F}}$ pp. 103, 365
$\widehat{\mathcal{F}}^{-1}$ p. 141
$\widetilde{\mathcal{F}}$ pp. 91, 363
$\widetilde{\widetilde{\mathcal{F}}}$ p. 115
$\mathcal{F}_A$ p. 227
$\mathcal{F}_{\mathrm{owa}}$ p. 226
$\mathcal{F}_{\mathrm{glb}}$ p. 153
$\mathcal{F}_K$ p. 248
$\mathcal{F}_L$ p. 367
$\mathcal{F}_M$ p. 247
$\mathcal{F}_{\mathrm{lub}}$ p. 154
$\mathcal{F}_P$ p. 247
$\mathcal{F}_R$ p. 363
$\mathcal{F}_S$ p. 227
$\mathcal{F}_Z$ p. 248
$\mathcal{F}_\xi$ p. 224
$\mathcal{F}^\sigma$ p. 150
$\mathcal{F}_\varphi$ p. 269
$\mathcal{F}_\psi$ p. 263
$\mathcal{F}_\Omega$ p. 240
$\mathcal{F}_\omega$ p. 243
$f \sqsubseteq f'$ p. 271
f^{-1} p. 102
$\widehat{f}$ p. 101
$\hat{\hat{f}}$ p. 103
$\breve{f}$ p. 184
$\widetilde{f}$ p. 91

$f^{\diamond}$ p. 204
$f^{\sharp}$ p. 195
$f^{\flat}$ p. 195
f^* p. 115
f_0^* p. 195
f_1^* p. 195
f_*^0 p. 195
$f_*^{\frac{1}{2}}$ p. 195
$f_*^{1\uparrow}$ p. 195
f_A p. 268
$f|_A$ restriction of f to A
$f_{Q,X_1,\ldots,X_n}$ p. 268
$f^{(n)}_{Q,X_1,\ldots,X_n}$ p. 268
FE-count pp. 49, 412
$\mathrm{FE}^{(1)}_{\mathrm{abs}}$ p. 412
FG-count pp. 47, 406
FG-count(X_2/X_1) p. 407
$\mathrm{FG}^{(1)}_{\mathrm{abs}}$ p. 406
$\mathrm{FG}^{(2)}_{\mathrm{abs}}$ p. 407
$\mathrm{FG}^{(1)}_{\mathrm{prp}}$ p. 407
$\mathrm{FG}^{(2)}_{\mathrm{prp}}$ p. 408
g_α p. 400
H^+ p. 311
H^- p. 311
H_A p. 310
H_S p. 48
$\mathbb{H}$ p. 197
h p. 78
Hist_X p. 310
I p. 18
$\mathrm{I}(Q)$ p. 121
$\mathrm{id}_{\mathbf{2}}$ p. 112
$\mathrm{id}_{\mathbf{I}}$ p. 112
$\mathrm{Im}\, f$ image of f
inf infimum
J^* p. 291
$\widehat{J}$ p. 293
j^* p. 291
$\widehat{\jmath}$ p. 293
j_{pk} p. 283
$\mathbb{K}$ p. 238
$\mathbb{L}$ p. 243
$L(Q, V, W)$ p. 208
$\ell(j)$ p. 309
ℓ_V p. 302
less than k p. 59
M^I p. 24
M^T p. 24
$\mathcal{M}$ p. 192
$\mathcal{M}^*$ p. 200
$\mathcal{M}_*$ p. 201
$\mathcal{M}_{\mathcal{B}}$ p. 194
$\mathcal{M}_{CX}$ p. 201
$\mathcal{M}_S$ p. 200
$\mathcal{M}_U$ p. 199
$\mathrm{m}_{\frac{1}{2}}$ p. 152
many p. 371
many more than p. 333
max maximum
$\mathrm{med}_{\frac{1}{2}}$ p. 152
min minimum
more than p. 62
more than k p. 59
more than p percent p. 61
most pp. 61, 356
N p. 39
$\mathbb{N}$ cardinals $0, 1, 2, \ldots$
$\mathrm{N}(Q)$ p. 121
$\mathrm{NV}(A)$ p. 265
no p. 59
$\mathbb{O}$ p. 168
often p. 353
only p. 117
orness p. 400

$\mathrm{OWA}^{(1)}_{\mathrm{prp}}$ p. 400
$\mathrm{OWA}^{(2)}_{\mathrm{prp}}$ p. 401
$\mathbb{P}$ p. 304
$\mathbb{P}^*$ p. 304
$\mathcal{P}(E)$ p. 11
$\breve{\mathcal{P}}(E)$ p. 185
$\widetilde{\mathcal{P}}(E)$ p. 18
$\mathrm{pol}(p)$ p. 305
Prop p. 35
Q pp. 57, 71, 359
$Q \,\widetilde{@}\, Q'$ p. 369
$Q \preceq_c Q'$ p. 163
$Q \leq_{(U,V)} Q'$ p. 129
$Q \sim_{(X_1,\dots,X_n)} Q'$ p. 135
$Q \,\widetilde{\wedge}\, Q'$ p. 154
$Q \,\widetilde{\vee}\, Q'$ p. 154
Q^+ p. 167
Q^- p. 167
$Q\neg$ p. 93
$Q\widetilde{\Box}$ pp. 94, 364
$Q\cup$ pp. 96, 364
$Q\cup^k$ p. 143
$Q\cap$ p. 123
$Q\triangle A$ p. 122
$Q\triangleleft A$ p. 123
$Q \,\widetilde{\triangleleft}\, A$ p. 172
Q_f p. 91
$\widetilde{Q}^L_{V,W}$ p. 208
$\widetilde{Q}^U_{V,W}$ p. 208
$Q\beta$ p. 116
Q_γ p. 190
$Q\tau_i$ p. 117
$\langle Q \rangle$ p. 142
$\widetilde{Q}$ pp. 31, 66, 360
$\widetilde{Q} \,@\, \widetilde{Q}'$ p. 369
$\widetilde{Q} \preceq_c \widetilde{Q}'$ p. 163
$\widetilde{Q} \leq_{(U,V)} \widetilde{Q}'$ p. 129
$\widetilde{Q} \,\widetilde{\wedge}\, \widetilde{Q}'$ p. 154
$\widetilde{Q} \,\widetilde{\vee}\, \widetilde{Q}'$ p. 154
$\widetilde{Q}\widetilde{\neg}$ p. 93
$\widetilde{Q}\widetilde{\Box}$ pp. 94, 364
$\widetilde{Q}\widetilde{\cup}$ pp. 96, 364
$\widetilde{Q}\widetilde{\cup}^k$ p. 143
$\widetilde{Q}\widetilde{\cap}$ p. 123
$\widetilde{Q} \,\widetilde{\triangle}\, A$ p. 122
$\widetilde{Q}\triangleleft A$ p. 123
$\widetilde{Q}_z$ p. 274
$\widetilde{Q}\beta$ p. 116
$\langle \widetilde{Q} \rangle$ p. 142
$\breve{Q}$ p. 186
$\mathcal{Q}$ p. 357
$\mathbb{Q}$ the set of rational numbers
$q(j)$ p. 282
$q^{\max}(\ell, u)$ p. 283
$q^{\min}(\ell, u)$ p. 283
$\mathbb{R}$ the set of real numbers
$\mathbb{R}^+ = [0, \infty)$ the set of non-negative reals
R_j p. 300
$R_{Q,X_1,\dots,X_n}$ p. 153
$R_\gamma(X_1, \dots, X_n)$ p. 299
$R_\gamma^{\Phi_1,\dots,\Phi_K}(X_1, \dots, X_n)$ p. 299
$\mathcal{R}(f)$ p. 182
$\mathrm{R}(Q)$ p. 121
r_+ p. 261
$r_+(A)$ p. 262
$r_+(D)$ p. 261
$r_+(f)$ p. 269
$[\textbf{rate} \geq r]$ p. 61
$[\textbf{rate} > r]$ p. 61
$[\textbf{rate} \leq r]$ p. 61
$[\textbf{rate} < r]$ p. 61
$[\textbf{rate} = r]$ p. 61
$[r_1 \leq \textbf{rate} \leq r_2]$ p. 61

rel many$_{\rho,\tau}$ p. 72
RES p. 183
S p. 33
$(S)\int X\,dQ$ p. 212
$S \sqsubseteq S'$ p. 240
$S \preceq_c S'$ p. 249
$S \Subset S'$ p. 249
$S^{\Box}$ p. 254
$S^{\sharp}$ p. 240
$S^{\flat}$ p. 240
$S^{\ddagger}$ p. 241
S_j p. 293
$S(p)$ p. 302
$S(P)$ p. 304
$S_{Q,X_1,\ldots,X_n}$ p. 238
$s \sqsubseteq s'$ p. 244
$s \trianglelefteq s'$ p. 245
$s(z)$ p. 239
$s_{Q,X_1,\ldots,X_n}$ p. 239
$s^{\ddagger}$ p. 246
$s_*^{\top,0}$ p. 246
$s_1^{\top,*}$ p. 246
$s_*^{\bot,0}$ p. 246
$s_1^{\bot,*}$ p. 246
$s_*^{\le\frac{1}{2}}$ p. 246
$s_*^{\ge\frac{1}{2}}$ p. 246
same number of p. 333
$\mathrm{SC}^{(1)}_{\mathrm{abs}}$ p. 395
$\mathrm{SC}^{(2)}_{\mathrm{abs}}$ p. 395
$\mathrm{SC}^{(1)}_{\mathrm{prp}}$ p. 395
$\mathrm{SC}^{(2)}_{\mathrm{prp}}$ p. 395
some p. 59
sometimes p. 353
spp(X) p. 135
sup supremum
$\mathrm{T}_\gamma(X)$ p. 187
$\mathcal{T}^*$ p. 295
$\mathcal{T}(X)$ p. 185
$\mathcal{T}(x)$ p. 184
$\mathcal{T}_\gamma(X)$ p. 188
$\mathbb{T}$ p. 223
$t_\gamma(x)$ p. 187
the$_{\mathrm{sg}}$ p. 62
the$_{\mathrm{pl}}$ p. 62
trp$_{a,b,c}$ p. 352
type(p) p. 305
$\mathcal{U}$ pp. 75, 361
$U_{m'}$ p. 316
$U(Q,V,W)$ p. 208
$u(j)$ p. 309
u_V p. 302
$\mathbb{V}$ p. 302
v_0 p. 321
X p. 18
$\mathbb{X}$ p. 269
$X \subseteq X'$ p. 98
$X \preceq_c X'$ p. 163
$X_{\ge\alpha}$ p. 180
$X_{>\alpha}$ p. 180
$X^{[p]}$ p. 302
$X^{\langle v\rangle}$ p. 301
$X^{\min}$ p. 185
$X^{\max}$ p. 185
$X_\gamma^{\min}$ p. 188
$X_\gamma^{\max}$ p. 188
$|X|_\gamma^{\min}$ p. 284
$|X|_\gamma^{\max}$ p. 284
$x \preceq_c y$ p. 151
xor exclusive or/antivalence
Y_j p. 282
$Y^{(\ell)}$ p. 296
$(Y_1,\ldots,Y_n)^*$ p. 294
$(Y_1,\ldots,Y_n) \sim (Y_1',\ldots,Y_n')$ p. 294
$(Y_1,\ldots,Y_n) \approx (Y_1',\ldots,Y_n')$ p. 299

$\mathcal{Z}$ p. 35

$\mathbb{Z}$ the set of integers

Z_V p. 302

$\mathcal{Z}^{(1)}_{\mathrm{abs}}$ pp. 35, 392

$\mathcal{Z}^{(2)}_{\mathrm{abs}}$ pp. 35, 392

$\mathcal{Z}^{(1)}_{\mathrm{prp}}$ pp. 35, 392

$\mathcal{Z}^{(2)}_{\mathrm{prp}}$ pp. 35, 392

z_+ p. 262

$z_+(A)$ p. 262

$z_+(f)$ p. 269

List of Figures

1.1 A possible membership function for "almost all" 34
1.2 Cloudiness in Southern Germany (guiding example) 42
1.3 "About ten percent of Southern Germany are cloudy" (Sigma-count) 44
1.4 "At least 60 percent of Southern Germany are cloudy" (Sigma-count) 44
1.5 "At least 60 percent of Southern Germany are cloudy" (OWA approach) 46
1.6 "At least 5 percent of Southern Germany are cloudy" (FG-count) 49
1.7 "The image region X is nonempty" (FE-count) 50

3.1 The Aristotelian square of "all" 95

5.1 Plot of the fuzzy median 152

7.1 α-cut as a function of α and $\mu_X(v)$ 181
7.2 Three-valued cut as a function of γ and $\mu_X(v)$ 188

11.1 Ranking for "As much as possible of Southern Germany is cloudy" 350
11.2 "Q-times cloudy in the last days": an example of image summarization 353

13.1 Coverage of quantificational phenomena: DFS vs. existing approaches 387

A.1 "About 10 percent of Southern Germany are cloudy" (Sigma-Count) 397
A.2 "At least 60 percent of Southern Germany are cloudy" (Sigma-Count) 399
A.3 "At least 60 percent of Southern Germany are cloudy" (OWA approach) 402
A.4 "More than 60 percent of Southern Germany are cloudy" (OWA approach) 403
A.5 "At least 5 percent of Southern Germany are cloudy" (FG-Count) 411
A.6 "Less than thirty percent of Southern Germany are cloudy" (FG-count) 412
A.7 "The image region X is nonempty" (FE-count) 414

List of Tables

1.1 Examples of NL quantification in various areas of everyday life . 2

1.2 Quantifying propositions and their translations into predicate logic . . . 6

1.3 The fields of fuzzy quantification . . . 22

1.4 Rescher's quantifiers . . . 24

1.5 "Less than 60 percent of Southern Germany are cloudy" (OWA approach) . . . 47

7.1 Computation of $L(Q, V, W)$ and $U(Q, V, W)$ . . . 209

7.2 Computation of $\widetilde{Q}^L_{V,W}(X)$ and $\widetilde{Q}^U_{V,W}(X)$ . . . 210

11.1 Algorithm for unary quantifiers in $\mathcal{F}_{\text{owa}}$ (floating-point arithmetics) . . . 314

11.2 Algorithm for unary quantifiers in $\mathcal{M}$ (floating-point arithmetics) 314

11.3 Algorithm for unary quantifiers in $\mathcal{M}_{CX}$ (floating-point arithmetics) . . . 315

11.4 Algorithm for unary quantifiers in $\mathcal{F}_{\text{owa}}$ (integer arithmetics) . . . 317

11.5 Algorithm for unary quantifiers in $\mathcal{M}$ (integer arithmetics) . . . 318

11.6 Algorithm for unary quantifiers in $\mathcal{M}_{CX}$ (integer arithmetics) . . 319

11.7 Examples of quantifiers of exception . . . 320

11.8 Algorithm for proportional quantifiers in $\mathcal{F}_{\text{owa}}$ (floating-point arithmetics) . . . 325

11.9 Algorithm for proportional quantifiers in $\mathcal{M}$ (floating-point arithmetics) . . . 326

11.10 Algorithm for proportional quantifiers in $\mathcal{M}_{CX}$ (floating-point arithmetics) . . . 327

11.11 Algorithm for proportional quantifiers in $\mathcal{F}_{\text{owa}}$ (integer arithmetics) 328
11.12 Algorithm for proportional quantifiers in $\mathcal{M}$ (integer arithmetics) 329
11.13 Algorithm for proportional quantifiers in $\mathcal{M}_{CX}$ (integer arithmetics) 330
11.14 Algorithm for cardinal comparatives in $\mathcal{F}_{\text{owa}}$ (floating-point arithmetics) 337
11.15 Algorithm for cardinal comparatives in $\mathcal{M}$ (floating-point arithmetics) 338
11.16 Algorithm for cardinal comparatives in $\mathcal{M}_{CX}$ (floating-point arithmetics) 339
11.17 Algorithm for cardinal comparatives in $\mathcal{F}_{\text{owa}}$ (integer arithmetics) 340
11.18 Algorithm for cardinal comparatives in $\mathcal{M}$ (integer arithmetics) 341
11.19 Algorithm for cardinal comparatives in $\mathcal{M}_{CX}$ (integer arithmetics) 342
11.20 Complexity of the algorithms for $\mathcal{F}_{\text{owa}}$, $\mathcal{M}$ and $\mathcal{M}_{CX}$: Main loop 348

A.1 "Less than 60 percent of Southern Germany are cloudy" (OWA approach) 406